# Agroforestry for Ecosystem Services and Environmental Benefits

# Agroforestry for Ecosystem Services and Environmental Benefits

Edited by Lester Bane

STATES
ACADEMIC PRESS
www.statesacademicpress.com

**States Academic Press,**
109 South 5th Street,
Brooklyn, NY 11249, USA

Visit us on the World Wide Web at:
www.statesacademicpress.com



ISBN: 978-1-63989-747-6



**Cataloging-in-Publication Data**

Agroforestry for ecosystem services and environmental benefits / edited by Lester Bane.
p. cm.
Includes bibliographical references and index.
ISBN 978-1-63989-747-6
1. Agroforestry. 2. Agroforestry systems. 3. Ecosystem services. 4. Agroforestry--Environmental aspects.
5. Forests and forestry--Environmental aspects. I. Bane, Lester.

S494.5.A45 .A375 2023
634.99--dc23

# Table of Contents

# Preface

Agroforestry is a type of land management system where trees or shrubs are planted around or surrounding the crops or pastureland. Such type of land management helps in producing a variety of products such as fruits, wood products, pharmaceuticals and other valuable products that can be traded commercially. It ensures long-term sustainability of agriculture. Some of its other benefits include improved farmer livelihoods through wealth creation, higher yields from staple food crops, carbon sequestration, increased biodiversity, and enhanced soil structure and health. Agroforestry helps blend forestry and farming, which helps restore the land by reducing soil erosion and regulating the water accessibility to restore the land. The practice of agroforestry leads to ecosystem services, which refers to the human benefits apart from an increase in farm produce. Agroforestry-based ecosystem services help in improving agricultural sustainability by ensuring better usage of fertilizers, enhanced ground cover, improved insect resistance, and weed avoidance. This book elucidates the concepts and innovative models around prospective developments with respect to agroforestry-based ecosystem services and environmental benefits of agroforestry. It will also provide interesting topics for research, which interested readers can take up.

This book is the end result of constructive efforts and intensive research done by experts in this field. The aim of this book is to enlighten the readers with recent information in this area of research. The information provided in this profound book would serve as a valuable reference to students and researchers in this field.

At the end, I would like to thank all the authors for devoting their precious time and providing their valuable contribution to this book. I would also like to express my gratitude to my fellow colleagues who encouraged me throughout the process.

**Editor**

1

# Agroforestry Benefits and Challenges for Adoption in Europe and Beyond

**Maya Sollen-Norrlin [1], Bhim Bahadur Ghaley [2] and Naomi Laura Jane Rintoul [1,*]**

[1] School of Psychology and Life Sciences, Canterbury Christ Church University, Canterbury CT1 1TE, UK; maya.sollen@gmail.com
[2] Department of Plant and Environmental Sciences, Faculty of Science, University of Copenhagen, 2630 Taastrup, Denmark; bbg@plen.ku.dk
* Correspondence: naomi.rintoul@canterbury.ac.uk

**Abstract:** Soil degradation is a global concern, decreasing the soil's ability to perform a multitude of functions. In Europe, one of the leading causes of soil degradation is unsustainable agricultural practices. Hence, there is a need to explore alternative production systems for enhanced agronomic productivity and environmental performance, such as agroforestry systems (AFS). Given this, the objective of the study is to enumerate the major benefits and challenges in the adoption of AFS. AFS can improve agronomic productivity, carbon sequestration, nutrient cycling, soil biodiversity, water retention, and pollination. Furthermore, they can reduce soil erosion and incidence of fire and provide recreational and cultural benefits. There are several challenges to the adoption and uptake of AFS in Europe, including high costs for implementation, lack of financial incentives, limited AFS product marketing, lack of education, awareness, and field demonstrations. Policies for financial incentives such as subsidies and payments for ecosystem services provided by AFS must be introduced or amended. Awareness of AFS products must be increased for consumers through appropriate marketing strategies, and landowners need more opportunities for education on how to successfully manage diverse, economically viable AFS. Finally, field-based evidence is required for informed decision-making by farmers, advisory services, and policy-making bodies.

**Keywords:** agroforestry; agronomic productivity; ecosystem services; agroecology; farmland management; sustainability

---

## 1. Introduction

Soil degradation is the "decline in soil quality or its capacity to produce economic goods or provide ecosystem services" [1] (p. 709). Soil degradation can result in soil compaction, salinity, nutrient loss, reduction in biodiversity, or contamination leading to unproductive soils, dependent on the crops grown, soil type, and management in an agro-ecosystem. Agro-ecosystems provide provisioning, regulating, supporting, and cultural ecosystem services, and amongst these services, provisioning services like food, fodder, fibre, and fuelwood have economic values due to their marketable values [2]. However, it is a challenge to add economic values to regulating, supporting, and cultural ecosystem services as the goods and services are non-marketable and the importance of these ecosystem services are location-specific, adding another complexity to valuation.

Soil degradation is a global concern that is most prominent in the tropics and subtropics. Soil degradation is a concern in Europe as there is a need to maintain the multifunctional role of soils for primary productivity, carbon sequestration, nutrient storage and regulation, water storage and cycling, and habitat for biodiversity [3]. The leading cause of soil degradation in Europe is the intensive conventional arable production system, where losses of up to 300 Tg C ($10^{12}$) year$^{-1}$ is reported [4]. The use of mechanical tillage causes adverse effects on soil structure and enhances the decomposition process, leading to nutrient loss through leaching and volatilization and imbalances in nutrient supply [5]. Furthermore, the use of pesticides has adverse effects on soil biodiversity [6]. Given the challenges of soil degradation due to conventional intensive production systems, there is need for production systems that enhance agronomic productivity and environmental performance.

Agroforestry, the practice of integrating trees and shrubs with crops or livestock production [7], is one such potentially sustainable way of achieving healthier soils for increased agricultural production [8,9] and environmental performance. Agroforestry systems (AFS) include silvoarable systems (combination of trees/shrubs with crops), silvopastoral (combination of trees with livestock), and agro-silvopastoral (combination of trees/shrubs with both crops and livestock), riparian buffer strips, and home gardens [10]. Besides provisioning services, viz., food, fodder, fibre, and fuelwood production, AFS provide several other ecosystem services, including regulation of nutrient cycling, carbon sequestration, habitat for biodiversity, erosion control, fire and flood control, and recreational and cultural services [11,12]. Hence, the objective of this study is to assess the beneficial effects of AFS on agronomic productivity, carbon sequestration, soil nutrient cycling, soil biodiversity, soil erosion and water retention, recreation and culture, fire and pollination, and to highlight the challenges faced in the adoption of AFS.

## 2. Benefits of AFS

### 2.1. *Agronomic Productivity*

The land equivalent ratio (LER) was first proposed by Mead and Willey [13] and is defined as the ratio of the area under sole cropping to the area under intercropping needed to give equal amounts of yield at the same management level [14]. It is the most common measure for comparing productivity in terms of biomass or other yields [15] and is a valuable tool to assess the production system when considering conversion of conventional agricultural land to AFS [16].

The evidence of higher productivity in agroforestry systems has been reported in different studies. The LER of a combined food and energy (CFE) production system in Denmark ranged from 1.14–1.34 [17], indicating that crop and tree yields produced in agroforestry required 14–34% less land or fewer resources in terms of light, water, nutrients, compared to monoculture. This is a significant advantage and provides avenues for ecological intensification to produce more with fewer inputs. A recent study investigated five agroforestry systems that differed in crop, tree, and grass species under diverse pedo-climatic zones and management regimes; LER values ranged between 1.36–2.00, demonstrating a higher productivity in a diversity of AFS [18]. Likewise, in the oldest and most well-documented agroforestry experimental site in Europe, silvoarable agroforestry systems had LER estimates of 1.3–1.6 [16]. Other studies have used modelling to determine the productivity of different agricultural systems. Over 30 years, it was estimated that 1.28 ha of separate forestry and arable systems would be needed to produce the crop and timber yields obtained in a silvoarable system [19]. In a study of 14 different agroforestry practices by Sereke et al. [20], the AFS were commonly more productive than separate forestry or arable systems (12 out of the 14 options, land equivalent ratio = 0.95–1.30).

Despite numerous studies demonstrating the higher productivity of AFS, these results contradict the views of farmers, who see AFS as unproductive and thus as financially unviable [20]. This view is reiterated in many other papers assessing farmers' opinions of barriers to AFS adoption, where farmers are often concerned with the financial risk [7,19,21,22]. Given that there is a need to increase food production significantly by 2050 [23], high-yielding agroforestry provides a viable option to contribute to the yield increase to meet the increase in food demand due to population growth and changes in food consumption patterns.

### 2.2. Soil Carbon Sequestration

Soil represents a large pool of carbon (C), consisting of about 1550 Gt of soil organic carbon (SOC) and 950 Gt of soil inorganic carbon (SIC) [24]. Soil organic carbon occurs mostly as a fraction of organic matter (OM), and its levels in soils are determined by a dynamic balance of inputs and outputs [24]. Carbon sequestration occurs when the C input to soil is higher than the C release [25]. Agricultural soils in Europe generally act as carbon sources, but have a high potential to become carbon sinks with changes in land management [24]. Because soil organic matter (SOM) contains 50–58% carbon, the loss of OM from soils in conventional agriculture contributes to C sources [26]. Agroforestry systems add significant amounts of SOM, resulting in multiple benefits such as improved soil moisture retention, greater soil nitrogen mineralisation, and, importantly, increased carbon sequestration, and thus climate change mitigation [27].

A field experiment of 17 arable agroforestry fields in Belgium revealed that trees increased carbon by 5.3 ton organic carbon (OC) $ha^{-1}$ in the plough layer, compared to treeless control plots [28]. Similarly, a Canadian field experiment recorded an increase in SOC of 6.2 ton OC $ha^{-1}$ in the top 0–20 cm of soil after 21 years of intercropping with poplar, compared to arable cropping without trees [29]. Inclusion of woody crops, viz., willow, can add up to 103 Mg carbon $ha^{-1}$ [30], inclusive of aboveground, belowground, and soil organic carbon. Kay et al. [31] identified priority areas across Europe where the implementation of agroforestry could be particularly effective and examined the impact of AFS on soil carbon. They determined that conversion of arable systems to AFS within these priority areas could lead to the sequestration of 2.1 to 63.9 million t C $a^{-1}$, depending on the type of agroforestry, which corresponds to 1.4% to 43.4% of European agricultural greenhouse gas (GHG) emissions. Aertsens, De Nocker and Gobin [25] estimated that implementation of agroforestry on arable land in the EU has the potential to sequester 2.75 t C $ha^{-1}$ $year^{-1}$ and a further 2.75 t C $ha^{-1}$ $year^{-1}$ on pasture land. This is 5–10 times the C sequestration potential of conventional agriculture [25]. Comparing the environmental footprint in terms of global warming potential, acidification, and eutrophication, agroforestry systems were found to be more environmentally friendly compared to the conventional arable production system [18].

Alley cropping is a common agroforestry practice in Europe [32]. In this system, crops are grown in strips in-between rows of trees and shrubs [33]. This is distinctly different from conventional farming systems where one plant species is grown as a monoculture [25]. In silvopastoral systems, livestock is grazed on grasslands with trees growing on them, compared to conventional pastures consisting of grassland monocultures [34]. These AFS increase carbon sequestration in several ways, compared to the conventional systems. Firstly, the woody biomass of trees is 46–51% C and thus represents a main pathway of C accumulation [35]. Deposition of vegetation onto the soil through litter fall and crop residues leads to increased incorporation of C as OM into the soil [35,36], and enhanced root production increases C in the soil [37]. Indeed, roots represent a more important component of C sequestration in AFS in deep soil horizons than herbaceous understory and crop plants, and thus can relocate C to below the plough layer and protect it from oxidation [37].

### 2.3. Nutrient Cycling

Nutrients in AFS generally are provided from prunings and litter fall from the trees, making the nutrients available for the crops in the alley [38]. However, in some AFS synthetic fertilisers are also used [39]. Nutrient availability and cycling have been shown to be greater and more efficient in AFS, compared to conventional agro-ecosystems [40,41]. This can be attributed to several different processes: the incorporation of nitrogen-fixing tree species or the use of leguminous crops in crop rotation or as cover crops leads to increased stocks of nitrogen (N) [42]. For example, 650 woody species in temperate regions can fix nitrogen from the atmosphere, and 32–58% of the total nitrogen required for the alley-cropped maize came from the nitrogen fixation by the adjacent alder trees [43]. In a combined food and energy system in Denmark, mineralized nitrogen was 108 kg N $ha^{-1}$ $year^{-1}$ for AFS, compared to 64 kg N $ha^{-1}$ $year^{-1}$ in a conventional wheat production system [2]. In addition,

Pardon et al. [28] recorded increased concentrations of nutrients in AFS compared to treeless control plots, where concentrations were greater by 35.5 mg kg $dm^{-1}$, 28.8 mg kg $dm^{-1}$, 16.2 mg kg $dm^{-1}$, and 5 mg kg $dm^{-1}$ for K, P, Mg, and NA, respectively.

In AFS, biomass production of the trees is faster than in forests as there is less competition amongst the trees [25]. This rapidly produced biomass can then be recycled back into the system, which improves SOM content and nutrient recycling [40]. Tree roots can reach and utilise nutrients that have leached to deep soil layers that crop roots cannot reach [40]. These nutrients are then recycled into the system through leaf litter and turnover of fine roots [40]. Thus, agroforestry has the potential to improve soil fertility through improved nutrient cycling and retention [25,40,41]. For example, the leaf litter from adjacent poplar trees in agroforestry systems doubled the availability of nitrogen in soils for uptake by crops in the alleys to 7 kg N $ha^{-1}$ $year^{-1}$, compared to the soils located 8 to 11 m away from the trees [44]. Trees can also add significant nitrogen by intercepting rainfall through stemflow (rainfall intercepted by leaves and branches) and throughfall (rainwater intercepted by the canopy) to the tune of 15.22 and 10.99 kg N $ha^{-1}$ $yr^{-1}$ in hybrid poplar and silver maple tree systems, respectively [45].

Despite the opportunities that AFS present, there is a great deal of variability amongst different types of AFS in different contexts in terms of species, soil type, and other land-use conditions [46]. The net productivity of any agroforestry system is the result of interactions, negative and positive, between the components in the system [43]. In fact, some long-term yields may not differ between AFS and monoculture systems [47]. If all plant species in the system have the same requirements over time and space for the same resource (water, nutrients, light), they will compete with each other [43], and this can result in reduced yields [48]. For example, shading by trees can have negative impacts on crops as sub-optimal light availability can induce stress responses [49]. However, if the different species in the system use different resources at different temporal and spatial scales, they can instead improve each other's environment and facilitate growth [43,49]. Therefore, selection of tree and crop species that have different resource requirements is an important factor when maximising the efficiency of the AFS [46,50]. The water-retaining properties of the soil will also impact the degree of competition for water in the system, with more competition in free-draining soils and vice versa [43]. Further, different management practices, such as pruning, root cutting, and distance between and orientation of tree rows, as well as whether fertilizers are added or not, all have an impact on the interactive outcome of the species in the system [38,39,42,43,49,50].

### 2.4. Soil Biodiversity

Soils are incredibly diverse: 1 g of soil has been estimated to contain 1 billion bacteria from tens of thousands of taxa, 200 m of fungal hyphae, and a vast variety of nematodes, mites, earthworms, and arthropods [51]. Soil organisms can be grouped by their ecological function; they are classed as decomposers, nutrient transformers, ecosystem engineers, and bio-controllers [52–54]. These functional assemblages contribute to four aggregated ecosystem functions: carbon transformations, nutrient cycling, soil structure maintenance, and population regulation, which through a variety of soil-based delivery processes generate and sustain soil health [55]. The diversity of soil organisms due to differences in litter quality and quantity results in a diversity of enzymes and enzymatic activity, supporting the critical soil functions of organic matter breakdown, nutrient availability, and recycling [56,57]. Agroforestry enhances the growth of arbuscular mycorrhizal fungi, which enhances litter decomposition, resulting in enhanced plant-available nutrients, and such positive effects have been demonstrated in field studies [58,59]. Marsden et al. [60] collated the results from 62 publications in order to examine the effect of agroforestry on soil fauna abundance and/or diversity in temperate climates. Of the papers that examined AFS compared to arable systems, the effects of AFS were mostly positive (70% of the data sets), and only two studies found a negative impact of AFS on soil fauna biodiversity or abundance. Similarly, a meta-analysis by Torralba et al. [46] found an overall positive effect of agroforestry on biodiversity, although the benefits of AFS are context-related. However, studies that examined the abundance or diversity of specific groups of organisms provided clearer information on the impacts of AFS on the soil ecosystem.

Decomposers and nutrient transformers (e.g., microorganisms such as algae, bacteria, fungi, and microarthropods) play a key role in breaking down OM [53,61,62], which then releases nutrients and contributes to nutrient cycling [53,63]. They fragment decomposing litter and increase availability to microbes. The symbiotic relationship between nitrogen-fixing bacteria and some plants (mainly legumes and some tree species) is well established [64]. Co-operative microbial activities such as biological nitrogen fixation (BNF) processes can be exploited as a low-input biotechnology in sustainable AFS [64,65]. In addition to this, mycorrhizal fungi in the soil take up phosphorous—particularly in P-deficient soils—and recycle it back into the system [66–68]. The microbial biomass also represents the living part of SOM, and living microbial cells usually represent 1–5% of the total organic C in soils [65]. Although a relatively small proportion of soil C, the microbial biomass C still represents a global pool of about 23 Pg [69]. Thus, it seems a likely assumption that the increased abundance of soil microbes in AFS contributes to increased C storage. The high input of OM to the soil in AFS also contributes to the diversity and abundance of microorganisms as OM provides an energy source for the microbes [62].

Ecosystem engineers (e.g., earthworms and ants) alter the physical structure and porosity of the soil by creating burrows, chambers, and castings, and influence nutrient and energy flow by ingesting organic matter and mixing with the soil [52,53,70]. In a combined food and energy system in Denmark, the earthworm count in shelter belts was 148 $m^{-2}$, compared to 63 $m^{-2}$ in a conventional barley production system [2]. In a study of 13 agroforestry sites in France, abundance and biomass of earthworms were higher in AFS than in control plots, as were the soil organic carbon stocks [64]. Similarly, a North American study found that earthworm density was higher within AFS than within monocultures [71].

Bio-controllers (e.g., microfauna and macrofauna such as nematodes and collembolans) act as herbivores or predate on other soil organisms, regulating the activity of the soil microbial community and thus affecting nutrient flow [52–54]. Intensified agriculture has been shown to be detrimental to soil biodiversity, and a higher diversity of soil microorganisms has been observed in AFS compared to conventional agriculture systems [40,41]. In addition, according to Barrios et al. [55], AFS had a higher abundance of earthworms, beetles, centipedes, millipedes, termites, ants, collembola, mites, and non-parasitic nematodes than monocultures; the only group that had a higher abundance in monoculture was parasitic nematodes. Thus, AFS can increase both abundance and diversity of soil biota. Similarly, Mupeyo et al. [72] found that AFS can reduce parasitic nematode fecundity and thus have a positive impact on livestock. Banerjee et al. [73] quantified bacterial 16S ribosomal RNA (rRNA) genes in soil samples from AFS (woodland or hedgerows) and controls (either grassland or agricultural land) and found that there was a higher abundance of soil bacteria in AFS. In addition, Bainard et al. [66] found greater fungi richness in an agroforestry system than in a monocrop, and several taxa that were found in the AFS were not present in a monocrop system. However, AFS have been shown to increase in fungi-to-bacteria ratio [74,75], which suggests that AFS have varying effects on different groups of organisms.

Differences in soil biota diversity and abundance can be attributed partly to the increased plant diversity of AFS compared to monocultures in conventional systems as plant diversity is positively linked to soil microbial diversity [65]. For example, higher soil invertebrate diversity is found near the trees compared to the alleys, due to high diversity of leaf litter resulting in high organic matter accumulation [71]. In addition, AFS structural complexity may alter soil biota. For example, leaves intercept rainfall and transpire water taken up by roots, canopy cover provides shade, and pruned leaves and branches provide soil cover and nutrient inputs. This, in turn, affects the temperature, moisture, erosion, and nutrient content of the soil and indirectly affects the soil biota by "fostering conditions for increased biological activity" [55].

Torralba et al. [46] argued that this capacity of agroforestry to provide food, shelter, habitat, and other resources is one of the main reasons why many agroforestry areas are protected under the

Natura 2000 Directive and are frequently recorded as High Nature Value farmlands. In addition, almost a quarter of the natural habitat types listed in Annex I of the Directive refer to silvopastures. The EU Commission's Europe 2020 strategy for a "resource efficient" Europe highlights the necessity of protecting, valuing, and restoring biodiversity and ecosystem services. However, more research on the effects of AFS is required; most studies focus on a single species or taxonomic group and many organisms are under-researched, particularly for temperate AFS [60].

### 2.5. Soil Erosion and Water Retention

AFS have several documented benefits to reduce the non-point source pollution of ground water and aquatic bodies due to reduction and filtration of surface runoff and enhanced infiltration into the soil to recharge the aquifers [76–78]. For example, a combined food and energy system recorded a water-holding capacity of 411 mm ha$^{-1}$ compared to a conventional wheat production system with a water-holding capacity of 283 mm ha$^{-1}$ [2]. Similarly, soil erosion prevention in a combined food and energy system was 3.3 Mg ha$^{-1}$ year$^{-1}$, compared to equivalent soil loss due to soil erosion in conventional wheat production [2]. Agroforestry systems improve soil stability and prevent erosion through several processes. Increased ground cover from leaf litter and "tree mulch" intercepts rainfall, decreases the velocity of runoff water, reduces evapotranspiration, and limits soil crusting [79]. Increased OM improves the water-retaining properties and improves stability of the soil via improved aggregation and soil structure [79]. Moreover, trees and hedges provide a permeable barrier that slows down and intercepts water runoff [80]. Water infiltration rates were shown to be up to 60 times higher in areas planted with young trees than in adjacent grazed pastures [81]. Trees are also used to reduce wind erosion; globally, 300 million hectares of farmland are protected through windbreaks and shelterbelts [82]. In a meta-analysis by Torralba et al. [46] comparing agroforestry, conventional monoculture, and forested sites in Europe, agroforestry had a positive impact on erosion control.

Mulches, cover crops, and alley cropping help reduce soil temperature and soil loss [83]. For example, soil loss was reduced by up to 80% as a result of alley cropping [84]. Similarly, a study by Budelman [85] showed that mulches from trees or woody shrubs can decrease the soil temperature by up to 6.6 °C and increase soil moisture content by up to 4.6%. Soil evaporation is also regulated through shading [80]. The shading effect creates a humidity and temperature buffer that can protect the system from weather extremes such as torrential rain and heat waves [86,87]. Thus, AFS have the potential to both mitigate and adapt to increased rainfall and erosion and increasing temperatures, which justifies the argument for their wider implementation in the EU.

### 2.6. Recreation and Culture

Agroforestry systems can increase the diversity and improve the aesthetics of the landscape, which is more appealing for visitors compared to monoculture [12]. Agroforestry provides cultural services such as heritage value, recreation and ecotourism, knowledge systems, spiritual interactions, aesthetic values, and conserves species, habitats, and landscapes [11,12,88]. However, current measurements of cultural services are inadequate as they are the least researched ecosystem service category in agroforestry papers [11,89].

AFS represent management practices based on local or indigenous knowledge that has evolved over time; they are the cultural, social, economic, and ecological heritage of the people [89,90]. Thus, Vera [91] stated that "it is only by knowing the wilderness that we can understand our cultural landscape". Recreation and ecotourism services are also provided by agroforestry sites on a regular basis for many people. Common examples are wild foraging, hunting, wildlife observation, walking or hiking, fishing, use of off-road recreation vehicles, horseback riding, and camping [92]. One survey found that 92.4% of participants could name at least one recreational service to household members or outsiders, and landowners were providing more than four of these services on average [92]. Moreover, McAdam et al. [12] noted that activities such as hunting, fishing, mountain biking, and horseback riding can diversify income for farmers, while the public can benefit from improved health and enjoyment from agroforestry through sports and wildlife watching. Similar opportunities were described for AFS in Italy by Pardini [93]. Well-known landscapes have a particularly high value for ecotourism, such as those in the New Forest in England, the cork oak systems of Spain and Portugal, or the wood pastures of the Alps [94].

Nature-based tourism is a rapidly expanding sector within tourism in Europe [95]. Thus, AFS can improve the rural economy in Europe if management through agritourism and local engagement is supported [96]. Since the aesthetic and recreation values are crucial to the success of ecotourism, there is a need for a market-based contractual mechanism to promote recreation and ecotourism that ensures a high aesthetic and recreational value [95]. Fortunately, some of the fundamental features of AFS have a high aesthetic value. For example, Dudek [97] created a valuation tool for forests based on their features that places a higher aesthetic value on forests with a high tree stand diversity, high diversity of flora in undergrowth, and more canopy layers. Thus, the promotion of AFS would go hand-in-hand with the promotion of forests that have a high recreation and ecotourism value; these systems are an important part of the culture and heritage in Europe, thus they should be preserved and sustainably managed [98].

### *2.7. Pest and Disease Control*

It is estimated that 37% of USA crop yields are lost to pests [99] and 20–40% of global crops are lost to disease [100]. Furthermore, outbreaks of pests and diseases are expected to increase in the future [101]. Much of the current pest control measures, particularly in developed countries, are based on the application of chemical pesticides to crops. In the USA alone, USD 11 billion was spent on 480 million pounds of pesticides in 2008 [102]. Similarly, in the USA, invasive plant pathogens cause an estimated USD 21 billion in crop losses each year due to diseased crops [103]. However, chemical pest and disease control is not an ecologically sustainable or economically viable solution. Instead, agricultural systems that promote the natural enemies of pests are a sustainable solution: they can reduce yield loss of crops without the negative environmental consequences that result from chemical use and are more economically viable [104,105]. In fact, pest control by natural enemies was valued at USD 13 billion per year in the USA [106].

In a meta-analysis by Pumariño et al. [107], pest abundance and plant damage were lower in agroforestry treatments for perennial crops, but annuals were unaffected. This was attributed mostly to the effects of shading by trees, although other studies have attributed this to increased landscape complexity in AFS [105,108]. In addition to trees, hedges, boundary plantations, and windbreaks create a physical barrier, especially if pests or pathogens are transported by wind [109,110]. A large number of studies have shown that AFS—particularly those with high crop diversity—generally have a higher abundance of natural enemies of pests [104,105,107,108,110–112] and increased competition between pest and non-pest species [108,112]. In addition to reducing pests, AFS have been shown to reduce the environmental effects of pesticides. Pesticide runoff is lower in AFS, and microbial populations in these systems can break down pesticides [113].

Chaplin et al. [104] found that despite the increase in natural enemies, pest abundance and plant damage showed no relationship to landscape complexity such as AFS rather than monocultures. Some studies have also reported fewer herbivores and less crop damage in AFS [111]. However, the data are limited; for example, in one meta-analysis >80% of papers were on just two crop types [107]. More diverse research is required to get a better understanding of the effects of different AFS on pests, particularly in temperate zones [114]. Moreover, Iverson et al. [115] suggested that appropriate AFS management is key in providing a "win-win" relationship between crop yield and biocontrol; this is not fully taken into account in most studies. Smith, Pearce and Wolfe [112] suggested that good AFS management can involve strategies such as providing sources of adult parasitoid food; providing sites for mating, oviposition and resting; ensuring intermittent distribution of host plants, thus making it harder for pests to find the plants; using "trap-crop" species protecting other crops from herbivore attack; and using plants acting as a repellent to the herbivore [112]. Stamps and Linit [108] and Iverson et al. [115] agreed that decreased host plant apparency may have beneficial effects for pest management.

Although there is evidence to suggest that plant diseases were significantly reduced by agroforestry [107,109], far less research has been carried out investigating the effects on plant diseases of AFS compared to plant pests. In a meta-analysis on the effects of AFS on pests and diseases by Pumariño et al. [107], out of the 40 studies included, only two papers were on plant disease and both were for tropical AFS. Beule et al. [116] agreed that studies on the effects of AFS on disease control—particularly in temperate zones—are lacking.

As with pests, appropriate management of AFS is essential for disease control. Beule et al. [116] suggested that various methods of diversification of the AFS can suppress plant diseases. These include growing a mixture of crop varieties, crop rotation, intercropping, and alley cropping. This view was supported by Tilman et al. [117], who suggested that modern intensive agro-ecosystems are more vulnerable to crop damage by pests and diseases due to oversimplification of these systems [117]. Like pest management, reducing host plant density can reduce crop disease prevalence, particularly of soil-borne and splash-dispersed pathogens. Soil microbiology can have a positive impact where antagonist organisms can suppress disease [109,116]. In addition, higher tree density results in microclimatic effects and reduced wind speed, which can reduce dispersal of pathogens. Studies also found that shading by trees in temperate AFS reduced pest and disease incidence, which led to better disease resilience [109,118]. However, in a review of tropical AFS, Schroth et al. [119] found that the intensity of attacks by pests and diseases in these systems varies with the pest or disease and with its ecological requirements.

### *2.8. Fire Prevention*

In 2018 alone over 137,000 ha of land was lost in Europe due to wildfires. Of all the recorded fires in Europe, the Middle East, and North Africa, the largest proportion of burnt land was in forest and other wooded land (58%), followed by other natural land (32%), agriculture (9%), and artificial surfaces (1%) [120]. Thus, wildfires are particularly problematic in forests.

Wildfires require fuel, oxygen, and heat in order to burn and spread. Unmanaged forests typically contain high fuel loads, increasing the risk of large, rapidly spreading wildfires [121]. However, many AFS have adopted a silvopastoral approach, which incorporates heavy thinning and pruning to create "firebreaks". The land is then grazed by livestock to reduce shrub regrowth, and thus limits fire risk by depriving the fire of adequate fuel [121,122]. This is particularly common in Mediterranean countries such as France and Spain and has been adopted for decades in some cases, and some government policies actively encourage this system, such as in Catalonia [122]. In addition, AFS involve the incorporation of litter and deadwood pools and, therefore, a lower litter depth. As a result, there are fewer flammable forests due to a lower burn temperature, lower maximum flame height, and reduced flame propagation [121].

The EU Rural Development Programme recognised that agroforestry reduces fire risk and some measures to promote agroforestry centred around this benefit [32]. However, these policy measures were insufficient: although the measures promoted both afforestation and reforestation, payments covered loss of income for afforestation but not reforestation. Furthermore, no payments were provided to promote agricultural activity after trees were planted, and so most of the forestland was not cleared in order to reduce forest fires [123].

### *2.9. Pollination*

Approximately 85% of flowering plants depend on more than 100,000 animal species—mostly insects—for pollination [124,125]. This is particularly important for agricultural crops, where animal pollinators affect two thirds of the 1500 global crop species and are essential for up to 30% of all food production [126]. In Europe, 84% of the 264 crops studied depend on animal pollination to some extent. In a more recent global study, the level of dependence on animal-mediated pollination was calculated for 107 crops, where animal pollination was beneficial for two thirds of crops studied. Furthermore, animal pollination was classed as of high or essential importance for over 40% of the crops studied [127].

Intensively managed agriculture has been recognised as a major cause of loss of global biodiversity [128] due to the conversion of natural systems into agricultural land and agricultural intensification on land already used for agriculture [129]. Animal pollinators are in decline globally, and estimates suggest that up to 40% of invertebrate pollinators are at risk of extinction [130]. The global annual economic value of insect pollination is approximately EUR 153 billion. This is the equivalent of 9.5% of the total economic value of the global agricultural output of all food crops for human consumption [131]. Agroforestry systems have a significant role to play in pollination, but very few studies have quantified the role of agroforestry in pollination. One such study quantified and valued pollination at USD 47 $ha^{-1}$ $year^{-1}$ in combined food and energy agroforestry in Denmark [132]. Despite the significant negative impact on the agriculture industry, the decline in pollinators is mainly due to intensive agriculture [133]. Varah et al. [134] stated that conventional farming methods focus on providing just one ecosystem service—productivity—by reducing environmental complexity by growing intensive monocultures with better economies of scale. By focussing on productivity, many other ecosystem services suffer.

Bentrup et al. [135] provided a detailed review of the effect of temperate agroforestry practices and woody plants on insect pollinators and pollination, focussing on the services they provide pollinators. They found that trees are sources of resin for bees and nectar of high nutritional value and sugar content. Similarly, woody species provide early season pollen and nectar and are hosts to important larvae. The varied woodland structure and reduced disturbance in AFS provide sites for ground nesting, cavity nesting, and overwintering. In addition, the lower exposure to pesticides and runoff in AFS reduces the harmful effects to pollinator populations. The findings were supported by a field study of four silvopastoral and two silvoarable AFS in the UK by Potts et al. [133], where butterfly diversity was significantly higher for AFS than for the control plots (agricultural monocultures growing the same crop as the agroforestry treatment but without the trees). A review by Varah et al. [136] found that temperate AFS provide greater pollination service than monocultures: AFS had twice as many solitary bees and hoverflies, and the species richness of solitary bees was approximately 10.5 times greater in AFS.

Models based on a Swiss field experiment predict enhanced wild bee habitat quality (particularly for cavity-nesting species) and tend to predict increased provision of pollination services by wild bees in landscapes with higher proportions of AFS [137]. In addition, a meta-analysis found that despite a higher abundance of pollinators in AFS compared to arable systems, the abundance of natural enemies and pests was 24% greater, and arthropod herbivore/pest abundance was 25% lower in AFS, suggesting the issue is more complex than simply understanding the abundance or diversity of pollinators [138].

## 3. Opportunities and Challenges for Agroforestry in Europe

In the EU, approximately 20 million ha of land is occupied by AFS, and the majority (90%) are silvopastoral systems [32] located in the Mediterranean area [139]. Aertsens, De Nocker and Gobin [25] estimated that there are still 90 million ha of arable land and 50 million ha of pasture land in the EU that have the potential to be used as agroforestry. AFS provide multiple benefits, viz., provision of ecosystem services, product diversity, enhanced biodiversity and mitigation of GHG emissions and climate resilience, and these multiple functionalities [31,60,140] need to be taken into account in assessing the AFS. AFS provide a good compromise between the input-intensive monoculture for high productivity and nature-based or conservation agriculture with lower productivity, for production of food and non-food products. Due to the diversity of crop and tree species in AFS, yields are more stable in cases of adverse climatic conditions [141,142], reducing the risk exposure for the farmers. Further, there is better nutrient cycling due to the complementarity in time and space in resource use between the crop and tree species, enhancing the resource use and cycling [143] within the production system. AFS are not without challenges due to various reasons. If AFS need to be optimized for productivity [144], there is a need to balance the crop and tree population, spacing and cropped area in order to optimize the complementarity between the species. There is a need for on-farm demonstrations and robust field-based evidence of AFS in the relevant socioeconomic settings to match the local demand for produce with the supply from the AFS. These activities need to be matched with agriculture advisors to advise the farmers and policy changes in order to influence the policy-making bodies to bring forth and communicate the multiple benefits of AFS from local to global scale. Hence, we highlight some of the key barriers to implementation, including the costs for the implementation of agroforestry systems, lack of financial incentives, marketing, and awareness/education.

### 3.1. High Costs for Implementation

High costs are common barriers to implementation of new farming practices and production techniques [145]. In the last 10–15 years, the EU has provided several voluntary and mandatory policy instruments (Table 1) for financial support to farmers to promote agroforestry implementation [32,146,147]. Low profitability is still a major concern of farmers and farm managers considering implementation of agroforestry [147,148]. AFS are labour-intensive and the high cost of labour in Europe can be a considerable deterrent for people considering adopting AFS [149]. Also, complementarity in temporal and spatial nutrient, water, and radiation use amongst the AFS components is the main reason for increased yields, and if not managed well, the competition amongst the AFS components can decrease yields and cause financial risk [22,148]. Garcia de Jalon et al. [19] modelled and valued the environmental impacts of arable, forestry, and agroforestry systems in Europe. They found that AFS were the second most profitable system after arable, where the financial equivalent annual value with grants was EUR 559 $ha^{-1}$ $yr^{-1}$ for arable and EUR 364 $ha^{-1}$ $yr^{-1}$ for AFS. The forestry system had a much lower financial equivalent annual value of EUR 194 $ha^{-1}$ $yr^{-1}$ with grants. However, in a 20-year field study of combined food and energy production systems in Denmark, calculations on agronomic productivity, costs of production, and revenue flows showed that agroforestry systems had higher net present values compared to monocrops [17]. Further, sensitivity analysis was carried out with different discount rates between 0–10%, and agroforestry systems were most insensitive to the discount rate and provided a stable and consistent net present value [17]. This study demonstrated the need for more field-based studies to demystify the notion that agroforestry systems are economically unviable.

**Table 1.** Overview of EU policy instruments relevant to agroforestry systems (AFS).

| Policy | Relevance to AFS |
|---|---|
| Common Agricultural Policy (CAP) | Payments to farmers fulfilling 13 Statutory Management Requirements related to environment, climate change, and good agricultural condition land. |
| European Biodiversity Strategy to 2020 | Aims to halt deterioration and improve the status of species and habitats covered by EU nature legislation. Linked to CAP and Natura 2000. |
| Natura 2000—Habitats and Birds Directives | Aims to protect threatened species and associated habitats. Network of special protection areas (SPAs) and sites of Community importance (SCIs). Several EU funds available. |
| European Strategy for Sustainable Development | Broad strategy involving many sectors. One objective is to manage natural resources responsibly and protect habitats and ecosystems, as well as halt the loss of biodiversity. |
| European Climate Change Programme (ECCP) | EU strategy to implement the Kyoto Protocol. Measures to improve energy efficiency and reduce greenhouse gas emissions. |
| European Forest Strategy | Aims to ensure sustainable and balanced forest management. |

### 3.2. Inadequate Financial Incentives

Financial incentives as compensation for provision of ecosystem services from agroforestry are lacking [148]. Payment for ecosystem services is an approach where land stewards such as farmers receive compensation for protecting and restoring ecosystems [150]. Although the EU promotes agroforestry through Pillars I and II of the Common Agricultural Policy (CAP, Table 1), this applies only for establishing new AFS and not those already existing [147], a disincentive to maintaining the AFS for those already practising AFS. Furthermore, the EU Emission Trading system does not currently allow trading with carbon credits for land-use or forestry systems [151]. Thus, farmers cannot benefit from selling carbon credits [151] despite the high C sequestration potential of AFS [25,37]. Subsidies and payments for ecosystem services to landowners who already practice agroforestry might mitigate some of the financial risk and encourage farmers and farm managers to maintain AFS in the long term [147].

Of the 13 Statutory Management Requirements under CAP, the first four—protection of water against pollution caused by nitrates; conservation of wild birds; conservation of natural habitats and of wild flora and fauna; and food and feed law [32]—are most relevant to AFS. The Natura 2000 sites are supported through the CAP Pillars 1 and 2, and other EU funds such as the Programme for the Environment and Climate Action and European Structural Funds, viz., Regional Development Fund, Cohesion Fund, and Social Fund [32]. Because of the multifunctional benefits of AFS, AFS are potential candidates to meet several national- and EU-level agricultural policy measures, viz., Nitrate Directive, Water Framework Directive, Greening Measures, Birds and Habitat Directive, Agri-Environmental Schemes, and EU2030 Climate and Energy Framework. Recognizing the role of AFS in meeting these measures can provide a significant boost to adoption of AFS by the farmers, which will bring them economic benefits in the short run and environmental benefits at the EU and global scale [152].

### *3.3. Products and Marketing*

Agroforestry produces a diversity of food and non-food products, providing multiple income streams for the farming households or farm manager. This diversity contributes to stable incomes as loss in market value of one product can be compensated by better prices of other products. AFS can provide products such as timber, crops, fruits, nuts, mushrooms, forages, livestock, biomass, Christmas trees, and herbal medicine [113]. Moreover, Wilson and Lovell [113] stated that "a diverse portfolio of products would allow revenue streams to be spread out over the short-term (crops, forage, livestock, mushrooms, certain fruits like currants), medium-term (nuts, fruits such as apples or persimmons, biomass, medicinal plants), and long-term (lumber, increased property value)". Income stability is an insurance against complete loss of income from monoculture with adverse climatic conditions, which is a clear benefit from adoption of agroforestry [17]. Although the complexity of managing AFS is a concern, the diversity of produce is seen by landowners as a benefit of land conversion to AFS, as long as these are financially viable [19,22,153]. However, in a survey of landowners and professionals, the lack of markets and lack of market information were the second highest concerns when considering adoption to AFS [154].

High quality products that have traditionally been products of AFS are of particularly high value, for example, Iberian pig ham from Iberian dehesas or reindeer meat in the boreal forest and subarctic tundra zone [88,155], and so landowners should focus on more marketable goods such as these. New needs for natural and high quality products derived from AFS need also to be explored, such as tannins for tanning leather and antioxidants or gluten-free flours [156]. Increasing the portfolio of AFS products, coupled with improved marketing of agroforestry products, [147] is essential. Increased awareness amongst the general public can create incentives for consumers to pay premium prices for agroforestry products and thus provide further incentives for farmers and farm managers [22,147]. For example, certification schemes, such as that of the Rainforest Alliance, have been demonstrated to improve the financial stability of farmers who adopted AFS with shaded coffee plantations in Latin America, which increased coffee yields and provided additional profit [157]. Similar schemes for agroforestry products from Europe might increase awareness amongst the public and provide a potential financial incentive for farmers to adopt AFS.

### *3.4. Education and Awareness*

Many traditional AFS in Europe disappeared in the 20th century due to intensification and mechanisation of agriculture [33], and the disappearance of these traditional systems in Europe resulted in a loss of the knowledge base amongst farmers [139]. This lack of knowledge and advisory service support in the implementation and management of agroforestry systems is a significant barrier to implementation [158]. The choice of tree and crop species and knowledge of their complementarity in resource use influence the profitability of AFS [50], and education and information sharing can be useful in helping farmers choose species that together enhance yields without compromising the main crop [50,147,148]. For example, there is an increasing demand for high quality hardwood timber production in Europe [149], which could pose a potential opportunity for profit from adding trees to agricultural lands. Education programmes and exchange opportunities amongst farmers can be useful tools to achieve successful implementation of profitable AFS [147]. Currently, awareness amongst farmers of the benefits of agroforestry is lacking and to overcome the barriers to implementation of AFS in Europe, raising awareness and experiential learning through exchange of experiences amongst farmers and the general public is necessary [22].

### *3.5. Field Demonstrations*

Local dissemination activities may consist of activities such as workshops, role-play activities, and focus groups, although the most promising dissemination activity is the use of field demonstrations. Among other challenges, field demonstrations on agroforestry systems are very few and scattered. Hence, there is a clear need to augment the robust scientific evidence on the multiple benefits of agroforestry with on-the-ground demonstrations. The demonstrations can serve as the training platform for the farmers, advisory services, and policy-making bodies and to teach students in the real environments about the multifunctionality of agroforestry systems. In such demonstration sites, training of trainers courses can be organized to provide insights into the costs and revenues generated and preparation of business plans, in order to convince the current and the potential farmers of the economics of agroforestry adoption and its beneficial effects on the environment.

When communication—particularly in a multilingual situation—may be an issue, role-play can precede the field demonstration to ensure that those running the demonstration have the appropriate technical knowledge and confidence that are essential to a successful field demonstration. Moreover, the field day can help to identify management problems, confirm accuracy of data that has been collected, ensure that landowners' felt needs are properly addressed, expose landowners to new management options and give them an opportunity to evaluate these, and to select more landowners for future on-site testing/field demonstrations. Thus landowners will have more ownership of the problem and recognise their important role in the overall research and dissemination process. Extension agents, too, own the problem and are taking part in finding the solution [159]. Workman et al. [151] came to similar conclusions: in a survey of landowners and extension professionals, the lack of demonstrations and lack of familiarity with the practices were cited as major obstacles to the adoption of agroforestry practices. Adesina and Chianu [160] found that the limited amount of contact with extension agents (e.g., at demonstration days) was a barrier to AFS implementation for landowners. However, Arbuckle et al. [161] found that only 16% of AFS landowners had attended a field demonstration in the last two years, whilst double the number of landowners sought advice from a professional in the past two years. Thus, the issue is twofold: (i) only a third of farmers are seeking professional advice on their AFS, and (ii) field demonstrations are not appealing and/or not readily available.

## **4. Conclusions**

In Europe, there is a great potential for AFS to improve agronomic productivity, carbon sequestration, nutrient cycling, soil biodiversity, water retention, and pollination; to reduce soil erosion and fire; and to provide recreational and cultural benefits. However, there are several challenges due to the high costs of implementation, lack of financial incentives, inadequate product marketing and pricing, lack of awareness and education, and lack of field demonstrations. These challenges need to be addressed in order to encourage farmers to adopt AFS.

**Author Contributions:** M.S.-N. wrote the preliminary draft of the paper. N.L.J.R. made significant edits to the draft, expanding on original text, adding new sections where necessary and providing many additional references throughout. B.B.G. made significant improvements to the draft and improved the scientific content of the manuscript. All authors have read and agreed to the published version of the manuscript.

**Acknowledgments:** The support from SUSTAINFARM (grant agreement no. 652615) and WATERFARMING (grant agreement no: 689271) projects for revision to improve the scientific content of the manuscript is acknowledged.

## References

1. Lal, R. Managing Soils and Ecosystems for Mitigating Anthropogenic Carbon Emissions and Advancing Global Food Security. *Bioscience* **2010**, *60*, 708–721. [CrossRef]
2. Ghaley, B.B.; Vesterdal, L.; Porter, J.R. Quantification and valuation of ecosystem services in diverse production systems for informed decision-making. *Environ. Sci. Policy* **2014**, *39*, 139–149. [CrossRef]
3. Ghaley, B.B.; Rusu, T.; Sandén, T.; Spiegel, H.; Menta, C.; Visioli, G.; O'Sullivan, L.; Gattin, I.T.; Delgado, A.; Liebig, M.A.; et al. Assessment of Benefits of Conservation Agriculture on Soil Functions in Arable Production Systems in Europe. *Sustainability* **2018**, *10*, 794. [CrossRef]
4. Janssens, I.A.; Freibauer, A.; Schlamadinger, B.; Ceulemans, R.; Ciais, P.; Dolman, A.J.; Heimann, M.; Nabuurs, G.-J.; Smith, P.; Valentini, R. The carbon budget of terrestrial ecosystems at country-scale—A European case study. *Biogeosci. Discuss.* **2005**, *1*, 167–193. [CrossRef]
5. Blanco-Canqui, H.; Lal, R. Mismanagement of Cultivated Lands. In *Principles of Soil Conservation and Management*; Springer: Dordrecht, The Netherlands, 2010; p. 10. [CrossRef]
6. Usman, M.; Ibrahim, F.; Oyetola, S.O. Sustainable agriculture in relation to problems of soil degradation and how to amend such soils for optimum crop production in Nigeria. *Int. J. Res. Agric. Food Sci.* **2018**, *4*, 1–17.
7. Smith, J.; Pearce, B.D.; Wolfe, M.S. A European perspective for developing modern multifunctional agroforestry systems for sustainable intensification. *Renew. Agric. Food Syst.* **2012**, *27*, 323–332. [CrossRef]
8. Ilany, T.; Ashton, M.S.; Montagnini, F.; Martínez, C. Using agroforestry to improve soil fertility: Effects of intercropping on Ilex paraguariensis (yerba mate) plantations with Araucaria angustifolia. *Agrofor. Syst.* **2010**, *80*, 399–409. [CrossRef]
9. Mbow, C.; Smith, P.; Skole, D.; Duguma, L.A.; Bustamante, M.M.C. Achieving mitigation and adaptation to climate change through sustainable agroforestry practices in Africa. *Curr. Opin. Environ. Sustain.* **2014**, *6*, 8–14. [CrossRef]
10. Mosquera-Losada, M.; Santiago-Freijanes, J.; Rois-Díaz, M.; Moreno, G.; Herder, M.D.; Aldrey-Vázquez, J.; Ferreiro-Domínguez, N.; Pantera, A.; Pisanelli, A.; Rigueiro-Rodríguez, A. Agroforestry in Europe: A land management policy tool to combat climate change. *Land Use Policy* **2018**, *78*, 603–613. [CrossRef]
11. Fagerholm, N.; Torralba, M.; Burgess, P.J.; Plieninger, T. A systematic map of ecosystem services assessments around European agroforestry. *Ecol. Indic.* **2016**, *62*, 47–65. [CrossRef]
12. McAdam, J.H.; Burgess, P.J.; Graves, A.; Rigueiro-Rodríguez, A. Classifications and Functions of Agroforestry Systems in Europe. In *Agroforestry in Europe: Current Status and Future Propspects*; Rigueiro-Rodríguez, A., McAdam, J., Mosquera-Losada, M.R., Eds.; Springer: Dordrecht, The Netherlands, 2009; pp. 21–41. [CrossRef]
13. Mead, R.; Willey, R.W. The Concept of a 'Land Equivalent Ratio' and Advantages in Yields from Intercropping. *Exp. Agric.* **1980**, *16*, 217–228. [CrossRef]
14. FAO. *Guidelines: Land Evaluation for Rainfed Agriculture, Soils Bulletin 55*; FAO: Rome, Italy, 1985.
15. Malezieux, E.; Crozat, Y.; Dupraz, C.; Laurans, M.; Makowski, D.; Ozier-Lafontaine, H.; Rapidell, B.; de Tourdonnet, S.; Valentin-Morison, M. Mixing plant species in cropping systems: Concepts, tools and models: A review. *Agron. Sustain. Dev.* **2009**, *29*, 43–62. [CrossRef]
16. Lovell, S.T.; Dupraz, C.; Gold, M.; Jose, S.; Revord, R.; Stanek, E.; Wolz, K.J. Temperate agroforestry research: Considering multifunctional woody polycultures and the design of long-term field trials. *Agrofor. Syst.* **2017**, *92*, 1397–1415. [CrossRef]

17. Xu, Y.; Lehmann, L.M.; De Jalón, S.G.; Ghaley, B.B. Assessment of Productivity and Economic Viability of Combined Food and Energy (CFE) Production System in Denmark. *Energies* **2019**, *12*, 166. [CrossRef]
18. Lehmann, L.M.; Borzecka-Walker, M.; Żyłowska, K.; Pisanelli, A.; Russo, G.; Ghaley, B.B. Environmental Impact Assessments of Integrated Food and Non-Food Production Systems in Italy and Denmark. *Energies* **2020**, *13*, 849. [CrossRef]
19. García De Jalón, S.; Burgess, P.J.; Graves, A.; Moreno, G.; McAdam, J.; Pottier, E.; Novak, S.; Bondesan, V.; Mosquera-Losada, M.; Crous-Duran, J.; et al. How is agroforestry perceived in Europe? An assessment of positive and negative aspects by stakeholders. *Agrofor. Syst.* **2018**, *92*, 829–848. [CrossRef]
20. Sereke, F.; Graves, A.R.; Dux, D.; Palma, J.; Herzog, F. Innovative agroecosystem goods and services: Key profitability drivers in Swiss agroforestry. *Agron. Sustain. Dev.* **2014**, *35*, 759–770. [CrossRef]
21. Trozzo, K.E.; Munsell, J.F.; Chamberlain, J.L. Landowner interest in multifunctional agroforestry Riparian buffers. *Agrofor. Syst.* **2014**, *88*, 619–629. [CrossRef]
22. Rois-Díaz, M.; Lovrić, N.; Lovric, M.; Ferreiro-Domínguez, N.; Mosquera-Losada, M.; Herder, M.D.; Graves, A.; Palma, J.; Paulo, J.A.; Pisanelli, A.; et al. Farmers' reasoning behind the uptake of agroforestry practices: Evidence from multiple case-studies across Europe. *Agrofor. Syst.* **2017**, *92*, 811–828. [CrossRef]
23. Hall, C.; Dawson, T.P.; MacDiarmid, J.I.; Matthews, R.; Smith, P. The impact of population growth and climate change on food security in Africa: Looking ahead to 2050. *Int. J. Agric. Sustain.* **2017**, *15*, 124–135. [CrossRef]
24. Lal, R. Soil Carbon Sequestration Impacts on Global Climate Change and Food Security. *Science* **2004**, *304*, 1623–1627. [CrossRef] [PubMed]
25. Aertsens, J.; De Nocker, L.; Gobin, A. Valuing the carbon sequestration potential for European agriculture. *Land Use Policy* **2013**, *31*, 584–594. [CrossRef]
26. Mondelaers, K.; Aertsens, J.; Van Huylenbroeck, G. A meta-analysis of the differences in environmental impacts between organic and conventional farming. *Br. Food J.* **2009**, *111*, 1098–1119. [CrossRef]
27. Ghaley, B.B.; Porter, J.R. Ecosystem function and service quantification and valuation in a conventional winter wheat production system with DAISY model in Denmark. *Ecosyst. Serv.* **2014**, *10*, 79–83. [CrossRef]
28. Pardon, P.; Reubens, B.; Reheul, D.; Mertens, J.; De Frenne, P.; Coussement, T.; Janssens, P.; Verheyen, K. Trees increase soil organic carbon and nutrient availability in temperate agroforestry systems. *Agric. Ecosyst. Environ.* **2017**, *247*, 98–111. [CrossRef]
29. Bambrick, A.D.; Whalen, J.K.; Bradley, R.L.; Cogliastro, A.; Gordon, A.M.; Olivier, A.; Thevathasan, N.V. Spatial heterogeneity of soil organic carbon in tree-based intercropping systems in Quebec and Ontario, Canada. *Agrofor. Syst.* **2010**, *79*, 343–353. [CrossRef]
30. Bazrgar, A.; Ng, A.; Coleman, B.; Ashiq, M.W.; Gordon, A.; Thevathasan, N.V. Long-Term Monitoring of Soil Carbon Sequestration in Woody and Herbaceous Bioenergy Crop Production Systems on Marginal Lands in Southern Ontario, Canada. *Sustainability* **2020**, *12*, 3901. [CrossRef]
31. Kay, S.; Graves, A.; Palma, J.H.; Moreno, G.; Roces-Díaz, J.V.; Aviron, S.; Chouvardas, D.; Crous-Duran, J.; Ferreiro-Domínguez, N.; De Jalón, S.G.; et al. Agroforestry is paying off—Economic evaluation of ecosystem services in European landscapes with and without agroforestry systems. *Ecosyst. Serv.* **2019**, *36*, 100896. [CrossRef]
32. Mosquera-Losada, M.R.; Santiago Freijanes, J.J.; Pisanelli, A.; Rois, M.; Smith, J.; den Herder, M.; Moreno, G.; Malignier, N.; Mirazo, J.; Lamersdorf, N.; et al. *Extent and Success of Current Policy Measures to Promote Agroforestry across Europe, Deliverable 8.23 for EU FP7 Research Project: AGFORWARD 613520*. 2016. Available online: https://www.agforward.eu/index.php/en/extent-and-success-of-current-policy-measures-to-promote-agroforestry-across-europe.html?file=files/agforward/documents/Deliverable8.23Extent_and_Success_of_Current_Policy_Measures_8_Dec_2016.pdf (accessed on 25 August 2020).
33. Quinkenstein, A.; Wöllecke, J.; Böhm, C.; Grünewald, H.; Freese, D.; Schneider, B.U.; Hüttl, R.F. Ecological benefits of the alley cropping agroforestry system in sensitive regions of Europe. *Environ. Sci. Policy* **2009**, *12*, 1112–1121. [CrossRef]
34. Giraldo, C.; Escobar, F.; Chara, J.; Calle, Z. The adoption of silvopastoral systems promotes the recovery of ecological processes regulated by dung beetles in the Colombian Andes. *Insect Conserv. Divers.* **2010**, *4*, 115–122. [CrossRef]
35. Kim, D.-G.; Kirschbaum, M.U.; Beedy, T.L. Carbon sequestration and net emissions of CH4 and N2O under agroforestry: Synthesizing available data and suggestions for future studies. *Agric. Ecosyst. Environ.* **2016**, *226*, 65–78. [CrossRef]

36. Negash, M.; Kanninen, M. Modeling biomass and soil carbon sequestration of indigenous agroforestry systems using CO2FIX approach. *Agric. Ecosyst. Environ.* **2015**, *203*, 147–155. [CrossRef]
37. Mosquera-Losada, M.R.; Freese, D.; Rigueiro-Rodríguez, A. *Carbon Sequestration in European Agroforestry Systems*; Springer: Dordrencht, The Netherlands, 2011; pp. 43–59. [CrossRef]
38. Dossa, E.L.; Fernandes, E.C.M.; Reid, W.S.; Ezui, K. Above- and belowground biomass, nutrient and carbon stocks contrasting an open-grown and a shaded coffee plantation. *Agrofor. Syst.* **2008**, *72*, 103–115. [CrossRef]
39. Tscharntke, T.; Clough, Y.; Bhagwat, S.A.; Buchori, D.; Faust, H.; Hertel, D.; Holscher, D.; Juhrbandt, J.; Kessler, M.; Perfecto, I.; et al. Multifunctional shade-tree management in tropical agroforestry landscapes—A review. *J. of App. Eco.* **2011**, *48*, 619–629. [CrossRef]
40. Nair, P.K.R. Agroforestry Systems and Environmental Quality: Introduction. *J. Environ. Qual.* **2011**, *40*, 784–790. [CrossRef] [PubMed]
41. Rivest, D.; Lorente, M.; Olivier, A.; Messier, C. Soil biochemical properties and microbial resilience in agroforestry systems: Effects on wheat growth under controlled drought and flooding conditions. *Sci. Total Environ.* **2013**, *463*, 51–60. [CrossRef] [PubMed]
42. Jose, S. Agroforestry for ecosystem services and environmental benefits: An overview. *Agrofor. Syst.* **2009**, *76*, 1–10. [CrossRef]
43. Jose, S.; Gillespie, A.; Pallardy, S. Interspecific interactions in temperate agroforestry. *Agrofor. Syst.* **2004**, *61–62*, 237–255. [CrossRef]
44. Thevathasan, N.; Gordon, A. Ecology of tree intercropping systems in the North temperate region: Experiences from southern Ontario, Canada. *Agrofor. Syst.* **2004**, *61–62*, 257–268. [CrossRef]
45. Zhang, P. The Impact of Nutrient Inputs from Stemflow, Throughfall, and Litterfall in a Tree-Based Temperate Intercrop-Ping System, Southern ON, Canada. Master's Thesis, Department of Environmental Biology, University of Guelph, Guelph, ON, Canada, 1999.
46. Torralba, M.; Fagerholm, N.; Burgess, P.; Moreno, G.; Plieninger, T. Do European agroforestry systems enhance biodiversity and ecosystem services? A meta-analysis. *Agric. Ecosyst. Environ.* **2016**, *230*, 150–161. [CrossRef]
47. Swieter, A.; Langhof, M.; Lamerre, J.; Greef, J.M. Long-term yields of oilseed rape and winter wheat in a short rotation alley cropping agroforestry system. *Agrofor. Syst.* **2018**, *93*, 1853–1864. [CrossRef]
48. Gao, L.; Xu, H.; Bi, H.; Xi, W.; Bao, B.; Wang, X.; Bi, C.; Chang, Y. Intercropping Competition between Apple Trees and Crops in Agroforestry Systems on the Loess Plateau of China. *PLoS ONE* **2013**, *8*, e70739. [CrossRef] [PubMed]
49. Dufour, L.; Metay, A.; Talbot, G.; Dupraz, C. Assessing Light Competition for Cereal Production in Temperate Agroforestry Systems using Experimentation and Crop Modelling. *J. Agron. Crop. Sci.* **2012**, *199*, 217–227. [CrossRef]
50. Palma, J.; Graves, A.; Burgess, P.J.; Van Der Werf, W.; Herzog, F. Integrating environmental and economic performance to assess modern silvoarable agroforestry in Europe. *Ecol. Econ.* **2007**, *63*, 759–767. [CrossRef]
51. Wagg, C.; Bender, S.F.; Widmer, F.; Van Der Heijden, M.G.A. Soil biodiversity and soil community composition determine ecosystem multifunctionality. *Proc. Natl. Acad. Sci. USA* **2014**, *111*, 5266–5270. [CrossRef] [PubMed]
52. Kibblewhite, M.; Ritz, K.; Swift, M. Soil health in agricultural systems. *Philos. Trans. R. Soc. B Biol. Sci.* **2007**, *363*, 685–701. [CrossRef] [PubMed]
53. Coleman, D.C.; Crossley, D.; Hendrix, P.F. *Fundamentals of Soil Ecology*, 2nd ed.; Elsevier BV: Burlington, NJ, USA, 2004.
54. Havlicek, E.; Mitchell, E.A.D. Soils Suppressing Biodiversity. In *Interactions in Soil: Promoting Plant Growth*; John, D., Jennifer, A.K., Eds.; Springer: Dordrecht, The Netherland, 2014; pp. 95–118.
55. Barrios, E.; Sileshi, G.W.; Shepherd, K.D.; Sinclair, F. Agroforestry and Soil Health: Linking Trees, Soil Biota, and Ecosystem Services. In *Soil Ecology and Ecosystem Services*; Diana, H.W., Richard, D.B., Valerie, B.-P., Jeffrey, E.H., Hefin, J.T., Eds.; Oxford University Press (OUP): Oxford, UK, 2012.
56. Udawatta, R.P.; Kremer, R.J.; Adamson, B.W.; Anderson, S.H. Variations in soil aggregate stability and enzyme activities in a temperate agroforestry practice. *Appl. Soil Ecol.* **2008**, *39*, 153–160. [CrossRef]
57. Lacombe, S.; Bradley, R.L.; Hamel, C.; Beaulieu, C. Do tree-based intercropping systems increase the diversity and stability of soil microbial communities? *Agric. Ecosyst. Environ.* **2009**, *131*, 25–31. [CrossRef]

58. Chifflot, V.; Rivest, D.; Olivier, A.; Cogliastro, A.; Khasa, D. Molecular analysis of arbuscular mycorrhizal community structure and spores distribution in tree-based intercropping and forest systems. *Agric. Ecosyst. Environ.* **2009**, *131*, 32–39. [CrossRef]
59. Schädler, M.; Brandl, R.; Kempel, A. "Afterlife" effects of mycorrhization on the decomposition of plant residues. *Soil Biol. Biochem.* **2010**, *42*, 521–523. [CrossRef]
60. Marsden, C.; Martin-Chave, A.; Cortet, J.; Hedde, M.; Capowiez, Y. How agroforestry systems influence soil fauna and their functions—A review. *Plant Soil* **2019**, 1–16. [CrossRef]
61. Freedman, A.J.; Tan, B.; Thompson, J.R. Microbial potential for carbon and nutrient cycling in a geogenic supercritical carbon dioxide reservoir. *Environ. Microbiol.* **2017**, *19*, 2228–2245. [CrossRef] [PubMed]
62. Dollinger, J.; Jose, S. Agroforestry for soil health. *Agrofor. Syst.* **2018**, *92*, 1–7. [CrossRef]
63. Nair, P.K.R.; Buresh, R.J.; Mugendi, D.N.; Latt, C.R. Nutrient cycling in tropical agroforestry systems: Myths and science. In *Agroforestry in Sustainable Agricultural Systems*; Buck, L.E., Lassoie, J.P., Fernandez, E.C.M., Eds.; CRC Press: Boca Raton, FL, USA, 1999; pp. 1–31. [CrossRef]
64. Barea, J.M.; Pozo, M.J.; Azcón, R.; Azcón-Aguilar, C. Microbial co-operation in the rhizosphere. *J. Exp. Bot.* **2005**, *56*, 1761–1778. [CrossRef]
65. De Araújo, A.S.F.; Leite, L.F.C.; Iwata, B.D.F.; Lira, M.D.A.; Xavier, G.R.; Figueiredo, M.D.V.B. Microbiological process in agroforestry systems. A review. *Agron. Sustain. Dev.* **2011**, *32*, 215–226. [CrossRef]
66. Bainard, L.D.; Klironomos, J.N.; Gordon, A.M. Arbuscular mycorrhizal fungi in tree-based intercropping systems: A review of their abundance and diversity. *Pedobiologia* **2011**, *54*, 57–61. [CrossRef]
67. Rintoul, N. Arbuscular mycorrhizal associations in plant nutrition and health. *CAB Rev. Perspect. Agric. Veter-Sci. Nutr. Nat. Resour.* **2016**, *11*, 1–16. [CrossRef]
68. Muchane, M.N.; Sileshi, G.W.; Gripenberg, S.; Jonsson, M.; Pumariño, L.; Barrios, E. Agroforestry boosts soil health in the humid and sub-humid tropics: A meta-analysis. *Agric. Ecosyst. Environ.* **2020**, *295*, 106899. [CrossRef]
69. Xu, X.; Thornton, P.E.; Post, W.M. A global analysis of soil microbial biomass carbon, nitrogen and phosphorus in terrestrial ecosystems. *Glob. Ecol. Biogeogr.* **2012**, *22*, 737–749. [CrossRef]
70. Cardinael, R.; Hoeffner, K.; Chenu, C.; Chevallier, T.; Béral, C.; Dewisme, A.; Cluzeau, D. Spatial variation of earthworm communities and soil organic carbon in temperate agroforestry. *Biol. Fertil. Soils* **2018**, *55*, 171–183. [CrossRef]
71. Price, G.W.; Gordon, A.M. Spatial and temporal distribution of earthworms in a temperate intercropping system in southern Ontario, Canada. *Agrofor. Syst.* **1998**, *44*, 141–149. [CrossRef]
72. Mupeyo, B.; Barry, T.; Pomroy, W.E.; Ramírez-Restrepo, C.; Lopez-Villalobos, N.; Pernthaner, A. Effects of feeding willow (Salix spp.) upon death of established parasites and parasite fecundity. *Anim. Feed Sci. Technol.* **2011**, *164*, 8–20. [CrossRef]
73. Banerjee, S.; Baah-Acheamfour, M.; Carlyle, C.N.; Bissett, A.; Richardson, A.E.; Siddique, T.; Bork, E.W.; Chang, S.X. Determinants of bacterial communities in Canadian agroforestry systems. *Environ. Microbiol.* **2015**, *18*, 1805–1816. [CrossRef] [PubMed]
74. Beuschel, R.; Piepho, H.-P.; Joergensen, R.G.; Wachendorf, C. Correction to: Similar spatial patterns of soil quality indicators in three poplar-based silvo-arable alley cropping systems in Germany. *Biol. Fertil. Soils* **2018**, *55*, 15–16. [CrossRef]
75. Beule, L.; Corre, M.D.; Schmidt, M.; Göbel, L.; Veldkamp, E.; Karlovsky, P. Conversion of monoculture cropland and open grassland to agroforestry alters the abundance of soil bacteria, fungi and soil-N-cycling genes. *PLoS ONE* **2019**, *14*, e0218779. [CrossRef]
76. Dosskey, M.G. Toward Quantifying Water Pollution Abatement in Response to Installing Buffers on Crop Land. *Environ. Manag.* **2001**, *28*, 577–598. [CrossRef]
77. Udawatta, R.P.; Krstansky, J.J.; Henderson, G.S.; Garrett, H.E. Agroforestry Practices, Runoff, and Nutrient Loss. *J. Environ. Qual.* **2002**, *31*, 1214–1225. [CrossRef]
78. Udawatta, R.P.; Garrett, H.E.; Kallenbach, R.L. Agroforestry and grass buffer effects on water quality in grazed pastures. *Agrofor. Syst.* **2010**, *79*, 81–87. [CrossRef]
79. Sepúlveda, R.B.; Carrillo, A.A. Soil erosion and erosion thresholds in an agroforestry system of coffee (Coffea arabica) and mixed shade trees (*Inga* spp. and *Musa* spp.) in Northern Nicaragua. *Agric. Ecosyst. Environ.* **2015**, *210*, 25–35. [CrossRef]

80. Mutegi, J.K.; Mugendi, D.N.; Verchot, L.V.; Kung'U, J.B. Combining napier grass with leguminous shrubs in contour hedgerows controls soil erosion without competing with crops. *Agrofor. Syst.* **2008**, *74*, 37–49. [CrossRef]
81. Carroll, Z.; Bird, S.; Emmett, B.; Reynolds, B.; Sinclair, F. Can tree shelterbelts on agricultural land reduce flood risk? *Soil Use Manag.* **2006**, *20*, 357–359. [CrossRef]
82. Nair, P.K.R.; Garrity, D. *Agroforestry—The Future of Global Land Use*; Springer: Dordrecht, The Netherlands, 2012.
83. Lin, B.B. The role of agroforestry in reducing water loss through soil evaporation and crop transpiration in coffee agroecosystems. *Agric. For. Meteorol.* **2010**, *150*, 510–518. [CrossRef]
84. Sajjapongse, A.; Zhu, Q.; Chen, Y.; Wang, H. Development of sustainable agriculture on sloping lands in China. In Proceedings of the 12th ISCO Conference, Bejing, China, 26–31 May 2002; pp. 335–341.
85. Budelman, A. The performance of selected leaf mulches in temperature reduction and moisture conservation in the upper soil stratum. *Agrofor. Syst.* **1989**, *8*, 53–66. [CrossRef]
86. Schwendenmann, L.; Veldkamp, E.; Moser, G.; Hölscher, D.; Köhler, M.; Clough, Y.; Anas, I.; Djajakirana, G.; Erasmi, S.; Hertel, D.; et al. Effects of an experimental drought on the functioning of a cacao agroforestry system, Sulawesi, Indonesia. *Glob. Chang. Biol.* **2010**, *16*, 1515–1530. [CrossRef]
87. Moberg, F. Jordbruket vid ett Vägskäl: Hur får vi Maten och Planetens Resurser att Räcka Till Alla? *Svenska Sammanfattning av Studien IAASTD*. 2011. Available online: https://www.naturskyddsforeningen.se/sites/default/files/dokument-media/jordbruket_vid_ett_vagskal_-_iaastd.pdf (accessed on 25 August 2020).
88. Moreno, G.; Aviron, S.; Berg, S.; Crous-Duran, J.; Franca, A.; De Jalón, S.G.; Hartel, T.; Mirck, J.; Pantera, A.; Palma, J.; et al. Agroforestry systems of high nature and cultural value in Europe: Provision of commercial goods and other ecosystem services. *Agrofor. Syst.* **2017**, *92*, 877–891. [CrossRef]
89. Ispikoudis, I.; Sioliou, K.M. Cultural aspects of silvopastoral systems. In *Silvopastoralism and Sustainable Management*; Mosquera-Losada, M.R., McAdam, J., Rigueiro-Rodríguez, A., Eds.; CABI Publishing: Wallingford, CT, USA, 2005. [CrossRef]
90. Papanastasis, V.P.; Mantzanas, K.; Dini-Papanastasi, O.; Ispikoudis, I. Traditional Agroforestry Systems and Their Evolution in Greece. In *Advances in Agroforestry*; Rigueiro-Rodríguez, A., McAdam, J., Mosquera-Losada, M.R., Eds.; Springer: Berlin, Germany, 2008; pp. 89–109. [CrossRef]
91. Vera, F.W.M. *Grazing Ecology and Forest History*; CABI: Wallingford, UK, 2000.
92. Barbieri, C.; Valdivia, C. Recreation and agroforestry: Examining new dimensions of multifunctionality in family farms. *J. Rural Stud.* **2010**, *26*, 465–473. [CrossRef]
93. Pardini, A. Agroforestry Systems in Italy: Traditions Towards Modern Management. In *Advances in Agroforestry*; Rigueiro-Rodróguez, A., McAdam, J., Mosquera-Losada, M.R., Eds.; Springer: Dordrecht, The Netherlands, 2009; Volume 6, pp. 255–267.
94. Smith, J. *Agroforestry: Reconciling Production with Protection of the Environment*; The Organic Research Centre: Berkshire, MA, USA, 2010.
95. Mäntymaa, E.; Ovaskainen, V.; Juutinen, A.; Tyrväinen, L. Integrating nature-based tourism and forestry in private lands under heterogeneous visitor preferences for forest attributes. *J. Environ. Plan. Manag.* **2017**, *61*, 724–746. [CrossRef]
96. Pantera, A.; Papadopoulos, A.; Papanastasis, V.P. Valonia oak agroforestry systems in Greece: An overview. *Agrofor. Syst.* **2018**, *92*, 921–931. [CrossRef]
97. Dudek, T. Influence of selected features of forests on forest landscape aesthetic value—Example of se poland. *J. Environ. Eng. Landsc. Manag.* **2018**, *26*, 275–284. [CrossRef]
98. Isted, R. Wood pasture and parkland; overlooked jewels of the English countryside. In *S Silvopastoralism and Sustainable Management Mosquera-Losada*; Rigueiro-Rodríguez, M.R., McAdam, A.J., Eds.; CABI Publishing: Wallingford, CT, USA, 2005. [CrossRef]
99. Pimentel, D.; Acquay, H.; Biltonen, M.; Rice, P.; Silva, M.; Nelson, J.; Lipner, V.; Giordano, S.; Horowitz, A.; D'Amore, M. Environmental and economic costs of pesticide use. *BioScience* **1992**, *42*, 750–760. [CrossRef]
100. Savary, S.; Ficke, A.; Aubertot, J.-N.; Hollier, C. Crop losses due to diseases and their implications for global food production losses and food security. *Food Secur.* **2012**, *4*, 519–537. [CrossRef]
101. Frison, E.A.; Cherfas, J.; Hodgkin, T. Agricultural Biodiversity Is Essential for a Sustainable Improvement in Food and Nutrition Security. *Sustainability* **2011**, *3*, 238–253. [CrossRef]

102. Fernandez-Cornejo, J.; Nehring, R.F.; Sinha, E.N.; Grube, A.; Vialou, A. Assessing Recent Trends in Pesticide Use in US Agriculture. In Proceedings of the Annual Meeting of the Agricultural and Applied Economics Association (AAEA), Milwaukee, WI, USA, 26–28 July 2009.
103. Rossman, A.Y. The impact of invasive fungi on agricultural ecosystems in the United States. *Biol. Invasions* **2008**, *11*, 97–107. [CrossRef]
104. Chaplin, R.E.; O'Rourke, M.E.; Blitzer, E.J.; Kremen, C. A meta-analysis of crop pest and natural enemy response to landscape complexity. *Ecol. Lett.* **2011**, *14*, 922–932. [CrossRef] [PubMed]
105. Bianchi, F.; Booij, C.; Tscharntke, T. Sustainable pest regulation in agricultural landscapes: A review on landscape composition, biodiversity and natural pest control. *Proc. R. Soc. B Biol. Sci.* **2006**, *273*, 1715–1727. [CrossRef] [PubMed]
106. Losey, J.E.; Vaughan, M. The Economic Value of Ecological Services Provided by Insects. *Bioscience* **2006**, *56*, 311. [CrossRef]
107. Pumariño, L.; Sileshi, G.W.; Gripenberg, S.; Kaartinen, R.; Barrios, E.; Muchane, M.N.; Midega, C.; Jonsson, M. Effects of agroforestry on pest, disease and weed control: A meta-analysis. *Basic Appl. Ecol.* **2015**, *16*, 573–582. [CrossRef]
108. Stamps, W.T.; Linit, M.J. Plant diversity and arthropod communities: Implications for temperate agroforestry. *Agrofor. Syst.* **1997**, *39*, 73–89. [CrossRef]
109. Ratnadass, A.; Fernandes, P.; Avelino, J.; Habib, R. Plant species diversity for sustainable management of crop pests and diseases in agroecosystems: A review. *Agron. Sustain. Dev.* **2011**, *32*, 273–303. [CrossRef]
110. Sileshi, G.; Schroth, G.; Rao, M.; Girma, H. Weeds, Diseases, Insect Pests, and Tri-Trophic Interactions in Tropical Agroforestry. In *Ecological Basis of Agroforestry*; Batish, D.R., Kohli, R.K., Jose, S., Singh, H.P., Eds.; CRC Press: London, UK, 2008; pp. 73–94.
111. Letourneau, D.K.; Armbrecht, I.; Rivera, B.S.; Montoya-Lerma, J.; Carmona, E.J.; Daza, M.C.; Escobar-Ramírez, S.; Galindo, V.; Gutiérrez, C.; López, S.; et al. Does plant diversity benefit agroecosystems? A synthetic review. *Ecol. Appl.* **2011**, *21*, 9–21. [CrossRef]
112. Smith, J.; Pearce, B.D.; Wolfe, M.S. Reconciling productivity with protection of the environment: Is temperate agroforestry the answer? *Renew. Agric. Food Syst.* **2012**, *28*, 80–92. [CrossRef]
113. Wilson, M.; Lovell, S.T. Agroforestry—The Next Step in Sustainable and Resilient Agriculture. *Sustainability* **2016**, *8*, 574. [CrossRef]
114. Martin-Chave, A.; Béral, C.; Capowiez, Y. Agroforestry has an impact on nocturnal predation by ground beetles and Opiliones in a temperate organic alley cropping system. *Biol. Control* **2019**, *129*, 128–135. [CrossRef]
115. Iverson, A.; Marín, L.E.; Ennis, K.K.; Gonthier, D.J.; Connor-Barrie, B.T.; Remfert, J.L.; Cardinale, B.J.; Perfecto, I. REVIEW: Do polycultures promote win-wins or trade-offs in agricultural ecosystem services? A meta-analysis. *J. Appl. Ecol.* **2014**, *51*, 1593–1602. [CrossRef]
116. Beule, L.; Lehtsaar, E.; Rathgeb, A.; Karlovsky, P. Crop Diseases and Mycotoxin Accumulation in Temperate Agroforestry Systems. *Sustainability* **2019**, *11*, 2925. [CrossRef]
117. Tilman, D.; Cassman, K.G.; Matson, P.A.; Naylor, R.; Polasky, S. Agricultural sustainability and intensive production practices. *Nature* **2002**, *418*, 671–677. [CrossRef]
118. Artru, S.; Garré, S.; Dupraz, C.; Hiel, M.-P.; Blitz-Frayret, C.; Lassois, L. Impact of spatio-temporal shade dynamics on wheat growth and yield, perspectives for temperate agroforestry. *Eur. J. Agron.* **2017**, *82*, 60–70. [CrossRef]
119. Schroth, G.; Krauss, U.; Gasparotto, L.; Aguilar, J.A.D.; Vohland, K. Pests and diseases in agroforestry systems of the humid tropics. *Agrofor. Syst.* **2000**, *50*, 199–241. [CrossRef]
120. San-Miguel-Ayanz, J.; Durrant, T.; Boca, R.; Libertà, G.; Branco, A.; de Rigo, D.; Ferrari, D.; Maianti, P.; Artes Vivancos, T.; Pfeiffer, H.; et al. *Forest Fires in Europe, Middle East and North Africa*; Publications Office of the European Union: Luxembourg, 2019.
121. Damianidis, C.; Santiago-Freijanes, J.J.; Herder, M.D.; Burgess, P.; Mosquera-Losada, M.R.; Graves, A.; Papadopoulos, A.; Pisanelli, A.; Camilli, F.; Rois-Díaz, M.; et al. Agroforestry as a sustainable land use option to reduce wildfires risk in European Mediterranean areas. *Agrofor. Syst.* **2020**, 1–11. [CrossRef]
122. Dupraz, C.; Lawson, G.J.; Lamersdorf, N.; Papanastasis, V.P.; Rosati, A.; Ruiz-Mirazo, J. Temperate agroforestry: The European way. In *Temperate Agroforestry Systems, 2nd ed*; Gordon, A.M., Newman, S.M., Coleman, B., Eds.; CABI: Wallingford, UK, 2018; pp. 98–152.

123. Santiago-Freijanes, J.; Pisanelli, A.; Rois-Díaz, M.; Aldrey-Vázquez, J.; Rigueiro-Rodríguez, A.; Pantera, A.; Vityi, A.; Lojka, B.; Ferreiro-Domínguez, N.; Mosquera-Losada, M. Agroforestry development in Europe: Policy issues. *Land Use Policy* **2018**, *76*, 144–156. [CrossRef]
124. Ingram, M.; Nabhan, G.C.; Buchmann, S.L. Impending pollination crisis threatens biodiversity and agriculture. *Tropine* **1996**, *7*, 1.
125. Ollerton, J.; Winfree, R.; Tarrant, S. How many flowering plants are pollinated by animals? *Oikos* **2011**, *120*, 321–326. [CrossRef]
126. Kremen, C.; Williams, N.M.; Thorp, R.W. Crop pollination from native bees at risk from agricultural intensification. *Proc. Natl. Acad. Sci. USA* **2002**, *99*, 16812–16816. [CrossRef] [PubMed]
127. Klein, A.; E Vaissière, B.; Cane, J.H.; Steffan-Dewenter, I.; Cunningham, S.A.; Kremen, C.; Tscharntke, T. Importance of pollinators in changing landscapes for world crops. *Proc. R. Soc. B Biol. Sci.* **2006**, *274*, 303–313. [CrossRef] [PubMed]
128. Ormerod, S.J.; Marshall, E.; Kerby, G.; Rushton, S. Meeting the ecological challenges of agricultural change: Editors' introduction. *J. Appl. Ecol.* **2003**, *40*, 939–946. [CrossRef]
129. Attwood, S.J.; Maron, M.; House, A.P.N.; Zammit, C. Do arthropod assemblages display globally consistent responses to intensified agricultural land use and management? *Glob. Ecol. Biogeogr.* **2008**, *17*, 585–599. [CrossRef]
130. IPBES. *The Assessment Report of the Intergovernmental Science-Policy Platform on Biodiversity and Ecosystem Services on Pollinators, Pollination and Food Production*; Potts, S.G., Imperatriz-Fonseca, V.L., Ngo, H.T., Eds.; Secretariat of the Intergovernmental Science-Policy Platform on Biodiversity and Ecosystem Services: Bonn, Germany, 2016. [CrossRef]
131. Gallai, N.; Salles, J.-M.; Settele, J.; Vaissière, B.E. Economic valuation of the vulnerability of world agriculture confronted with pollinator decline. *Ecol. Econ.* **2009**, *68*, 810–821. [CrossRef]
132. Costanza, R.; D'Arge, R.; De Groot, R.; Farber, S.; Grasso, M.; Hannon, B.; Limburg, K.; Naeem, S.; O'Neill, R.V.; Paruelo, J.; et al. The value of the world's ecosystem services and natural capital. *Nature* **1997**, *387*, 253–260. [CrossRef]
133. Potts, S.G.; Biesmeijer, J.C.; Kremen, C.; Neumann, P.; Schweiger, O.; Kunin, W.E. Global pollinator declines: Trends, impacts and drivers. *Trends Ecol. Evol.* **2010**, *25*, 345–353. [CrossRef]
134. Varah, A.; Jones, H.; Smith, J.; Potts, S.G. Enhanced biodiversity and pollination in UK agroforestry systems. *J. Sci. Food Agric.* **2013**, *93*, 2073–2075. [CrossRef]
135. Bentrup, G.; Hopwood, J.; Adamson, N.L.; Vaughan, M. Temperate Agroforestry Systems and Insect Pollinators: A Review. *Forests* **2019**, *10*, 981. [CrossRef]
136. Varah, A.; Jones, H.; Smith, J.; Potts, S.G. Temperate agroforestry systems provide greater pollination service than monoculture. *Agric. Ecosyst. Environ.* **2020**, *301*, 107031. [CrossRef]
137. Kay, S.; Kühn, E.; Albrecht, M.; Sutter, L.; Szerencsits, E.; Herzog, F. Agroforestry can enhance foraging and nesting resources for pollinators with focus on solitary bees at the landscape scale. *Agrofor. Syst.* **2020**, *94*, 379–387. [CrossRef]
138. Staton, T.; Walters, R.J.; Smith, J.; Girling, R. Evaluating the effects of integrating trees into temperate arable systems on pest control and pollination. *Agric. Syst.* **2019**, *176*, 102676. [CrossRef]
139. Mosquera-Losada, M.; Moreno, G.; Pardini, A.; McAdam, J.H.; Papanastasis, V.; Burgess, P.J.; Lamersdorf, N.; Castro, M.; Liagre, F.; Rigueiro-Rodríguez, A. Past, Present and Future of Agroforestry Systems in Europe. In *Advances in Agroforestry*; Nair, P., Garrity, D., Eds.; Springer: Dordrencht, The Netherlands, 2012; Volume 9, pp. 285–312.
140. Marais, Z.E.; Baker, T.; O'Grady, A.P.; England, J.R.; Tinch, D.; Hunt, M.A. A Natural Capital Approach to Agroforestry Decision-Making at the Farm Scale. *Forest* **2019**, *10*, 980. [CrossRef]
141. Ofori, D.; Gyau, A.; Dawson, I.K.; Asaah, E.; Tchoundjeu, Z.; Jamnadass, R. Developing more productive African agroforestry systems and improving food and nutritional security through tree domestication. *Curr. Opin. Environ. Sustain.* **2014**, *6*, 123–127. [CrossRef]
142. Mantino, A.; Volpi, I.; Micci, M.; Pecchioni, G.; Bosco, S.; Dragoni, F.; Mele, M.; Ragaglini, G. Effect of Tree Presence and Soil Characteristics on Soybean Yield and Quality in an Innovative Alley-Cropping System. *Agronomy* **2019**, *10*, 52. [CrossRef]
143. Isaac, M.E.; Borden, K.A. Nutrient acquisition strategies in agroforestry systems. *Plant Soil* **2019**, *444*, 1–19. [CrossRef]

144. Rahman, S.A.; Baral, H.; Sharma, R.; Samsudin, Y.B.; Meyer, M.; Lo, M.; Artati, Y.; Simamora, T.I.; Andini, S.; Leksono, B.; et al. Integrating bioenergy and food production on degraded landscapes in Indonesia for improved socioeconomic and environmental outcomes. *Food Energy Secur.* **2019**, *8*, e00165. [CrossRef]
145. Long, T.B.; Blok, V.; Coninx, I. Barriers to the adoption and diffusion of technological innovations for climate-smart agriculture in Europe: Evidence from the Netherlands, France, Switzerland and Italy. *J. Clean. Prod.* **2016**, *112*, 9–21. [CrossRef]
146. Panagos, P.; Imeson, A.; Meusburger, K.; Borrelli, P.; Poesen, J.; Alewell, C. Soil Conservation in Europe: Wish or Reality? *Land Degrad. Dev.* **2016**, *27*, 1547–1551. [CrossRef]
147. Hernández-Morcillo, M.; Burgess, P.J.; Mirck, J.; Pantera, A.; Plieninger, T. Scanning agroforestry-based solutions for climate change mitigation and adaptation in Europe. *Environ. Sci. Policy* **2018**, *80*, 44–52. [CrossRef]
148. Graves, A.R.; Burgess, P.J.; Liagre, F.; Dupraz, C. Farmer perception of benefits, constraints and opportunities for silvoarable systems. *Outlook Agric.* **2017**, *46*, 74–83. [CrossRef]
149. Eichhorn, M.P.; Paris, P.; Herzog, F.; Incoll, L.D.; Liagre, F.; Mantzanas, K.; Mayus, M.; Moreno, G.; Papanastasis, V.P.; Pilbeam, D.J.; et al. Silvoarable Systems in Europe—Past, Present and Future Prospects. *Agrofor. Syst.* **2006**, *67*, 29–50. [CrossRef]
150. Milder, J.C.; Scherr, S.J.; Bracer, C. Trends and Future Potential of Payment for Ecosystem Services to Alleviate Rural Poverty in Developing Countries. *Ecol. Soc.* **2010**, *15*, 6. [CrossRef]
151. Van Vooren, L.; Reubens, B.; Broekx, S.; Pardon, P.; Reheul, D.; Van Winsen, F.; Verheyen, K.; Wauters, E.; Lauwers, L. Greening and producing: An economic assessment framework for integrating trees in cropping systems. *Agric. Syst.* **2016**, *148*, 44–57. [CrossRef]
152. Schulte, R.; Bampa, F.; Bardy, M.; Coyle, C.; Creamer, R.E.; Fealy, R.; Gardi, C.; Ghaley, B.B.; Jordan, P.; Laudon, H.; et al. Making the Most of Our Land: Managing Soil Functions from Local to Continental Scale. *Front. Environ. Sci.* **2015**, *3*, 1–14. [CrossRef]
153. Graves, A.R.; Burgess, P.J.; Liagre, F.; Pisanelli, A.; Paris, P.; Moreno, G.; Bellido, M.; Mayus, M.; Postma, M.; Schindler, B.; et al. Farmer Perceptions of Silvoarable Systems in Seven European Countries. In *Agroforestry in Europe: Current Status and Future Propspects*; Rigueiro-Rodríguez, A., McAdam, J., Mosquera-Losada, M.R., Eds.; Springer: Dordrecht, The Netherlands, 2009. [CrossRef]
154. Workman, S.W.; Bannister, M.E.; Nair, P. Agroforestry potential in the southeastern United States: Perceptions of landowners and extension professionals. *Agrofor. Syst.* **2003**, *59*, 73–83. [CrossRef]
155. Augère-Granier, M.L. Agroforestry in the European Union. 2020. Available online: https://www.europarl.europa.eu/RegData/etudes/BRIE/2020/651982/EPRS_BRI(2020)651982_EN.pdf (accessed on 25 August 2020).
156. Moreno, G.; Franca, A.; Pnto Correia, M.T.; Godinho, S. Multifunctionality and dynamics of silvopastoral systems. *Options Méditerr.* **2014**, *109*, 421–436.
157. Perfecto, I.; VanderMeer, J.; Mas, A.; Soto-Pinto, L. Biodiversity, yield, and shade coffee certification. *Ecol. Econ.* **2005**, *54*, 435–446. [CrossRef]
158. Gómez, J.A.; Guzmán, G.; Giráldez, J.V.; Fereres, E. The influence of cover crops and tillage on water and sediment yield, and on nutrient, and organic matter losses in an olive orchard on a sandy loam soil. *Soil Tillage Res.* **2009**, *106*, 137–144. [CrossRef]
159. Yamoah, C.F.; Grosz, R. Linking on-station research with on-farm testing: The case of agroforestry and organic matter-based cropping systems for the Rwanda farming systems improvement project. *Agroforest. Syst.* **1988**, *6*, 271–281. [CrossRef]
160. Adesina, A.A.; Chianu, J. Determinants of farmers' adoption and adaptation of alley farming technology in Nigeria. *Agrofor. Syst.* **2002**, *55*, 99–112. [CrossRef]
161. Arbuckle, J.G.; Valdivia, C.; Raedeke, A.; Green, J.; Rikoon, J.S. Non-operator landowner interest in agroforestry practices in two Missouri watersheds. *Agrofor. Syst.* **2008**, *75*, 73–82. [CrossRef]

# 2

# Agroforestry in the Andean Araucanía: An Experience of Agroecological Transition with Women from Cherquén

**Santiago Peredo Parada [1,2,*], Claudia Barrera [1,2], Sara Burbi [3,*] and Daniela Rocha [1]**

1 Laboratory of Agroecology and Biodiversity (LAB), Group of Agroecology and Environment (GAMA), University of Santiago de Chile, Santiago 7501015, Chile; cpazbarrera@gmail.com (C.B.); danielarochacorrea@gmail.com (D.R.)

2 Environment and Society Program, Pablo de Olavide University, 41013 Sevilla, Spain

3 Centre for Agroecology, Water and Resilience, Ryton Gardens Campus, Coventry University, Coventry CV8 3LG, UK

* Correspondence: santiago.peredo@usach.cl (S.P.P.); sara.burbi@coventry.ac.uk (S.B.)

**Abstract:** Agroforestry is a practice used for the establishment of integrated production systems as an economic alternative. In Chile, the most significant experiences have been developed with rainfed farmers in the central zone, where the arboreal component is the predominant one. This study analyses the agroecological transition process of a group of women from the Andean foothills of southern Chile in the establishment of an agroforestry system based on rosehip. The field work was developed in 4 stages: (1) problem survey and definition of strategy; (2) identification of an alternative market; (3) perception of the data collection work and; (4) implementation of a demonstration unit; which included (a) workshops and meetings for discussion, reflection, and feedback on what had been done and to agree on the actions to be implemented; and (b) the development of different activities to implement the actions agreed in the workshops and meetings. The results of this research show that agroecological projects open doors to the participation, visibility, and valorization of women's work while moving towards sustainable agroforestry systems integrated into the landscape.

**Keywords:** agroecology; peasant women; rosehip; local knowledge; socio-ecological transition

## 1. Introduction

Since its origins in Latin America, agroecology has been oriented "to agricultural development more sensitive to the complexities of local agriculture ... with an emphasis on the reproduction of the family" [1] (p. 33), [allowing] "That the communities help themselves to achieve a collective improvement of rural life at the local level" [2] (p. 182). Agroecological research, therefore, is "research that responds to the needs, aspirations and knowledge of local people" [1] (pp. 33, 34) and, to date, has generated sufficient knowledge and evidence for the establishment, among others, of biodiverse production systems in which various components are integrated for multiple purposes in which production is compatible with conservation of agroecosystems and landscapes.

An example of this is agroforestry, which can be considered an, "integrated land use that is particularly adapted to marginal areas and low-input systems" [3] (p. 229). Agroforestry is recognized for its potential particularly in marginal areas, where small-scale farmers cannot afford to adopt modern technologies requiring large financial investments. As a result, agroforestry can be particularly beneficial to small farmers who have very limited or no social or political power and those, "who have been overlooked by agricultural research" [3] (p. 230). Among the challenges these systems face, it is important to consider that agroforestry systems are ecosystem-specific and in certain low-quality soils the choice of appropriate plant species may be limited. Therefore, most agroforestry systems need a few years before the first returns can be observed.

For the Chilean context, the agroforestry initiatives developed by the National Forestry Institute of Chile (INFOR), under the Ministry of Agriculture, with small and medium-sized agricultural producers—owners of more than 8 million hectares in the country who allocate a low percentage to establish forest plantations (less than 2 ha between 2005 and 2010)—have considered, in their approach, the cultural identity and life system of such producers, the conservation of natural resources and the permanence of peasant families in the countryside, thus, avoiding migration to the city, as a central element of development rural [4]. The most accepted agroforestry systems by rainfed landowners have been silvopastoral systems (44.4%) and windbreaks (43.7%) [5] (p. 297).

Most resources in forestry research have been directed towards integrating trees into agricultural systems; however, agroforestry also plays an important role in forest conservation, for example as an alternative for the management of invasive plants, such as rosehip.

Rosehip is worldwide known for its various properties and multiple uses of its fruit and seed [6]. Considered as a non-timber forest product of high commercial value [7]—Chile being the largest exporter—the harvesting of its fruit, for subsequent sale, is not without controversy since they are obtained from invasive plants that compete with other native plants and invades cattle meadows decreasing its value. This condition has generated a contradictory situation, since on the one hand, harvesting is promoted to strengthen the food industry of this product and, on the other hand, the elimination of the plant is encouraged to stop the degradation of native forests and the areas' livestock.

The economic importance of rosehip for many peasant and indigenous families in southern Chile justifies its cultivation under an agroforestry system integrated into the forest as a use strategy that attempts to reconcile both positions. With the establishment of the crop, the pressure on native forests would be lessened through the controlled production of rosehip in places where other types of crops are not feasible.

The development of agroforestry in Chile has involved the participation of peasants or farmers in the identification, design, and execution of research activities [4] (p. 22) and the perspectives of different social actors, from producers to the community and State institutions, who have recommended the association between farmers [8]. Furthermore, it has been recognized that the implementation of an agroforestry system depends, obviously, on the interest and commitment of the producer, as well as the professionals' assessment of the sector as a determining factor in the advancement of agroforestry systems, due to the fact that they are immersed in local problems experienced by small farmers [9].

However, there are few records of agroecological-based agroforestry experiences in the mountain ranges of Araucanía developed by peasant women from the cultivation of the rosehip. Therefore, the objective of this work is to determine the aspects that facilitate and hinder an agroecological transition in initiatives with the aforementioned characteristics.

### *1.1. Agroforestry Systems in Chile*

The small and medium silvopastoral owners of Chile live, mainly, between the regions of O'Higgins in the north and Los Lagos in the south, where the greatest industrial silvopastoral development is concentrated [5]. Its farms are located in rainfed areas with preferably forest and livestock aptitude soils. They dedicate a large part of their soils to extensive livestock farming with sheep and cattle, and to annual crops, such as wheat and barley [5]. Its farms, especially rain-fed ones, due to the pressures submitted to soils with VI and VII use capacity (livestock and forestry) are eroded as a result of over-cutting and the low protection of arboreal or shrubby vegetation resources. This situation has caused that 49.1% of the soils of the national territory, equivalent to 36.8 million hectares, are found with some degree of erosion in some regions, such as Coquimbo in the north with 84% of the soils eroded, Valparaíso with 55%, and O'Higgins with 52% [10].

The most significant agroforestry experiences developed in Chile are those established by INFOR, since the beginning of the 20th century, which has oriented its work to smallholders (forest farmers). Agroforestry can be developed with these small landowners as systems and technologies for land use and natural resources in which woody species (trees, shrubs, palms, etc.) are deliberately used under a system of comprehensive management with agricultural crops and/or animal production, in some form of spatial arrangement or temporal sequence, which can provide interesting economic alternatives [4]. The woody component is included with the agricultural uses of the same farm in order to achieve a synergism between all its productive components that result in an improvement in both the productivity and sustainability of the system, as well as promoting various environmental and noncommercial benefits.

The inclusion of trees in the traditional agricultural systems of Chilean agriculture is not only aimed at recovering the rural landscape, improving its beauty and aesthetics [4], but also helps to recover soils and water systems, and to deliver, through the introduction of woody components, an additional economic alternative that can improve the profitability of the agricultural business.

In the Chilean context, the silvopastoral systems that best reflect the cultural reality and the economic and environmental needs of the forest peasantry [4] (pp. 22–24), are:

a. Silvopastoral system with tamarugo (*Prosopis tamarugo*), a native and endemic species of northern Chile, particularly in the Pampa del Tamarugal located 70 km east of the city of Iquique in the Tarapacá Region of northern Chile.
b. Silvopastoral system with forage woody species from Chile, such as *Atriplex* spp. and *Acacia saligna* in the semi-arid zone of northern Chile. In the central-south zone, meanwhile, there is the case of *Chamaecytus proliferus*, better known as tagasaste.
c. Silvopastoral system with hawthorn (*Acacia caven*) in the Maule Region. This management involves maintaining a tree cover that gives protection to the meadow, with a canopy cover between 40–50%, which will allow to improve the pasture production, improve animal production, and to give protection to the soil, especially in those hillside situations where removing the protective tree cover causes erosion and soil loss.
d. Silvopastoral system with poplar (*Populus* spp.) whose main advantage is that it grows in fertile soils that, with irrigation, allows the development of a natural or artificial meadow (in the first seven years of rotation) capable of sustaining an important livestock mass for the owners. In the central zone of Chile (Maule Region), this system combines poplar with corn cultivation in the inter-row during the first three years, and then pastures its plantations throughout the rotation. In the southern zone (Los Lagos Region), meanwhile, studies have been carried out in which the poplar has been combined with different forage species. This management experience has also been used for the recovery of degraded or unproductive soils in arid soils. Small and medium producers use it especially in fertile or humid soils, since it allows them to use these soils in a productive way. Poplars are also used massively in windbreaks to protect grasslands and crops.
e. Silvopastoral system with radiata pine (*Pinus radiata*) in the coastal dryland of the O'Higgins Region. Systems better evaluated in terms of profitability and, in addition, from the social point of view allow: to obtain periodic income as a result of annual crops and animal production, greater roots of the owners in their lands, avoiding migration to the cities, which causes marginal poverty and changes in agricultural culture.
f. Silvopastoral system with pine contorta (*Pinus contorta*) in the Aysén Region, Chile, is the extensive livestock activity, based on land covered with native grasslands of regular to poor quality, with low productivity for the production of meat and/or wool, and with eroded soils due to overcutting and low protection with perennial vegetation on them.

Within the traditional agroforestry practices used in Chile, it has been possible to identify those that, based on the combination of crops with livestock, trees can play a passive or active role, depending on the edaphoclimatic conditions [11]. Where there are water and/or soil limitations, trees are found in smaller proportions and associated with crops and forage. In areas with better environmental conditions, trees play an important role in maintaining productivity, helping to improve family income. Traditional practices are common and only differ in their spatial arrangements depending on environmental conditions and sociocultural characteristics. Traditional practices include the following:

a. *Protein banks*: These are forage shrub species cultivated on smooth and steep slopes, on platforms or terraces, usually in demarcated and protected plots.
b. *Grazing on farmland*: Grazing is done in rows or *melgas* planted on soft slopes and plains. Each area, shaped like a terrace, is separated by fences made with rocks and/or cobblestones and live fences made with bushes. After the harvest, animals are left to graze these areas to allow the regeneration of soil fertility.
c. *Family gardens*: This is one of the most common practices in the Chilean highlands. Family gardens are located on land near the farmer's house or outside family farms on small plots known as *hijuelas* or planks. Generally, square or triangular in shape, the rows are built with a combination of mud and rocks that serve to regulate water flows. Fruit trees are planted to provide food and wood, as well as shade and protection from low temperatures and frost. Crops are grown in the inter-rows, with live fences to regulate the fallow, the rotation of the crops, and the application of guano.
d. *Windbreaks and live fences*: Used mainly in the Norte Grande region of Chile, trees are located along the boundaries of farms as a form of demarcation and to protect crops from the wind.
e. *Mixed cultivation of trees and grasslands*: This is used in areas where irrigation is possible for the cultivation of grasslands. Trees or shrubs provide forage, usually leguminous plants, which are also used for their fruits and for firewood.

Agroforestry practices in agricultural communities are particularly important as a form of land tenure inherited from the colonial era. They are located in the Coquimbo regions and are associated with areas with severe environmental limitations.

f. *Non-irrigated farmland*: These are plots that the community gives to a member for cultivation. Their size depends on the family's labor capacity. They are demarcated with live fences, usually cactus. After being used for several years, usually for the cultivation of coriander, anise, and wheat, they are used for livestock grazing, mainly goats. The more degraded lands have been turned into forage forests using species such as *Atriplex nummularia* and/or *Acacia saligna.*
g. *Protective and productive forest areas*: These are found especially in ravines where natural forests already exist or trees are planted to protect animals from the sun or rain, for firewood, or to protect water resources.
h. *Spinal grazing*: This is the most common agroforestry practice in the central area and has existed throughout history in the Central Zone of Chile. It is carried out with hawthorn (*Acacia caven*) steppes. Hawthorn is a nitrogen-fixing leguminous tree that provides a natural canopy under which goats and cattle can graze on the wide variety of high-quality forage that grows below it. However, it is also used to produce charcoal, which, along with excessive grazing, has put the land at risk of erosion.

The establishment of agroforestry units together with small forest producers is not without limitations and difficulties. For the experiences in Chile, the following barriers are worth mentioning [12] (p. 264):

1. Lack of interest from private operators to participate in the establishment of agroforestry systems, due to the low economic outlook, as the density in the establishment of the forestry component has to be reduced and the management of the whole system is complex.

2. Producers' limited knowledge and valuation of integrated systems of agroforestry production, especially the medium- and long-term management of the system.

3. Limited knowledge from the institutions that manage mechanisms and incentives to promote agroforestry. There is a disparity of criteria and approaches between regions and areas, and a lack of knowledge of the proposals available.

4. Lack of uniformity of approval criteria is required, as well as the management and design of the agroforestry systems to be presented according to the context, the tools and the incentives available.

5. Limited technical capacity for the correct application of the proposed agroforestry systems, which needs to be improved to ensure an adequate relationship between the agroforestry production system and the site conditions.

*1.2. Rosehip: Harvesting and Sale of an Invasive Species Valued as Non-Timber Forest Products (NTFP)*

In Chile, as in many other countries in the world, NTFPs in both forests and agroforestry systems have great social significance, due to the strong reliance on unskilled labor for their collection and processing. Additionally, the largest collection and use of NFTPs is carried out in rural communities, mainly peasant and low-income communities [13]. In agroforestry systems, the collection and selling of NTFP represents 18.92% of the total activity. This activity (especially harvesting tasks) is mostly done by women (49.35%) and has an important contribution of young labor (22.08% of workers are under 18 years of age). Moreover, this activity represents 19.11% of total family income for the small owners that collect and sell the product [14].

NTFPs have experienced significant and sustained growth in the last 20 years, and this has been reflected in significant advances in new commercial ventures destined for the national and international market and, to a lesser extent, in research and development actions focused on value aggregation. Until 1969, rosehip was eliminated by farmers, as it was considered a fairly invasive weed. That year, CIDERE BIOBIO, a development institution in the region of Bio Bio, found a productive destination becoming a new export item for the country: dehydrated rosehip. The first exports of this fruit were made in 1970 [15].

However, the promising economic development of the export process contrasts with the low production levels achieved in the other areas that make up the production model, with gaps that put its sustainability at risk. The main issues are due to a lack of information on the quantification of production and consumption, and of sustainable extraction methods and techniques, the absence of management plans, as well as insufficient information on value chains and marketing processes. Additionally, market failures are often associated with information asymmetries, monopsonies, moral hazard, and adverse selection. These combined with high transaction costs, territorial dispersion, and excessive fragmentation in the value chain, are all factors that generate problems that should be addressed by public policies [16] (pp. 357, 358).

Land owners can agree with groups of rosehip pickers on a fee to access the property and carry out the harvest. However, the purchase price for pickers does not reflect the effort and work generated by this activity, with intermediaries being the ones gaining the most, i.e., 20–30% of the purchase price [17]. For this reason, the same pickers process the rosehip for family consumption and for sale in urban centers or municipal markets, allowing them to sell a product with greater added value at a better price [18]. Rosehip pickers have then two possibilities of selling their product: to primary buyers, which is handled as a formal marketing process where the pickers sell their production at a previously agreed value; and to local or private consumers (direct sale) for the production of jam or homemade infusions, products that can be consumed by themselves or sold in local markets or fairs. This last option is significantly less adopted than the export-related marketing dynamic. Thus, it is necessary to eliminate the link between pickers and intermediaries within the value chains, facilitating direct negotiation between the picker and the final buyer with an agreement on the price that benefits both.

However, it must be recognized that adopting the fair-trade framework allows us to expand the concept of trade, commonly considered as a strictly technical and profit-maximizing activity, by adding essential elements of sustainable development [19]. This value chain presents interesting opportunities to raise quality requirements, prices, continuity of supply, and more generally, best practices in national and international trade [20].

Some experiences of commercial cultivation of rosehip have been developed: the COESAM company has approximately 1700 plants and vertically integrates the entire marketing chain: production, processing, and export [8], as well as a research field characterized by a monoculture production logic in which in vitro propagation has been promoted [21], fertilization [22,23] and weed control [24] with inputs of synthetic origin. This system then obviously ceases to be a NTFP and cannot be considered an agroforestry system. However, due to its low profitability, this system is not largely adopted [8], with more than 90% of the rosehip marketed in the country of wild origin, which reiterates its status as NTFP.

In the municipality of Melipeuco (along with those of Loncoche and Toltén in the Araucanía Region), specifically in the Llaima area, NGOs and Mapuche associations collaborated with public institutions (e.g., Programa Servicio País of the Chilean state, local municipalities, and regional universities) within the framework of an intra- and extraterritorial intervention whose objective was to contribute to the territorial self-management of Mapuche communities under a scheme that tends to balance the spheres of sustainable development [25]. This led to an associative commercialization initiative of non-timber forest products (NTFP), among them rosehip, which sought to link the various fruit pickers of the territory in an association and, thereby, generate a significant volume to sell the products at a better price. Within this initiative focused on the development of Mapuche communities in the Llaima area, this study was carried out to trial the establishment of an agroecology-based agroforestry system with women from the Cherquén area in the municipality of Melipeuco.

## 2. Materials and Methods

The experience documented in this study corresponds to one of the six groups of peasants and farmers in the municipality of Melipeuco, and members of the Prodesal Alto operational unit, with whom a participatory action research initiative was carried out with the purpose of designing endogenous development strategies based on socio-ecological potential (Figure 1). Three main reasons highlighted the need to undertake this study. Firstly, the women in the area identified the need to have control over their properties during any initiative undertaken involving their participation. Secondly, the women themselves expressed interest in participating in this study, that differs from other studies and initiatives in the area geared towards supplying a rosehip collection center. Thirdly, in the study area the design of agroforestry systems is not common, making the study even more of a novel experience in evaluating an agroforestry system based on shrub species considered invasive.

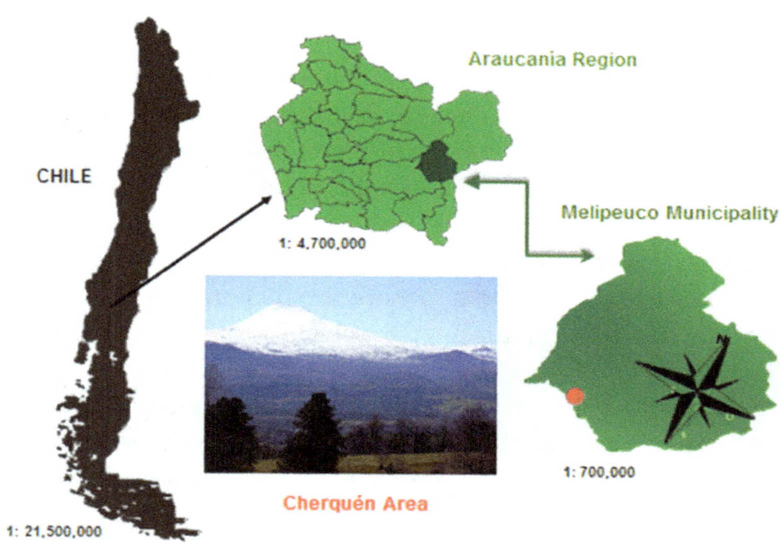


**Figure 1.** Study location.

The municipality of Melipeuco is located in the foothills of the Araucanía Region, 92 km east of the city of Temuco (regional capital). Its geographical location is between 38°35′–39°00′ S and 71°24′–71°52′ W. This municipality has the particularity of having parts of two Protected Areas, Conguillío National Park and China Muerta Natural Reserve, which occupy approximately 2/3 of the municipality. Located in the sub-basin of the Allipén river, it is naturally delimited by mountain ranges of the southern ends of the Cordillera de Las Raices to the north; the Nevados del Sollipulli and the Curacalco hills, to the south; the Cordillera de Los Truenos and Cordillera Colorado to the east and the Llaima volcano to the west.

It has a temperate rainy climate with temperatures ranging between 1 °C in the coldest months in winter and 23 °C in the hot summer months, with temperature differences between the valleys and the mountains. Precipitation varies between 1000 and 2500 mm annually with two dry months in the summer. Land uses are manly forests with 52.98% of the territory of the municipality, while agricultural activity occupies only 0.03% [26].

The main economic activity is agriculture (includes hunting, livestock, and forestry), which is carried out mainly in the valleys and is concentrated on soils with low fertility used by small landowners for self-sufficiency. Cattle farming is the most frequent activity, together with the exploitation of native forests, which historically has been practiced to obtain wood, firewood, and construction. In recent years, the collection of NWFP and tourist activity has been valued [27].

The Cherquén area is located to the southwest of the Melipeuco municipality. The group of women involved in this study lives in the highlands (Cherquén Alto). The collaboration evolved from an initiative created by the Women's Service of the Ministry of Planning (now replaced by the Ministry of Women and Gender Equity and the Ministry of Planning is called the Ministry of Social Development and Family) as an opportunity to create spaces for reflection of the working life of women. Within this collaborative space, each woman could define her work experience and her work project according to the labor market in her community, with the support offered by the program to improve her working conditions and employability. Their complementary objective was to develop social skills for work and to reinforce among the participants their identity as heads of household and as working women.

### 2.1. *Participatory and Needs-Based Research*

Understanding the dynamics that occur in agroecological systems requires integrative analytical approaches that allow the articulation of the various epistemic sources from which agroecology draws upon. Such methodological strategies are defined on the basis of the inclusion of the various actors who possess complementary forms of knowledge within the objectives outlined in a research study. A central element in this type of research, then, is participation in such a way that it gives the different actors a state of horizontality when defining a research strategy [28]. Participatory research is a methodological proposal with a defined action strategy, which involves its beneficiaries in the production of knowledge. It is a combination of research, education, learning, and action [29]. The purpose of this type of research [30] is the people themselves, and not the advancement of science, to know and analyze a reality in its three constitutive moments: the objective processes; the perception of these processes in people; and the experience within their structures. Participatory research is characterized as qualitative research in which quantitative elements can be included, but always within the context of a qualitative problem. Another fundamental characteristic of this type of research is the importance of working together with the various institutions that make up the social fabric of a territory. The researcher, as the subject of the investigation, embarks then together with the other actors, in the process of research, learning, and action, maintaining the scientific criteria of precision in observation and analysis, combining them with the interests of the community. Participation is visualized in the design and execution of the research, but also in the use of the results for the actions, by the subjects [31].

From the epistemological point of view, agroecology understands agroecosystems as a result of the co-evolutionary process between social and natural systems. It is important to consider that "environmental factors influence the suitability of particular aspects of social systems and, in turn, social systems influence the suitability of particular aspects of environmental systems" [32] (p. 171). Social systems can be subdivided into systems of knowledge, values, organization, and technologies, which co-evolve with environmental systems. Each of these systems is related to all the others and each one changes and influences the change of the others (Figure 2).

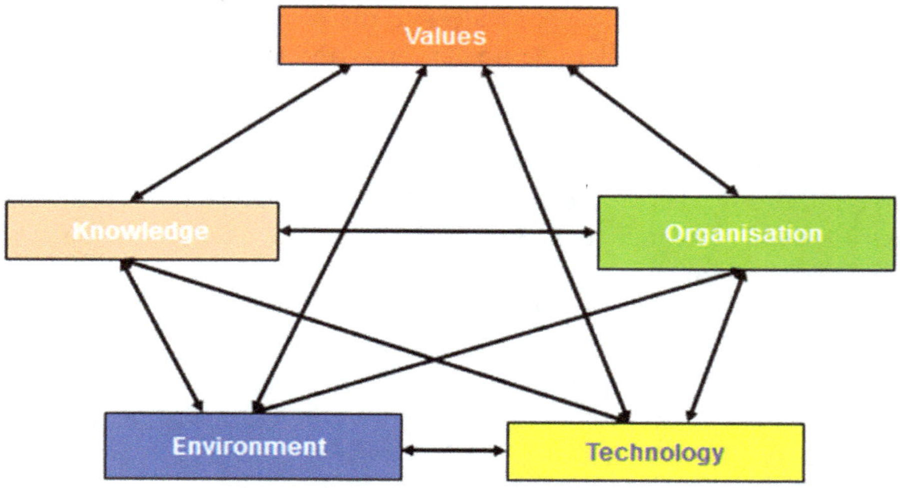


**Figure 2.** Elements that constitute the social and natural systems in coevolution. Source: [32].

The elements of a coevolutionary interaction adapted to agroecosystems [33] are as follows:

- The environment is understood as the biophysical environment in which the members of a group find local material resources for their subsistence.
- The values involved in the beliefs and worldviews that give an identity to a group aimed at articulating their life plans in relation to their environment and appropriate resources.
- Knowledge refers to the repertoire of abilities and skills with which a group appropriates "pieces" of its environment.
- Technology focuses on productive practices, including the different uses and management of natural resources (collection and exchange).
- The organization referring to all forms of collective action aimed at the sustainable management of its resources.

Table 1 illustrates how these principles were applied in this study.

**Table 1.** Constitutive elements of coevolution principles adapted and applied to this study.

| Factors | Adaptation | Operationalization | Application |
|---|---|---|---|
| Environment | Local resources | Understood as the biophysical environment in which the members of a group find local material resources for their subsistence. | Rosehip |
| Technology | Cultural techniques and practices | The production practices, including the various uses and management of natural resources (collection and exchange). | Agroecological management of the farm |
| Organization | – | Referred to all forms of collective action aimed at the sustainable management of its resources. | Workshops |
| Values | Life plans | Beliefs and worldviews that give an identity to a group aimed at articulating their life plans in relation to their environment and appropriate resources. | Control of the farm |
| Knowledge | Local Knowledge | Repertoire of abilities and skills with which a group appropriates "pieces" of their environment. | Production systems |

For the development of this participatory research, a consensual and planned work strategy was carried out with various local actors in the municipality of Melipeuco: peasants and farmers, leaders of municipal social and professional organizations, the Local Economic Development Unit, UDEL, and PRODESAL. This strategy included the women of Cherquén and consisted of the following elements [34]:

a. *Local initiatives*: These refer to the fact that any process that one wishes to implement, transition to or go through must emerge as an initiative of the farmers involved, since they will be the ones who will permanently demand and ensure compliance with the objectives set, in addition to deciding the pace of its achievement. As a result, there will be a greater commitment on their part, as they are managers of their own development process. The fact that the initiative arises from the peasants themselves is of real importance, since relationships of equality and exchange that are necessary in an agroecological process are generated more naturally and spontaneously, leading to increasing levels of involvement that translate into higher degrees of participation and prominence during the process.
b. *Accompaniment program*: The program must respond to the needs of all the actors involved and it must also consider their creative capacity. Actors are immersed in a socio-economic context where development programs are often conceived only as a productive project. Trained professionals are expected to provide support in the implementation of techniques that are not degrading natural resources. However, social relationships that occur in the community are also to be considered, as they allow the integration of the groups of peasants with the wider society. This process of accompaniment facilitates mechanisms of interpretation of state bureaucracies, thus, facilitating the self-management of peasant groups.
c. *Mobilization of the subjects of change*: This represents support that should be able to mobilize people in order to organize them either formally or informally. As a result, it could then be possible to channel the proposals for change designed by the peasants themselves.
d. *Choice of the technical proposal*: From the productive point of view, the choice of the technical proposal is decisive to support an agroecological process. It also constitutes a quantitative evaluation factor and a possible cause of the success or failure of a project. However, it should not only respond to a merely productive objective. Because of its agroecological nature, the technical proposal to be implemented should be one that at least considers the use of local resources, that respects biological cycles and the reproduction of these, and that does not generate more waste than the agroecosystem itself can recycle. Therefore, the choice of a suitable technical proposal is of great importance, mainly due to three aspects: (1) the agrarian nature of both the economy and the social relations of the peasant; (2) it must be in line with the agroecological nature as a complete process; and (3) the socio-economic importance that the productive dimension acquires in rural development projects. However, its implementation will have a beneficial impact on the achievement of agroecological objectives if it is implemented within a broader strategy that involves the parallel redesign of the property or farm, with limited financial investments initially, and where production is implemented based on the resources and inputs that are available locally.
e. *Presence of local markets*: The support and promotion of local fairs so that they become local market places based on the trust of the social networks established in the region, may constitute a concrete action that facilitates the establishment of autonomous relationships between the peasants and the market, where the latter constitutes the final destination of the produce rather than where factors affecting production are mobilized. Therefore, it can be suggested that a strategy that would allow timely access to a local market is to be preferred, as it has the potential to promote short circuits. Another potentially successful strategy could be focusing on organic production, considered also traditional, that is based on reduction of costs as a result of the use of local inputs (or the gradual independence from external inputs). This approach also includes the possibility of selling at a premium price, depending on the characteristics of the market,

which is expected to have a positive impact and support the economy of peasant families in spite of the influence of the large-scale national agri-food production.

f. *Empowerment of peasant farmers*: One of the key elements to establish a strategy of agroecological transition is the power that is believed in hands of the peasants involved in agroecological processes. The generation of any type of knowledge, whether theoretical and/or practical, methodological and/or technological, must be assimilated and appropriated by the peasants with whom researchers and other professionals are working and accompanying, even more so when this knowledge has been generated from the local culture, as a product of the interaction and dialogues of knowledge between peasants and technicians. The process of empowering peasant farmers includes the following aspects:

- Assimilation of techniques and technologies: all technologies and agricultural practices that are implemented to manage the natural resources of the agroecosystem, whether external or locally generated, must be incorporated by the farmer in a constant process of education, in order to grant autonomy and control over production processes. As a result, the adoption of new technologies must be understood as a process of assimilation based on local experiences.
- Adequate tools: the construction of adequate tools should be in line with peasants' approaches, so as to allow the interpretation of a technical language often not understood by peasant farmers. Despite having been generated on the basis of traditional and local knowledge, tools often end up being a one-way interpretation of agroecosystemic contexts, where the technician obtains greater (sometimes absolute) benefits from the information gathered. Therefore, the generation of cumbersome and complex methodological tools that only manage to accumulate information, hindering its interpretation and analysis, should be avoided.
- Access to information: access to clear and timely information continues to be an obstacle to the development of peasant groups when selling the surplus production that the support institutions themselves, both technical and financial, demand of them. Therefore, an empowerment strategy must be generated so that information regarding fair prices and local markets is managed and administered by the farmers themselves without the intervention of intermediaries or third parties.
- Decision-making power and negotiation: in order to facilitate development, peasant communities must achieve decision-making power and negotiating capacity when deciding the central elements of a program and directing the processes in which they are included. Agroecology can and should play a central role, since the methodologies used enhance peasants' participation in the projects that it supports. Participation must be understood as a negotiation between the different actors involved (internal and external), where the constant and permanent appropriation and assimilation of both methodological and technological elements are fundamental, generating decision-making powers, together with determination and control over the processes that are carried out autonomously, respectful and in line with peasants' logic and organizational structures.

Once the elements considered in the accompaniment strategy carried out at different stages were defined, the study was evaluated through the analytical categories designed to eliminate the gap between the theoretical agroecological proposal and its consequent conceptual transformation [35], considering that the discipline of agroecology "is being built in reciprocity to social and political movements and processes" [36] (p. 18). This framework of analysis has been proposed as a theoretical approach for the analysis of experiences of agroecological transition with groups of peasant women in the Chilean context [37]. The variables used were: localized knowledge; temporality, uses, and customs of the territory; socio-political power and relations; inductive practices from the local; and economic alternatives to the hegemonic model. Results will then be discussed in line with these variables.

### 2.2. *Study Phases*

The present study was developed in 4 stages:

(1) identification of the problem and definition of a strategy,
(2) search for an alternative market,
(3) perception of collection work and,
(4) implementation of a demonstrative unit.

Each stage consisted of: (a) workshops/meetings for discussion, reflection, and feedback on what was done and to agree on the actions to be implemented; and (b) the development of different activities for the implementations of the actions agreed upon in the workshops and meetings (Figure 3).

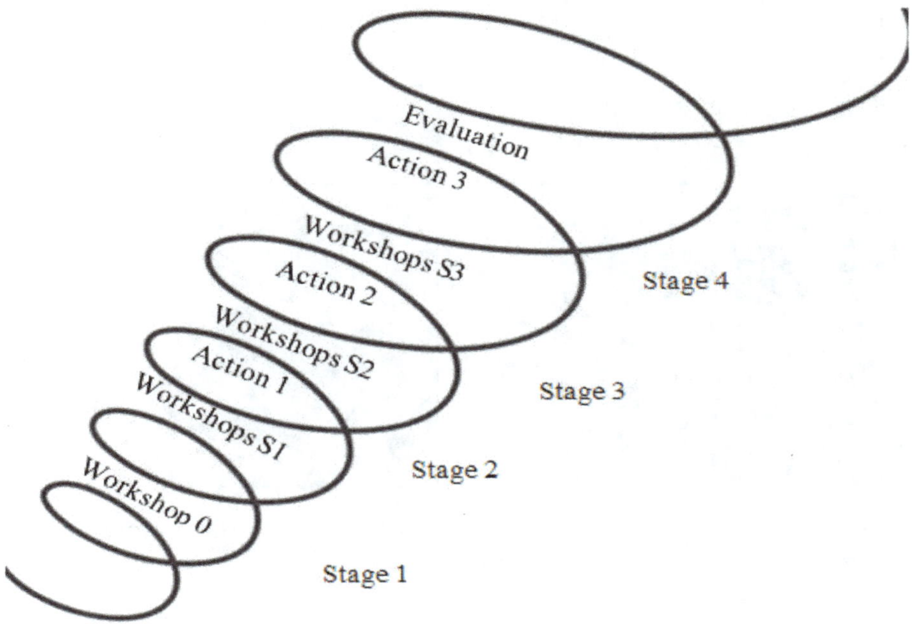


**Figure 3.** Stages of the study.

#### 2.2.1. Identification of the Problem and Definition of a Strategy

This was carried out with the Women's Group in the Cherquén area. The first discussion workshop was aimed at collectively identifying the problem and elaborating a dedicated action strategy.

#### 2.2.2. Search for an Alternative Market for the Sale of Rosehip

This consisted of a micro-market study to find out the purchasing conditions established by the two organizations that, at that time of the study, acquired rosehip from the direct sale from peasants and farmers. This information was shared in workshops for analysis and discussion about the possibilities of meeting the conditions and requirements of these organizations. Subsequently, the participants collectively agreed to determine their women's perceptions of the work of harvesting rosehip and the labor conditions.

#### 2.2.3. Perception of Collection Work and Conditions for Collective Work

A discussion group was held with the women members of the group, using a semi-structured approach. The information obtained was recorded and transcribed to be later categorized through content analysis. The results obtained were analyzed and discussed in discussion groups with the women, and it was agreed to carry out a single demonstration unit, of a small size, on the farm of one of the participants.

### 2.2.4. Implementation of the Demonstration Unit

The unit consisted of the establishment of an agroforestry system integrated in the forest landscape (park) in the Cherquén area with the establishment of rosehip bushes alternated with native prairie. Plant clumps were obtained from the collection sites usually used by the women participants, previously selected for site characteristics similar to the one selected for planting and attributes such as color and size of fruits. The seedlings were planted following topographical keylines at a distance of 1.5 m above the row and 3 m in the inter-row, where then guano was incorporated at the time of planting (Figure 4). Finally, a workshop was held where a general evaluation of the experience was made.

**Figure 4.** (**A**) Rosehip plantation: collection locations; (**B**) mugrones; (**C**) contour lines; (**D**) plantation of a rosehip mop, and incorporation of guano (manure).

## 3. Results

### *3.1. Identification of the Problem and Definition of a Strategy*

The problem was defined as a matter related to "combating the theft" of rosehip by pickers from other parts of the country who, without the authorization of the owners, entered the farms to collect rosehip fruits. The women of Cherquén, who carry out the harvest and collection work, expressed the possibility of establishing production systems in places closer to their homes that would allow them to regain control of their property, while at the same time making home care tasks compatible with their farming activity (e.g., feeding, upbringing). The definition of a collective strategy, then, consisted in that prior to carrying out any productive economic initiative, women wanted to be sure of their capacities and possibilities to carry out the activity. Therefore, the advantages and disadvantages of an associated rosehip sale, the conditions for an association, and the technical requirements of a rosehip culture were discussed. As a result, participants agreed to carry out a search for alternative rosehip buyers.

### *3.2. Identification of Alternative Channels for the Commercialization of Rosehip*

The background obtained from the interview with the managers of the two commercial entities and potential buyers are outlined in Table 2.

Results indicate the importance of having a minimum level of organization for the commercialization of rosehip and, thus, meet the requirements established by commercial entities. This information was share among and analyzed by the women, eliciting the following comments (Table 3).

**Table 2.** General description of commercial entities as buyers of products made from rosehip.

| Commercial Entities | Buyer 1 | Buyer 2 |
|---|---|---|
| Description | Organic farmers association that markets its members' products but is complemented by the offer of organic products produced by third parties. They are approx. 30 partners and within some of its objectives it promotes organic agriculture, shares knowledge with agroecological farmers. | Cooperative that emerges as a fair-oriented commercial initiative. Producers can be organizations of artisans and peasants or companies with good social policy. |
| Requirements (jam) | The products marketed are from partners or external producers. The elaborated product must have organic certification. They mark up the price of the products marketed by 35%. The freight is not included in this percentage, so it must be paid by the association or the producer, influencing the final price. Labelling must carry everything that the law stipulates. A certificate issued by the certifying entity is required. | To market the products, producers can be members of the cooperative or a producer related to the cooperative. The price of the products is marked up by no more than 20%. Products must have the approval of the Health Service. The freight is paid by the producer or the cooperative and the cost must be applied to the price of the product. The ideal solution is to deliver the product in small jars. |

**Table 3.** Main opinions and perceptions of women regarding the conditions established by the companies that buy rosehip.

| Participants | Comments |
|---|---|
| Participant 1 | *"For us it is super complicated, because imagine that they tell us how many kilos from 30,000 to more and for example I do not consider myself a collector and I believe that we are far from that, for the reason that I personally do not always go out to collect. A person who is a collector takes advantage of the season and leaves from 9 in the morning and returns at 6 in the afternoon, that is what a collector is. And we are more like nothing an entertainment or suddenly to make some extra pesos, but it is not the strong one like us as a family".* <br> *"We have no serious commitment to. It is that we cannot either, because suddenly we have to do one thing and others that things to do, the truth is that the harvest of rosehip is not the strength of us".* |
| Participant 2 | *"It would more than make oil deliver oil is not a bad idea, but it is more expensive to do a project, so we train, and get resources pa ' buy the machines must have a room where you work, where they meet the requirements of health and all that".* <br> *"Maybe we could organize ourselves as a Prodesal group and ask ourselves to help us in that because it is something innovative, it is something new, to be able to obtain resources, to see what possibilities there are of making a small group".* |
| Participant 3 | *"Of course I can tell you if it is okay, but we cannot assume you, well, personally, and I think that the girls are going to share with me. I can tell you if, if I am going to and if I don't comply there is the issue, that is if I commit I have to comply and it is not the idea if I say yes, but I do not comply, because one is supposed to take responsibility. And this is far from reality, 12,000–30,000 kilos. And the price is perhaps not bad either, if you sell it for $ 150 per kilo, you have $ 15,000 in a bag, but you don't collect all that."* <br> *"This is fine, but the issue is that for us too, or because pa ' start if we think of oil and the whole subject, think that we have to be trained for that, because this is not to say one wants to produce oil, if we do not know how we are going to produce it. If one had access to all things well but, unfortunately, we do not have access, unless we just sell the raw fruit, it is the only thing we can do".* |
| Participant 4 | *"And another thing, now, if we have a commitment, this commitment must be fulfilled, but what do we get out of making a commitment if on the radio they say that the musket season begins and all the people start to remove the muskets that there are? over her so how are we going to fight, telling them not to remove our muskets".* <br> *"We would lack the resources to start, because for example to start with oil we would have to be trained and from that we would already have to enter to organize ourselves, enter to request training". "If we had the opportunity, where we train and come up with a project to buy the machines and all would be good pa 'us, those opportunities must not lose".* |

The opinions and comments expressed by women highlight the difficulty of meeting the requirements of companies, specifically, with delivery volumes. They express a high sense of responsibility for the fulfilment of commitments between them, the convenience, and need to receive aid such as subsidies and training.

### *3.3. Determination of the Endogenous and Exogenous Factors that Impact on the Organization*

The background obtained from the group discussion held with the women of the community of Cherquén is presented in Table 4 below:

**Table 4.** Summary aggregate of the endogenous and exogenous factors that impact on the organization.

| Categories | Subcategories | Comments | Inferential Transcription |
|---|---|---|---|
| Perception of collection work | Sacrifice | *"When you go too far to collect"; "We must leave the house alone with the children"* | Long distances that must go, leaving neglected to the children. |
| | Priority | *"You have to stay longer in the house to do lunch, toilet and take care of the daughters"* | It prioritizes the tasks of the home and the care of the children. |
| Rosehip for sale | Price | *"It is not very fair, hard work sacrifice and little profit"* | Payment for the fruit of rosehip is not the proper for the work that requires the collection. |
| | Request | *"A direct delivery, without intermediaries"; "A person who makes the contacts for a direct delivery"* | Need for external support to achieve direct delivery of the product to final buyers. |
| Income source for farmers | - | *"It is a good source for people who need extra pesos, it is one more entry, as a support"* | Selling rosehip fruits generates extra revenues that support the economic costs of production. |
| Rating of the collection work | - | *"They do not value so much the sacrifice that people make to collect"; "They are interested in buying it no more, because they are the ones who win"* | Lack of interest of the intermediaries to know about what that implies the work of collection. |
| Peasant Organization | Initiative | *"There has been no thought of organizing to sell rosehip, if there was a better offer perhaps there would be interest and with direct buyers"* | Scarce interest of organization for the sale of rosehip fruits, but if the payment conditions improve, would be an incentive to join. |
| - | Experience Women's Group | *"We need to organize, is that people are difficult, it is difficult to agree"; "There is also a lack of more motivation from the board to do something"; "It will happen that outside support is lacking"; "I am the president and we have not thought organized nothing else"* | Lack of internal coordination within the group; little incentive from the board and need for external support to regroup; demonstration of poor leadership. |
| - | Experience with other organizations | *"They always cooperate, and the others don't"; "When some help, there they all arrive* | Lack of cooperation and unity within the organization. |

Results highlight a number of aspects that should improve with internal coordination as well as cooperation and cohesion within the group. These also include the difficulties in harvesting rosehip products and how such activity can be time consuming. All these factors affect a possible organization of the women for the collective sale of rosehip.

### *3.4. Establishment of Agroforestry Systems with Rosehip*

These were implemented through the planting of vine branches in contours on the property of one of the women (Figure 5).

**Figure 5.** (**A**) Agroforestry system with rosehip: plantation incorporated into the landscape; (**B**) rosehip plants in flower; (**C**) number of rosehip plants with fruits (red) and without fruits (blue).

The system is integrated into the landscape that is made up of laurifolio-type forest dominated by Patagonian oak (roble, *Nothofagus obliqua*), maqui (*Aristotelia chilensis*), lingue (*Persea lingue*), laurel (*Laurelia sempervirens*), palo santo (trevo, *Dasyphyllum diacanthoides*). Species density allows, on the one hand, to carry out handling (pruning) and harvesting tasks and, on the other hand, the growth of native species in the prairie and free movement of animals. The majority of established plants presented fruits at the time of this study.

## 4. Discussion

### *4.1. Local Knowledge*

Agroecology "is based on a set of knowledge and techniques that are developed by farmers and their experimentation processes" [36] (p. 6). In addition, agroecology proposes socio-ecological aspects [2] as an essential component of agroecological systems. As a result, it adopts the principle of pluri-epistemic knowledge, under which not only scientific knowledge is recognized for decision-making [38], but also the epistemologies of each culture are incorporated as generators of valid and useful knowledge for the sustainable development of the territory, leading to development options and approaches that are "consistent with its values and goals" [38] (p. 198). Therefore, it is essential to recognize the knowledge potential of traditional agricultural cultures, without excluding formal social and biological knowledge developed by conventional agricultural sciences, in order to improve agroecosystems with the contributions of traditional and modern knowledges [39].

The epistemological trajectory of agroecology, as well as the approach to the understanding of culture, fosters the dialogue between the unilinear and multilinear understandings of knowledge, recognizing the relevance of "transforming the dominant relationship between nature and culture or between nature and society, it is necessary to distance oneself from the relationship established by western logocentrism between nature and culture, or between nature and society" [40] (p. 146). Thus, knowledge is not only due to the recognition of the knowledge of the other in an equal way, but also to the responsibility that comes from managing information. Therefore, access to information is not due to a neutral motivation of knowledge, but rather to a commitment in which "it plays a role in restoring the ecological balance of agroecosystems" [41] (p. 67).

The joint work between the different actors involved in this study reveals the following aspects:

a. The importance of the knowledge of women and their families reflected in aspects related to their environment that allowed the correct choice of places to obtain the seedlings that were used in the rosehip plantation, as well as the criteria used for the selection and the attributes of the fruit. This is in line with what has been experienced with other groups of peasants from the Huechelepún area in the Melipeuco municipality that, because of the lack of scientific knowledge, use an endemic tree to feed their livestock [42,43].
b. Complementing the knowledge of professionals who are knowledgeable about tools and committed to the process, promoting important synergies when implementing technical proposals. The technicians' skills, such as the management of orchards and the implementation of the contour lines facilitate the application of knowledge for specific objectives. This highlighted the importance of the dialogue of knowledge and the local dimension applied within an agroecological approach, enriching and complementing the knowledge of different actors [44] in the improvement of agroecosystems.
c. The responsibility implied by the handling of the information, e.g., conditions demanded by the companies that buy rosehip contributed to decision-making for concrete actions. Accurate and timely information on these conditions, i.e., volume and price, allowed safeguarding the response capacities of women.

### *4.2. Temporality and Customs of the Territory*

Seasonal cycles are the relevant time for traditional peasant activities, which recognize the natural cycles in the process of food production. This looks different from the idea of agriculture based on the use of land, which adopts an industrial agriculture approach to the production of material goods and commodities: intensification of work and increasing production rates [45]. This study refers directly to the concept of territoriality that is the way the space beyond the geographical reference of the place is inhabited [46]. Thus, time and space, being categories widely analyzed by different theoretical approaches highlight the challenge in understanding approaches to land use and the customs of peasant communities [47]. As a result, with the recognition of the concept of territoriality, this study contains cultural and social signifiers that mark the relationship "between the natural environment and human activity ( . . . ) the reduction of the territory to a cartographic issue is a highly abstract simplification that does not respond to the verifiable empirical demands of the concept of human reality" [48] (p. 42). The territory has then a subjective character and can be considered "the space within which the sociocultural relations of a group take place" and as such it is [48] (p. 48). The complexity of the factors at play in this context are then associated with the geographical and environmental conditions, as well as with the social ties and the activities carried out during the study.

Similarly, time or the historical dimension, which is linked to knowledge, recognizes the transmission of rural knowledge orally from generation to generation, when contemporary knowledge is present, originating in a different temporality that has been implemented and changed according to the transformation of the environmental conditions of the field. Space and time determine how agricultural production, rest, and harvest are lived, experienced, and felt. It is the way in which one lives in the space and the subjectivity of a producer; the result of "a cultural baggage that depending on the scale, is projected from the community the producer belongs to: the nucleus or family unit, the rural community, the region and, finally, the ethnic or cultural group" [49] (p. 71). The agroecological approach tends to understand the productive cycles of the land in harmony with the traditional cycles marked by environmental, climatic, social, and cultural conditions, as opposed to modernity, characterized by "the dismemberment of the social life of the local context and its increasing determinations by translocal forces; and the space/time distancing or the separation of space and place" [40] (p. 68).

The agroecological experiment carried out in this study in relation to space and time highlighted how women dwell, experience, and feel their agricultural production line, rest, and harvest. Women recognized in the rosehip harvest a task that can complement income, rather than the main source of income. They also did not seem to want it to become the main source of income. The time allocated for the harvest is marginal because there are other priorities in agricultural work, family and domestic work. They did not identify themselves as pickers or gatherers and were not exclusively devoted to this activity. This collection activity corresponds then to ecological-economic exchanges developed with the natural environment and as a result, the establishment of rosehip productions (either as monoculture or in agroforestry systems) would mean a change in the current interaction with natural socio-economic environment [49]. Therefore, the scale and the agroecological management agreed among the women for the establishment of rosehip production on the farms would correspond to a projection of this interaction with their environment [49], which is attempted to replicate, rather than the current situation.

As mentioned above, the economy of the families in this territory is based on forestry (wood and firewood) and livestock activities. In this case, the establishment of a productive unit focused on the production of rosehip for commercial purposes represents a process of appropriation, objectification, and assimilation carried out by a group (or community) in order to create a commodity from a natural good, which must respect the ways of life of said community [50]. In the case of the rosehip in this territory, the secondary transformation of this resource into oil for cosmetic use has constituted an important element for the economy of these families. Productive concentration or hyper-specialization runs the risk of putting pressure on agroecosystems to the point of eroding the dissipative structures that they constitute (e.g., soil and biodiversity). This is common in productive development strategies based on quality certifications or seals, such as designations and indications of origin that focus on the product rather than on the socio-ecosystem. However, the holistic view of the agroecological approach focuses precisely on the socio-ecosystem where the participation of all actors is sought to establish common attributes that define an agri-food product that circulates through value chains and networks [51]. In this respect, diversification of socio-ecosystems and the products introduced in value chains can then represent an effective agroecological strategy with the aim to strengthen livelihoods, improve food security, and promote food sovereignty [52] (p. 44).

### *4.3. Sociopolitical Power and Relations*

In the framework of agro-industrial production, where the use of monoculture and the concentration of land ownership lead to the impoverishment and confinement of locally producing communities and small-scale farmers [53], the socio-political proposal of agroecology takes full relevance based on the concept of equality [41]. Without falling into the idealization and exoticization of the peasant subject, the agroecological perspective recognizes that "peasant communities are not exempt from disputes, social conflicts and internal hierarchies, with gender inequalities being the most evident" [45] (p. 11). However, there are evidences of the awareness of the importance of the kinship legacy, class-consciousness, cultural identity, and "the intragenerational awareness or rejection of any form of exploitation in a historical moment linked to a possible generational domination" [54] (p. 2).

Today we live in times when colonial relations are visible in our societies, in which, in turn, there is a concern for recovery by rediscovering and re-engaging with people's roots and history, valuation by cultural traits, and the need to reinvent or build relationships based on equality. In this line of thought, decolonial and de-patriarchal approaches are modern strategies "to understand the complexity of this global crisis that is closely linked to the dominant structural myopia paradigms and knowledge (pre)built around the conventional scientific thinking" [55] (p. 96). From a socio-political perspective, agroecology focuses on "reviewing and relating the approaches to social change, from the personal and micro to the institutional and macro" [56] (p. 246), with the aim that these views are then reflected in a social reality of a sustainable nature.

The proposal for socio-political relations within an agroecological framework fosters social diversity, the intensity of knowledge and autonomy in relation to independence with respect to external inputs. As a result, agroecology becomes a tool that supports the struggle of political movements that aspire to food sovereignty [57].

The relevance of agroecology in the fight for food sovereignty in the current context of unsustainability of food systems is then highlighted and not only glimpsed through the pressures exerted by the "large agri-food transnationals [that] make it impossible all agroecological dynamics of social transition" [56] (p. 246) and also hinder relations between producers, distributors, and consumers. On the other hand, in relation to food sovereignty, the relationship between producers and consumers is a fundamental point, while the right to healthy food, free from agrochemicals, for healthy communities represents the strong link between those who produce and those who consume. This relationship can be understood as "an extensive democratization of our sociocultural relationships with a view to achieving a sustainable social or socio-vital metabolism" [56] (p. 250).

This study has contributed to reconfiguring equality relationships in the following aspects:

a. Gender issues: although the greater female presence in rosehip harvesting has been highlighted, in reality that most projects of this type are planned and executed by men (professionals and farmers). In this study, the perceptions and opinions of the women involved have been collected, which are the fundamental input in the decision-making process.

b. Independence from the use of external inputs: both the valorization of local knowledge and the dialogue of knowledges established with professionals have highlighted the feasibility of establishing an agroecology-based agroforestry system with rosehip using local resources and without the application of inputs of synthetic origin.

c. Food autonomy: this was reflected at various stages during the study, from the moment in which the women have revealed the undervaluation that is carried out with respect to the work of harvesting the rosehip, to facing the monopsony and establishing a negotiation process without intermediaries.

d. Relations at the institutional level: participants reviewed these relations, since they are the ones that demand technical assistance from institutions, and organization and training on specific topics that can benefit their work.

### *4.4. Inductive Practices at Local Level*

This study adopts the perspective of agroecology as a holistic approach and a systemic strategy that has the potential to "stop the degrading and degrading forms of production of nature and society, causing the current ecological crisis" [57] (p. 42). This is a challenge that must necessarily be established through inductive practices at a local level, in order to contribute to the construction of sustainable societies.

The novelty of the agroecological approach is that it has the ability to recognize and rescue what was overwhelmed by modern agro-industrial systems, looking at memories of languages and practices locally, forms and knowledge that appear as a remedy to heal the current ecological and social crisis. In this scenario, the local dimension stands as the carrier of an endogenous potential that, through the articulation of peasant and scientific knowledge, advances towards the implementation of alternative agriculture systems that tend towards ecological and sociocultural biodiversity [57].

Currently, although the rural scene is not dichotomous, distinct opposite features of land work can be observed: one is based on exploitation and over-production that leads to ecological fatigue; and another achieves, from the knowledge of the recovery cycles of natural resources, a comprehensive production that does not respond to immediacy but is projected in the long term and that is variable depending on local determination and contexts. Thus, "traditional farmers have developed and inherited complex agricultural systems, adapted to local conditions; where successful native agricultural strategies are a tribute to the creativity of small farmers" [41] (p. 61) who develop their own technology, under pressure or need and with their own resources [58], since it is accepted that local cultures interact with their own local ecosystem, as well as with landscapes and local biodiversity, resulting in a complex array of specific interactions and synergies [49].

Undoubtedly, agroecology is capable of releasing ancestral knowledge and putting it into local practice. However, to use local resources correctly people need to have a cognitive system and knowledge that allows the connection between the tangible (material life) and the intangible (symbolic life) [59]. Language keeps local knowledge alive, since through orality the memory of the peoples is transmitted in space and time. In this sense, the agroecological approach must pay attention to the signs, symbols, concepts, and perceptions that make up the traditional cognitive system of the communities involved [60]. These principles underpin the importance of traditional knowledge, where "popular wisdom" carried out in rural practice is a source of observation and knowledge through dialogue with local knowledge established by agroecological approaches.

*4.5. Economic Alternatives to the Hegemonic Model*

Agroecology aims to promote the development of own, local economic systems that are based on reciprocity and that maintain a harmonious relationship with nature, with the firm intention of improving the quality of life within indigenous and sociocultural systems, and also achieving greater equality [61].

In contexts of social and environmental imbalance, like the one of this study, the agroecological perspective proposes to return to the traditional concepts of solidarity, reciprocity, and cooperation, in coherence with an economy that respects the production cycles of the land, the territory, maintains the ranges of production that are more suitable to the local context, and do not rely on agrochemicals. This position challenges, among other things, the current system to take distance from the anthropocentric bias of the western market societies and, by adopting a biocentric ethical approach, promote new forms of production and consumption that do not follow the logic of accumulation, but pay attention on basic needs [62]. The construction of these new economies, which are not subjugated to the dilemma between the environment and economic growth, requires a new social order that contemplates ecological sustainability, participatory democracy, and environmental rationality [41]. The definition of a counter-hegemonic economic alternative to capitalism must, on the one hand, be adapted to the ecological systems of the world and, on the other hand, be based on a new rationality that considers the value of the territory, the identity and autonomy of cultural diversity, and people's quality of life [38]. In this logic, the market spaces must maintain cycles of exchange and commercialization that favor the local environment [63], under the model of production of self-consumption and surplus. This is important to achieve greater equality because it ensures that agroecology does not become a discourse of goods of consumption of the elite and privileged urban middle classes, as it tends instead towards a more equitable model of healthy food distribution [64].

Although it is true that the study promoted the search for alternatives to the conventional purchasing companies, the possibilities of integrating rosehip in short value chains were scarce. This was not only because there are only two suitable companies, but also because of the aspirations expressed by women, who did not consider undertaking a larger-scale initiative. The labor that the harvest requires to them, the summer heat, the characteristics of the plant, the undervaluation of their labor, and the fact that they do not recognize themselves as pickers influenced the development of this study towards approaches for two different added-value products: on the one hand, the processing and sale of rosehip oil extracted from the fruit of the rosehip and, on the other, the production and sale of traditional rosehip jam. In both cases, what women proposed was the establishment of exchanges that value labor and the identity of their land and that, being extractive in nature, can transition to organic production respecting the cycles and uses of the territory, with benefits that can be distributed between the women themselves and the local communities.

*4.6. Barriers to and Drivers for Agroecological Transition*

Finally, the study highlighted key barriers to and drivers for agroecological transition. The main barriers are constituted by current socio-economic structures and gender issues, such as the lack of value chains adequately supporting the rosehip production activities and the lack of recognition of women's labor and contribution to the economy of rosehip plantations. In line with results from similar studies, these aspects are particularly relevant in rural contexts and areas further away from the main centers of consumption where there is a need for better integration of operations within the local value chains and commercial outlets [63,65] (p. 126).

On the other hand, the study reinforces that the integration of knowledge between the women involved in the study and the researchers, as well as all actors along the value chain constitute the key drivers for agroecological transitions. This is in line with similar studies that highlighted the relevance and role of local knowledge in agroecological transition processes, where the recovery of traditional agroecological knowledge should not only be considered the resuming of the cultivation of certain old varieties, but also the mobilization of "a collective knowledge linked to the management and use of said products that, inserted in an agroecological process of a territorial dimension, will contribute to strengthening social ties" [66] (p. 126). In this respect, this study also highlights the importance of knowledge integration in fostering social cohesion and identity, while at the same valuing cultural roots and diversity [28,67,68].

## 5. Conclusions

This study of an agroecological transition highlighted the potential that women and their families own, especially with regard to the establishment of the agroforestry system with rosehip using agroecological principles. The experience showed that the practices carried out in such agroecosystems projected in the rosehip plantation were successful and pertinent. Specifically, the use of vine branches as plant material to reproduce, use of guano (manure) as fertilizer, and integrating the system into the landscape evidenced that the practices developed by women and their families allow us to establish a culture of rosehip outside logic conventional industry that puts pressure on monoculture, the use of synthetic fertilizers and more complex propagation systems that have proven to be unnecessary for the requirements of rosehip production [8].

The study highlighted the suitability of open, reflective, and iterative methodological strategies or design in which the continuous exercise of feedback and self-criticism allows participants to reorient and redirect the processes towards the objectives. Dynamic approaches, such as the spiral one adopted in this study, are widely adopted in participatory research. Each loop corresponds to a phase that, like the activities and actions carried out in each one of them, can be repeated (iterated) to achieve the objectives and, thus, finish a cycle to start a new one, which, according to the becoming of their own activities and actions can take unexpected and diverse directions. In this study, it is worth noting the change in direction that the research took in phase three, due to the analysis of the results obtained from the discussion group, where participants agreed to establish an agroforestry rosehip system in only one of the farms as a demonstrative unit.

This type of research with communities effectively contributes to finding solutions collectively with the participation of the various actors involved. The establishment of agroforestry-based agroecological systems with rosehip by means of participatory approaches that really involve farmers from their design constitutes an interesting tool for the agroecological transition in rural areas. Further studies need to be carried out to deepen the knowledge about community-led approaches and how researchers can facilitate and support the redesign of socio-economic systems based on equality and food sovereignty of the communities involved.

**Author Contributions:** Conceptualization, S.P.P. and C.B.; methodology, S.P.P. and C.B.; data collection: S.P.P., C.B. and D.R.; formal analysis, S.P.P. and D.R.; writing—original draft preparation, S.P.P.; writing—review and editing S.P.P., C.B. and S.B.; funding acquisition, S.P.P., C.B. All authors have read and agreed to the published version of the manuscript.

**Acknowledgments:** The authors would like to thank the "Araucanía Tierra Viva" program and PRODESAL Alto of the Municipality of Melipeuco for financing; the professionals of the Local Economic Development Unit (UDEL) of the Municipality of Melipeuco, the families of the women protagonists of this work, and the group of local students for their collaboration; the Vice-Rectory for Research, Development and Innovation of the University of Santiago de Chile for financing the scholarly visit to the Laboratory of History of Agroecosystems of the Pablo de Olavide University in Seville, Spain.

## References

1. Altieri, M.A.; Yurjevic, A. La agroecologia y el desarrollo rural sostenible en América Latina. *Agroecol. Desarro.* **1991**, *1*, 25–36.
2. Altieri, M.A. *Agroecology: The Science of Sustainable Agriculture*; Westview Press: BL, Colorado, USA, 1995; p. 182.
3. Farrell, J.G.; Altieri, M.A. Sistemas agroforestales. In *Agroecología: Bases Científicas Para Una Agricultura Sustentable. Ed. Altieri.*; Nordan-Comunidad: Montevideo, Uruguay, 1999; pp. 229–243.
4. Sotomayor, A. Sistemas silvopastorales: Alternativa de producción integrada para un desarrollo sustentable de la agricultura en Chile. *Cienc. Investig. For.* **2010**, *16*, 19–51.
5. Sotomayor, A.; Barros, S. (Eds.) *Los Sistemas Agroforestales en Chile*; Instituto Forestal: Santiago, Chile, 2016.
6. Espinoza, T.; Valencia, E.; Quevedo, R.; Díaz, O. Importancia y propiedades físico química de la Rosa mosqueta (*R. canina, R. rubiginosa*): Una revisión. *Sci. Agropecu.* **2010**, *7*, 67–78. [CrossRef]
7. Tacón, A. *Cuadernos de Campo de Buenas Prácticas de Recolección Sustentable Para Productos Forestales No Madereros Prioritarios. Rosa Mosqueta (Rosa spp). Serie Cuadernos Para la Innovación Agraria*; Publicaciones FIA: Santiago, Chile, 2017; ISBN N° 978-956-328-221-4.
8. Salinas, M.; Tapia, M.; Albornoz, F.; Cartes, F. *Resultados y Lecciones en Cultivo de Rosa Mosqueta. Serie Experiencias en Innovación Para el Emprendimiento Agrario*; Ministerio de Agricultura: Santiago, Chile, 2008; ISBN 9789563280029.
9. Jofré, P.; Barrales, L.; Hernández, B. Estrategia participativa para implementación de sistemas agroforestales en pequeñas propiedades, Región de Los Ríos. *Cienc. Investig. For.* **2013**, *19*, 93–99.
10. Centro de Información en Recursos Naturales (CIREN). *Determinación de la Erosión Actual y Potencial de Los Suelos de Chile*; Publicación. N° 139; Centro de Información en Recursos Naturales (CIREN): Santiago, Chile, 2010; ISBN 9789567153916.
11. Rojas, P.; González, M.; Benedetti, S.; Yates, P.; Sotomayor, A.; Dube, F. Silvopastoral Systems in Arid and Semiarid Zones of Chile. In *Silvopastoral Systems in Southern South. America*; Peri, P.L., Dube, F., Varella, A., Eds.; Advances in Agroforestry; Spinger: Cham, Switzerland, 2016; Volume 11. [CrossRef]
12. Coda, R.; Sotomayor, A. La agroforestería una alternativa para la agricultura familiar campesina. *Cienc. Investig. For. INFOR. Santiago, Chile.* **2011**, *17*, 255–265.
13. Campos, J. *Productos Forestales No Madereros en Chile*; Dirección de Productos Forestales; Food and Agriculture Organisation (FAO) of the United Nations: Santiago, Chile, 1998.
14. Chung, P.; Sotomayor, A.; Alejandro, L. Diagnóstico del impacto de los productos forestales no madereros (PFnM) en el ingreso y alimentación de los pequeños propietarios en la Región del Bío Bío. *Cienc. Investig. For. INFOR. Santiago, Chile.* **2013**, *19*, 53–68.
15. Becerra, P. *Estudio de Prefactibilidad Técnico Económica Para la Obtención Industrial de Rosa Mosqueta. Facultad de Ingeniería, Ciencias y Administración*; Universidad de la Frontera: Temuco, Chile, 1998.
16. Valdebenito, G. Perspectivas de los Productos Forestales No Madereros (PFnM) en Chile. In *Los Sistemas Agroforestales en Chile*; Sotomayor, A., Barros, S., Eds.; INFOR: Santiago, Chile, 2016; pp. 357–404.
17. Joublan, J.; Berti, M. Rosa mosqueta, una alternativa agrícola para zonas de secano. *Rev. Agroanálisis* **1997**, *13*, 29–33.
18. Instituto Forestal. *Innovación Tecnológica y Comercial de Productos Forestales No Madereros (PFNM) en Chile*; INFOR: Santiago, Chile, 2003; p. 8.

19. Organización de las Naciones Unidas para la Agricultura ya la Alimentación (FAO). *El Comercio Justo: Una Alternativa Para la Agroindustria Rural de América Latina*; FAO: Santiago, Chile, 2002; p. 64.
20. Fundación de Apoyo a la Microempresa Rural de América Latina y El Caribe (FAMER). *El Comercio Justo: Historia, Metodologías, Oportunidades y Desafíos Para la Pequeña Agricultura*; FAMER: Santiago, Chile, 2006; p. 63.
21. Weldt, E. Establecimiento, multiplicación y enraizamiento in vitro de *Rosa canina* L. In *Tesis de Grado*; Universidad Austral de Chile: Valdivia, Chile, 2008.
22. Benavente, A. Fertilización en Rosa Mosqueta (*Rosa rubiginosa* L.) bajo condiciones de secano interior octava Región. In *Tesis de Grado*; Universidad de Concepción: Concepción, Chile, 2001.
23. González, C. Fertilización en rosa mosqueta (*Rosa rubiginosa* L.) bajo condiciones de secano. In *Tesis de Grado*; Universidad de Concepción: Concepción, Chile, 1999.
24. Gaete, F. Aplicación de glifosato y paraquat en rosa mosqueta (*Rosa rubiginosa* L.) cultivada. In *Tesis de Grado*; Universidad de Concepción: Concepción, Chile, 1999.
25. Muñoz, J.; Mardones, R. Autogestión e Intercambio Territorial Mapuche en la Comuna de Melipeuco (2003–2008): Una mirada desde el Desarrollo Local. In Proceedings of the Acta Científica XXIX Congreso de la Asociación Latinoamericana de Sociología, Santiago, Chile, 29 September–4 October 2013.
26. Centro de Información en Recursos Naturales (CIREN). *Delimitación de Microregiones, IX Región de la Araucanía*; Ministerio de Agricultura: Gobierno de Chile, Chile, 1994.
27. PLADECO. *Plan de Desarrollo Comunal 2012–2016; I. Municipalidad de Melipeuco*; PLADECO: Melipeuco, Chile, 2011.
28. Peredo, S.; Barrera, C. Agroecology, Local Knowledge and Participatory Research: Articulation of Knowledge for Sustainable Use of Plant Resources in Agroecosystems. In *Ethnobotany: Local Knowledge and Traditions*; Martínez, J.L., Muñoz, A., Rai, M., Eds.; Taylor & Francis; CRC Press: Boca Raton, FL, USA, 2019; pp. 19–33.
29. López de Ceballos, L. *Un Método Para la Investigación Acción*; Editorial Popular: Madrid, España, 1992.
30. Ardón, M. Métodos e instrumentos para la etnoecología participativa. *Etnoecología* **2002**, *6*, 129–143.
31. Guzmán, G.; Alonso, A. La investigación participativa en agroecología: Una herramienta para el desarrollo sustentable. *Ecosistemas* **2007**, *16*, 24–36.
32. Norgaard, R. Una sociología del medio ambiente coevolucionista. In *Sociología Del Medio Ambiente. Una Perspectiva Internacional*; Redclift, M., Woodgate, G., Eds.; Mc. Graw Hill: Madrid, Spain, 2002; pp. 167–178.
33. Peredo, S.; Acuña, B.; Barrera, C. Diálogos entre la teoría y prácticas agroecológicas para el establecimiento de estrategias de desarrollo rural sustentables: Investigación y acción a partir del potencial endógeno. In *Anuario de Antropología Lberoamericana*; Antropología Iberocamericana (AIBR): Madrid, Spain, 2019.
34. Peredo, S.; Barrera, C. Desarrollo Rural Endógeno: Condiciones para una transición agroecológica desde una experiencia de producción orgánica. Universidad Católica de Temuco. *CUHSO* **2002**, *6*, 71–90. [CrossRef]
35. Peredo, S.; Acuña-Jujihara, B.; Hurtado-Quiñones, A. Agroecología y antropología: Acercamientos para un encuentro transdisciplinario. In Proceedings of the VIII Congreso Chileno de Antropología, Colegio de Antropólogos, Arica, Chile, 11–15 November 2013.
36. Altieri, M.; Toledo, V. La Revolución Agroecológica en América Latina: Rescatar la naturaleza, asegurar la soberanía alimentaria y empoderar al campesino. In *El Otro Derecho 42*; Instituto Latinoamericano para una Sociedad y un Derecho Alternativos (ILSA): Bogotá, Colombia, 2010; pp. 163–202.
37. Peredo, S.; Barrera, C.; Acuña, B. Complejizando la mirada agroecológica: Reflexiones sobre los conceptos de agroecosistema y resciliencia para ampliar las iniciativas agroecológicas. USACH. *Sustentabilidades* **2019**, *10*, 137–159.
38. Sevilla-Guzmán, E.; Soler-Montiel, M. Agroecología y soberanía alimentaria: Alternativas a la globalización agroalimentaria. In *Patrimonio Cultural de la Nueva Ruralidad Andaluza, Junta de Andalucía*; Instituto Andaluz de Patrimonio Histórico, Ed.; Consejería de Cultura: Sevilla, España, 2010; pp. 190–217.
39. Caporal, F.R. Agroecología: Ciencia para agriculturas más sostenibles. *Am. Lat. Mov.* **2013**, *487*, 6–10.
40. Escobar, A. *Más Allá Del Tercer Mundo. Globalización y Diferencia*; Instituto Colombiano de Antropología e Historia: Bogotá, Colombia, 2005.
41. Gutiérrez, J.G.; Aguilera, L.I.; González, C.E. Agroecología y Sustentabilidad. *Rev. Cienc. Soc. Converg.* **2008**, *46*, 51–87.

42. Peredo, S.; Alvarez, R.; Barrera, C.; Parada, E. Nutritional value of *Dasyphyllum diacanthoides* (Less.) Carb. an endemic tree used as suplementary forage in agroforestry systems. *BIOAGRO* **2020**, *30*, 139–144.
43. Peredo, S.; Parada, E.; Alvarez, R.; Barrera, C. Propagación vegetativa por estacas de *Dasyphylum diacanthoides* mediante recursos endógenos. Una aproximación agroecológica. *Revista Boletín Latinoamericano y Caribeño de Plantas Medicinales y Aromáticas. BLACPMA* **2015**, *14*, 301–307.
44. Peredo, S.; Barrera, C. Usos etnobotánicos, estrategias de acción y transmisión cultural de los recursos florísticos en la localidad de Armerillo, Región del Maule (Chile). *BLACPMA* **2017**, *16*, 398–409.
45. Soler Montiel, M.; Rivera Ferré, M.G. *Agricultura Urbana, Sostenibilidad y Soberanía Alimentaria: Hacia Una Propuesta de Indicadores Desde la Soberanía Alimentaria*; X Congreso de la Federación Española de Sociología: Pamplona, Navarra, Spain, 2010; pp. 1–17.
46. Bello, Á. Espacio y territorio en perspectiva antropológica. El caso de los purhépechas de Nurío y Michoacán en México. *Rev. CUHSO* **2011**, 41–60. [CrossRef]
47. Ther Ríos, F. Antropología del territorio. *Polis* **2012**, *11*, 493–510. [CrossRef]
48. García, J.L. *Antropología del Territorio*; Taller de Ediciones Josefina Betancor: Madrid, España, 1976.
49. Toledo, V.M. *La Memoria Biocultural. La Importancia Ecológica de Las Sabidurías Tradicionales*; Icaria Editorial S.A.: Barcelona, España, 2008.
50. Peredo, S.; Barrera, C. Democratizando el consumo ecológico: Elementos para la acción y aprendizaje colectivo en procesos de investigación acción participativa. *Agroecología* **2018**, *13*, 57–69.
51. Peredo, S.; Barrera, C.; Acuña, B. De qué hablamos cuando hablamos de ampliar la Agroecología? Hacia una (re)configuración de sistemas agroalimentarios (re)localizados desde la democracia y la participación. *Revista Anual de Antropología Iberoamericana* **2019**. ISSN 2530-7843.
52. Anderzén, J.; Guzmán, A.; Luna-González, D.; Merrillb, S.; Caswella, M.; Méndez, V.; Hernández, R.; Mier y Terán Giménez, M. Effects of on-farm diversification strategies on smallholder coffee farmer food security and income sufficiency in Chiapas, Mexico. *J. Rural Stud.* **2020**, *77*, 33–46. [CrossRef]
53. Altieri, M.A.; Toledo, V.M. The agroecological revolution of Latin America: Rescuing nature, securing food sovereignty and empowering peasants. *J. Peasant Stud.* **2011**, *38*, 587–612. [CrossRef]
54. Sevilla Guzmán, E. La participación en la construcción histórica latinoamericana de la Agroecología y sus niveles de territorialidad. *Política Soc.* **2015**, *52*, 351–370, ISSN 1130-8001. [CrossRef]
55. Pérez-Neira, D.; Soler-Montiel, M. Agroecología y ecofeminismo para descolonizar y despatriarcalizar la alimentación globalizada. *Rev. Int. Pensam. Político* **2013**, *8*, 95–113.
56. Calle, Á.; Gallar, D.; Candón-Mena, J. Agroecología política: La transición social hacia sistemas agroalimentarios sustentables. *Rev. Econ. Crít.* **2013**, *16*, 244–277.
57. García-Roces, I.; Soler-Montiel, M.; Sabuco, I.; Cantó, A. Perspectiva ecofeminista de la Soberanía Alimentaria: La Red de Agroecología en la Comunidad Moreno Maia en la Amazonía brasileña. *Relac. Int.* **2014**, *27*, 75–96.
58. Sevilla-Guzmán, E. Sobre as perspectivas teórico metodológicas da Agroecología. *Redes St. Cruz Sul* **2017**, *22*, 13–30.
59. Altieri, M.A. Linking ecologists and traditional farmers in the search for sustainable agriculture. *Front. Ecol. Environ.* **2004**, *2*, 35–42. [CrossRef]
60. Barahona, R. Conocimiento campesino y sujeto social campesino. *Rev. Mex. Sociol.* **1987**, *49*, 167–190. [CrossRef]
61. Lamine, C.; Garçon, L.; Brunori, G. Territorial agrifood systems: A Franco-Italian contribution to the debates over alternative food networks in rural areas. *J. Rural Stud.* **2019**, *68*, 159–170. [CrossRef]
62. Cuellar, M.; Sevilla-Guzman, E. Aportando a la construcción de la soberanía alimentaria desde la agroecología. *Ecol. Política* **2009**, *5*, 43–52, ISSN 1130-6378.
63. Grando, S.; Carey, J.; Hegger, E.; Jahrl, I.; Ortolani, L. Short Food Supply Chains in Urban Areas: Who Takes the Lead? Evidence from Three Cities across Europe. *Urban Agric. Reg. Food Syst.* **2017**, 2. [CrossRef]
64. Vicente-Almazán Castro, L.; Herrera-Gil, M.; Escobar-Cruz, M. *Sistemas Alimentarios Territorializados en España. 100 Iniciativas Locales Para Una Alimentación Responsable y Sostenible*; Cerai-Carasso: Valencia, Spain, 2019.
65. Mier y Terán Giménez, M.; Giraldo, O.; Aldasoro, M.; Morales, H.; Ferguson, B.; Rosset, P.; Khadse, A.; Campos, C. Bringing agroecology to scale: Key drivers and emblematic cases. *Agroecol. Sustain. Food Syst.* **2018**, *42*, 637–665. [CrossRef]

66. Esplug, J.; López-García, D.; Clavet-Mir, L.; di Masso, M.; Pomad, A.; Tendero, G. Agroecología, conocimiento tradicional e identidades locales para la sostenibilidad y contra el despoblamiento rural. *Rev. PH* **2019**, *98*, 108–130.
67. Domené-Painenao, O.; Mier y Terán-Giménez, M.; Limón-Aguirre, F.; Rosset, P.; Contreras-Natera, M. Construcción territorial de agroecologías situadas: El Maestro Pueblo en Sanare, estado Lara-Venezuela. *Estud. Soc.* **2020**, *30*. [CrossRef]
68. López-García, D.; Pontijas, B.; González de Molina, M.; Delgado, M.; Guzmán, G.; Infante-Amate, I. Saltando de escala ... ¿hacia dónde? El papel de los actores convencionales en los sistemas alimentarios alternativos. *Ager* **2018**, 99–127. [CrossRef]

# 3

# Gendered Migration and Agroforestry in Indonesia: Livelihoods, Labor, Know-How, Networks

**Elok Mulyoutami** [1,*], **Betha Lusiana** [1] **and Meine van Noordwijk** [1,2]

1 World Agroforestry Centre (ICRAF), Bogor 16001, Indonesia; b.lusiana@cgiar.org (B.L.); m.vannoordwijk@cgiar.org (M.v.N.)

2 Plant Production Systems, Wageningen University, 6708 PB Wageningen, The Netherlands

* Correspondence: eloknco@gmail.com

**Abstract:** Migration connects land use in areas of origin with areas of new residence, impacting both through individual, gendered choices on the use of land, labor, and knowledge. Synthesizing across two case studies in Indonesia, we focus on five aspects: (i) conditions within the community of origin linked to the reason for people to venture elsewhere, temporarily or permanently; (ii) the changes in the receiving community and its environment, generally in rural areas with lower human population density; (iii) the effect of migration on land use and livelihoods in the areas of origin; (iv) the dynamics of migrants returning with different levels of success; and (v) interactions of migrants in all four aspects with government and other stakeholders of development policies. In-depth interviews and focus group discussions in the study areas showed how decisions vary with gender and age, between individuals, households, and groups of households joining after signs of success. Most of the decision making is linked to perceived poverty, natural resource and land competition, and emergencies, such as natural disasters or increased human conflicts. People returning successfully may help to rebuild the village and its agricultural and agroforestry systems and can invest in social capital (mosques, healthcare, schools).

**Keywords:** coffee; cocoa; Java; livelihoods; rural–urban; remittances; returning migrants; Sumatra; Sulawesi

## 1. Introduction

Contrary to the long-term attachment to place that prevails in "myths of origin" and cultural constructs of place-based identity [1,2], humans have a history of dispersal and migration [3], as reflected in our complex DNA and linguistic signatures [4]. Migration has been the demographic basis of the expansion of our species spreading to all parts of the world, adapting to a wide range of circumstances and learning how to cope with variability and diversity. Both cultural and genetic evidence suggests that human dispersal and migration were not a one-way process and that links to areas of origin were maintained through any means of communication and transport that was accessible in given periods of human history and development. Cultures have absorbed newcomers while migrants kept a cultural attachment to areas of origin, creating the rich "unity in diversity" fabric that characterizes many Asian countries [5]. Human dispersal and migration have had ethnobotanical consequences, with a large number of semi-domesticated trees and crops spreading along with humans as well as the knowledge of how to use them [6,7]. Working across Indonesia, anecdotes of how agroforestry and local forest management practices were inspired by experiences elsewhere during "circular migration" or were traced back to migrants from other parts of the country are commonly heard [8] but appear not to have been systematically analyzed. In the current "lockdown" response to the COVID-19 pandemic [9], both positive and negative aspects of such human movement call for a more nuanced analysis as part of the Sustainable Development Goals [10]. Migration can involve a radical decision to uproot and try one's luck elsewhere for a variety of social, political, economic, and environmental reasons [11,12] or be a more gradual process wherein temporary and circular migration precedes "permanent" migration at the individual or household level, with family members potentially following suit [13]. The decision to migrate circularly or permanently is usually taken step by step in response to perceived success or failure and interacting with external circumstances, such as policies in both the area of origin and area of the temporary new residence. While many studies have zoomed in on specific parts of the decisions to move and/or return, a holistic perspective requires a human life-cycle approach that is not easy to obtain experimentally but can be constructed by combining separately studied pieces of a larger puzzle.

Each year, around half a million Indonesians travel abroad to work, half of those to the Middle East. They are typically women from small cities or villages with primary education and limited work experience, hired to perform domestic work [14]. Econometric analysis of data on emigration rates of countries at different stages of economic development has revealed inverse U-shaped responses. With a GDP per capita at purchasing power parity of USD 3893 in 2018 [15], Indonesia as an emerging lower-middle-income country is approaching the income level at which international migration is expected to peak (USD 5–10,000). Although merely based on cross-sectional evidence, the "migration hump" is widely interpreted as a causal relationship. However, this interpretation is contested [16,17]. The fact that income growth increases the opportunity cost of migration and also eases liquidity constraints—two opposite forces at play—may explain the "hump". Further analysis of a migration matrix for all of Indonesia [18] found roles for ethnic networks in groups such as the Sundanese from West Java and the Buginese from South Sulawesi that reduced the fixed costs of international migration from rural areas of Indonesia and contributed to the recorded way income elasticity of migration varies depending on the exposure of the given type of landholdings to variability of rainfall and rice prices. Although these ethnic communities may be isolated from native ethnic populations when residing outside their historical homelands, they still have strong ties to the broader Sundanese and Buginese networks with the potential for connections to international labor markets. The choices involved in internal migration within large countries like Indonesia may, however, differ from those for international migration. A study in East Java found that individuals with access to water, health insurance, or markets, or those living in villages that have a large proportion of non-irrigated land being used for non-agricultural activities, were less likely to seek employment elsewhere [19].

Internal migration in Indonesia has been linked to productivity growth [20]. Between 1979 and 1988, the Transmigration Program relocated 2 million voluntary migrants (hereafter, transmigrants) from the Inner Islands of Java and Bali to newly created agricultural settlements in the Outer Islands (with Sumatra, Kalimantan, Sulawesi, and Papua being the largest). The success rate depended on the agro-ecological similarity between the areas of origin and migration targets [21], while diverse communities, rather than migrants from a single origin, could be linked to stronger integration and success in the new environment [22].

Migration decisions relate to gender-specific expectations and have gender-specific consequences for those who move, those who stay behind, and those who return. The latter can enrich local livelihoods with new knowledge, norms, and expectations, apart from the financial resources they may have. However, returning migrants will need to invest in social capital to earn their place back in the local community. In cases where males are primarily involved in migration to Africa and South Asia, effectson the source areas may be a "feminization" of agriculture [23,24], however, the opposite case—where females preferentially engage—is common in Asia as well. Distinct feminization of labor migration in Southeast Asia, participating in gender-segmented global labor markets, has significantly altered care arrangements, gender roles and practices, as well as family relationships within households [25]. Elsewhere, youth migration has become an issue [26]. What these all have in common is that selective migration of either gender or age segments leads to a reshuffling of roles in local livelihoods and households, with a gradual change in associated "identities". Burgers [27] described the consequences of female migration for the matrilineally inherited rice fields in Kerinci, Sumatra with a complex adaptive change to customary rules.

Anecdotal evidence, as reflected in journalists' accounts and literary reflections, suggests that the full spectrum of success to failure and exploitation exists. Policy responses that try to minimize risks of exploitation and trafficking need to be constructed without reducing opportunities for livelihood benefits, which is no easy task. Our analysis of case studies was aimed at obtaining a more complete understanding of the drivers and consequences of migration from densely populated rural parts of Java in Indonesia to overseas urban and rural target areas, using a gender lens. Gender-specific aspects can be expected to apply to the decision to move (for young males, young females, families); their roles and opportunities in the new areas of temporary residence interacting with local communities and businesses; their relations back home, including remittances, invitations to join, sharing of knowledge and experience; the consequences of their migration on those who stay behind; and the consequences in case they decide to return home with any assets they may have acquired, their new skills, and norms of behavior, as shown in Figure 1. Each of these aspects deserves a fully fledged study but, even with the limited evidence available, a system approach to the whole picture is needed to guide policy development and to inform public discourse, where strongly polarized opinions pro and con tend to dominate.

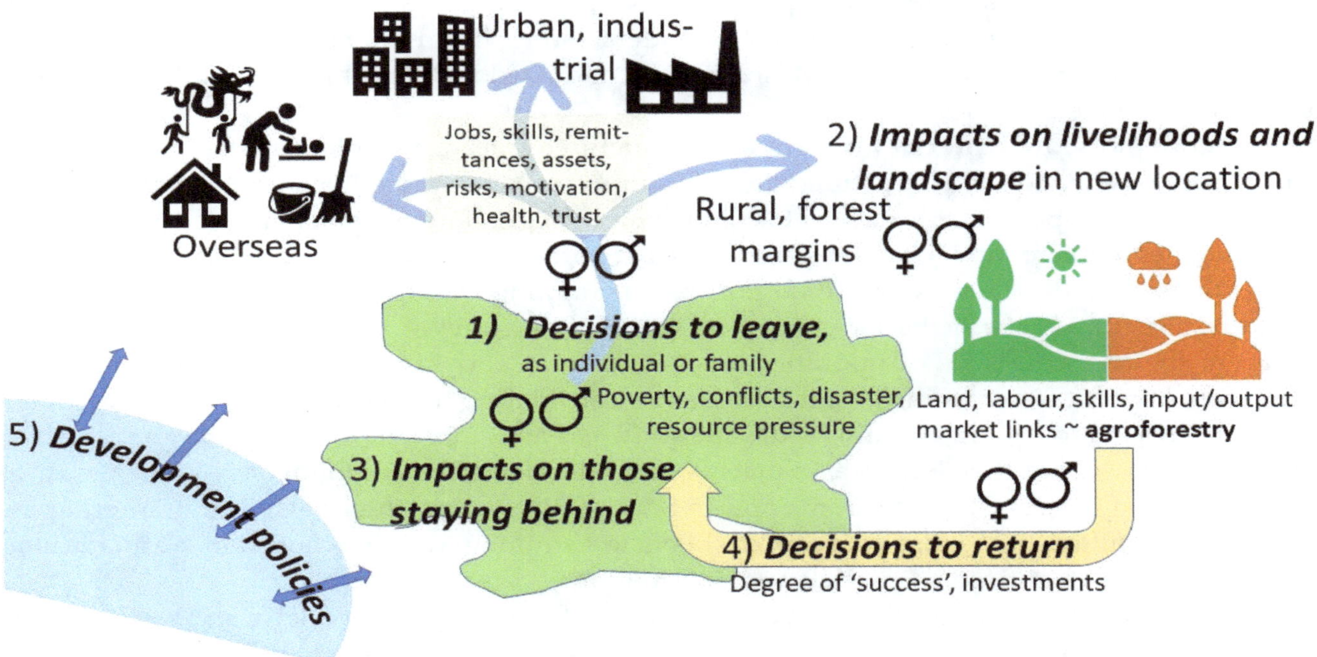


**Figure 1.** Gender-specific aspects before, during, and after (circular) migration decisions concerning land use in the area of origin.

While migration decisions—leaving and returning—are part of a large body of ethnographic studies, few studies have analyzed both the source and target areas of specific rural-to-rural migration patterns in Indonesia. We here describe two such cases, which allowed us partial answers to the following questions.

(1) What are the gendered patterns of movement concerning age and life histories in both source and target areas? Do gender norms of behavior influence land-use patterns differentially in source and target areas?

(2) How do (temporary) migrants compare the positive and negative aspects of home and temporary abodes?

(3) How do returnees reintegrate and modify land use, gender norms, and culture of the areas from which they originated?

We will first describe the two cases as such and then draw comparisons between them. The discussion will also touch on interactions with the government and other stakeholders concerning development policies.

## 2. Methods and Locations

### *2.1. Methods*

This paper is based on anthropological fieldwork by the first author analyzing gender aspects of social development across Indonesia and, in particular, in two cases where migration source areas (West Java and South Sulawesi) could be linked to specific migration target areas (Lampung and Southeast Sulawesi, respectively), as shown in Table 1. Details of some of the field studies synthesized here have been published elsewhere [28–30]. Their *post hoc* combination into a single study, based on emerging opportunities rather than prior design, is new here.

**Table 1.** Characteristics of migration in the study area.

| Study Area | Source | Area of Destination | Ethnicity | Type of Migration | Type of Work in the Destination Area |
|---|---|---|---|---|---|
| **West Java** | Ciamis | Sumberjaya, West Lampung | Sundanese | Temporary | Land-based and off farm |
| **Southeast Sulawesi** | Bone, Bulukumba, Soppeng, Sinjai, Wajo, Jeneponto, Maros in South Sulawesi | Kolaka and Konawe in Southeast Sulawesi | Bugis | Temporary | Land-based and off farm |
| | | | Bugis | Permanent | Land-based |

The two case studies were each explored in two phases. Phase 1 aimed at understanding the migration context in the origin areas; Phase 2 on understanding the in-migration pattern and challenge in receiving communities. The West Java study was conducted in 2016 while the Southeast Sulawesi study was conducted in 2013.

In-depth individual interviews, individual structured interviews through a household survey, participatory observation, and structured group discussions were the methods used to obtain primary data used for this paper. The primary data could explore migration phenomenon, networks, patterns, and challenges at micro- and meso-levels. Specific to Southeast Sulawesi, the migration network was deeply analyzed at these levels. Stratified purposive sampling was applied for each method, for the individual structured interviews and structured group discussions.

From the characteristics of migration in each area, as shown in Table 1, we identified each community household according to some typologies considering the status of migration (migrant or stayer), the reason for migration and its destination areas and, most importantly, the gender aspect. In the case study of South to Southeast Sulawesi, we interviewed 65 respondents—in the Southeast Sulawesi we only interviewed migrants and in South Sulawesi we interviewed both migrant and non-migrant families. In the case study of West Java to Lampung, we interviewed in West Java alone 120 respondents, categorized as migrant and non-migrant.

Snowball sampling was used in particular for in-depth individual interviews and social network analysis. In-depth interviews were conducted to understand the historical and social realities that described the established migration chain and network. Secondary data, such as the results of a population census by the Central Statistical Bureau (Biro Pusat Statistik: BPS) [31,32], related literature and documentation, were used to support an explanation of the migration phenomenon, mainly at meso- and macro-levels.

The results of the household surveys were analyzed using descriptive statistics and interpreted qualitatively combined with information from focus group discussions and in-depth interviews to explain migrants' and non-migrants' characteristics, social typology, and actors involved in the migration chains. T-test statistics were employed to determine the significant differences between males and females and differing situations of migration. The analysis of migration or social networks was conducted using NodeXL software, which can measure and visualize the relationships of actors.

### *2.2. Study Sites: Background and Context*

#### 2.2.1. Indonesia

Indonesia is an archipelagic nation with the world's fourth-largest population (around 260 million people). The nation has recently reached lower-middle-income status. The average annual population growth was 1.27% for 2005–2010 and is expected to decline to 0.82% for 2020–2025. In 2015, 52.6% of the population lived in urban areas. Urban growth is, at 4.1% $y^{-1}$, the highest in Southeast Asia. The sex ratio of recent migration has been 110.3 (males per 100 females).

In terms of human population density, the largest contrast is between (i) densely populated, volcanic Java and Bali, with fertile soils and wet-rice agricultural traditions; (ii) Sumatra with population densities around the national average and strong tree-crop traditions, with coffee in the mountains and rubber and oil palm in the lowlands; and (iii) the rest of the archipelago with lower human population densities, with the exception of South Sulawesi, which approximates Java. Gender roles in agriculture tend to vary with regional contexts [28,29]. In upland areas, farming systems have mostly shifted from rice swiddens to a reliance on tree crops, such as cocoa, coffee, and rubber [33] and oil palm in areas with suitable rainfall.

### 2.2.2. Case Study One: The South to Southeast Sulawesi Connection

Our first case study analyzed migration histories, patterns, and networks from South to Southeast Sulawesi and other areas in rural areas in Indonesia and abroad. Cocoa began to boom and experience a "golden age" in Lawonua around 1997–1998 following the rise of prices worldwide, owing to a decrease in production in the Ivory Coast that led to global shortages. Li [34] described the same condition in Central Sulawesi, which also experienced a large arrival wave from South Sulawesi during the same period. Indonesia became a promising candidate for "major cocoa producer" at the time.

Working in the context of a regional development program underway at the time, we found that Lawonua Village in Besulutu Sub-district, Kolaka District, Southeast Sulawesi, with ongoing cocoa expansion, was a destination area for migrants from South Sulawesi. In the area, we had contacts that allowed surveys to be undertaken. The migrant percentage and composition in Lawonua were similar to that of the sub-district as a whole. The flow of migration into the village had been continuous over the preceding few years, which assisted our study in tracing the identity of the migrant community at origin. The tracing was conducted in the context of creating a community profile, which included physical, social, and economic conditions. The tracing was not only conducted at the location of migration (their current place of living) but also included the conditions (physical, social, and economic) in their origin village.

From the tracing process in Lawonua, 60% of migrants (which consisted of 40% of the total village population) came from Kalobba Village in Tellu Limpoe Sub-district, Sinjai District, South Sulawesi, as shown in Figure 2. Kalobba, characterized by limited resources and medium agricultural technology, is classified as a "suburban" area with limited land resources owing to pressure from outsiders. Competition for land causes a fairly high number of outgoing migrations from this village.

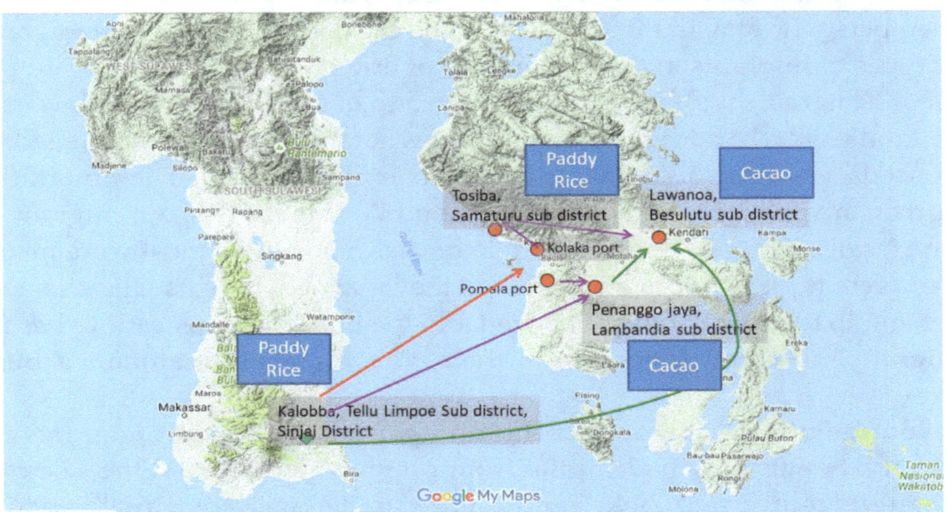


**Figure 2.** The source and destination of migrants in our first case study, in South (left) and Southeast Sulawesi (right), respectively.

2.2.3. Case Study Two: The West Java to Lampung Connection

Our second case study analyzed migration patterns, dynamics of migrant and stayer communities, migration decision making, and gender relations, with a focus on Ciamis District, West Java, where there is a close migration connection to coffee-growing landscapes in West Lampung District, Lampung, as shown in Figure 3. In Ciamis, the study was conducted in two villages in Panjalu Sub-district and two villages in Rajadesa Sub-district. Panjalu is in the northwest of the district and is the capital of Ciamis. Rajadesa is in the eastern part. Both villages are categorized as agricultural communities that rely heavily on farming as their source of livelihoods.

**Figure 3.** The source (right) and destination (left) of migrants in our second case study, West Java and Lampung.

## 3. Results

### *3.1. Migration Decision Making*

Within the community of origin, decisions to migrate or stay in the village were mainly due to economic opportunity. Many migrants chose to migrate based on the capital they had, the support provided by their extended family and neighbors who could reduce the need for money for taking care of family members who stay in the village, or even the cost of living in the migration destination. There were four types of migrants in the community of origin.

(1) Off-farm out-migrants without capital. This group of out-migrants had low–middle economic status and were landless or had low levels of land ownership. They usually tried to find work to meet their daily needs as well as accumulate capital to establish farms. Their migration destinations were mostly in urban areas nearby, both in Sulawesi and West Java. This out-migrant group generally consisted of some family members (either women or men); however, most were male. Most women who out-migrated were typically unmarried. When they married, they usually chose to stop working for money, preferring to take care of their children and the household instead. Other family members who did not migrate and chose to remain in their village usually maintained businesses and/or cultivated farms.

(2) Off-farm out-migrants with capital. This group of middle–up out-migrants owned medium-sized areas of land. They usually had capital from previous work, from the sale of crops, or from inheritance. This out-migrant group could feature entire families migrating to cities or other prospective rural areas, who would only return to their village of origin for holidays. Most of these out-migrants were male. If there were women who out-migrated, they usually went with other family members.

They were mostly interested in off-farm livelihood sources, e.g., selling secondhand iron in West Java, while in Lampung and Southeast Sulawesi, selling clothes and other household items. The success of previous out-migrants in this line of work, as well as the networks created, attracted others from their village to do the same.

(3) and (4) Land-based out-migrants with and without capital. Land-based migrants without capital consisted of low–middle economic status who were mostly landless or had limited land in the origin community. They moved to the destination as on-farm labor, mostly working on their relatives' land, with some developing patron–client relationships. Rajadesa Village members in West Java migrated to Sumberjaya, Lampung to support their relatives in growing coffee. In Sulawesi, the relationship between a landowner and their followers became one of the doors through which a large number of migrants arrived. Landowners (land-based out migrants with capital) had some funds to open cocoa plantations and recruit followers from their home village to clear the land, plant, and care for the cocoa. When the plantation began to produce—after five years—a sharing system was implemented. The landowners maintained relationships with their clients to support their economic activities as well as maintain the power of their networks. The patrons' clients, who generally came from lower socio-economic groups, had better income sources through the patron compared to those in their home village, which no longer attracted their attention. Furthermore, through the patron's support, their migration to the new area became lower in risk and the cost required to migrate was reduced. The reciprocal relationship (reciprocity), although it was not entirely symmetrical (it was often highly asymmetric), was still able to improve their income. A summarized excerpt from an interview that illustrates this pattern follows.

> *Around 1997, HS, a landlord from South Sulawesi, purchased a large amount of land in Southeast Sulawesi. HS [a patron] later recruited his men [clients] to manage his land. His men were given approximately four hectares each for cocoa plantations. During the first six months, HS's men were given a living allowance of approximately 20 kg of rice per month, salted fish, and a few other staples. After six months, HS's men subsisted on seasonal crops grown in their plantations. After producing cocoa—after about 5 years—the harvest was divided: one part for the landowner and another part for the workers. Nowadays, the land here still belongs to the landlord, the clients may get a small part of the land as theirs, though the other land was still owned by HS. [30]*

### *3.2. Social Network, The Instrument of Migration Decision Making*

As detailed elsewhere [30], three main network models reflected strategies used by migrants in their decision to migrate. The first model was a kinship-based network of either close relatives or immediate family, as well as distant relatives or extended family (53.84%). This strategy was commonly deployed by migrant communities who tended to be more mature, had sufficient capital to start migrating, as well as knowledge of cocoa or coffee cultivation as a requirement for planting. However, some landless people also decided to migrate as on-farm labor for other migrant communities who had established their farming practices in destination areas. Those migrants could be temporary, permanent, or seasonal. An excerpt from an interview that illustrates this pattern follows.

> Mr T, a resident of T, Lawonua Village, stated, *"I was visited by my uncle ... and he said, come on, move ... What are you doing staying in this village? You can plant a cocoa plantation there and the price of the land is cheap, not as expensive as here. Sell your plantation or cows here, it's enough to buy land there". [30]*

The second was a network set up to gain profit (44.12% of the total network). This network was either run in balance or not and built through a patron–client mechanism. A capital owner who later acts as a patron needs workers who are his inferiors as clients. The patron provides jobs and financial support, including the cost of migrating and supporting the clients' living needs in the early days of migration. These clients need the patron to improve the economic conditions of their families through managing their land as well as minimizing their migration risks. In this type of relationship, often the client's decision to migrate was not voluntary but forced owing to economic pressure and the vertical relationship with the patron.

The third network is a pattern of relationships that emerge owing to similarity of purpose. Generally, this pattern is characterized by identity, location of origin, and current residence similarities and was generally found in migrant communities who had been pioneer settlers in Southeast Sulawesi and Lampung. These groups built a network of neighborhood or identity similarity among community origin who had the same goal of increasing the number of plantations by expanding to villages that still had available land.

In the case of West Java, the first settlers in Lampung joined the transmigrant program in the early 1960s [35]. With their success in managing coffee systems apparent by the late 1980s, migrants joined from West Java originally as laborers and opened new coffee plots when they had access to capital [36].

These various relationship patterns often overlapped. A vertical relationship pattern, such as the patron–client relationship, could be reinforced by the patterns of kinship and neighborhoods, which were horizontal. For example, in the patron–client relationship, the kinship between the two often enlarged the client's decision to migrate not only because of economic need but also reluctance to reject an offer from a relative. Moreover, the overlapping relationship was also enforced by brokers, intermediaries, or ones who bridge the various groups of migrants from different regions to select land and encourage them to move. An intermediary or broker is an actor who can bridge and build the trust of the individual or group of individuals who initially were not interconnected. They facilitate social interaction, increase a community's economic activity, and minimize the risks of migration. On the other hand, the broker or intermediary is often associated with exploitation, transfer of risk into profits for intermediaries, and the accumulation of profit.

### *3.3. Type of In-Migrant*

Migrant characteristics in receiving communities vary, however, among (1) new migrants (first-time movers), (2) recurrent migrants (multiple movers), and (3) follow-up migrants (family movers), as shown in Table 2. New migrants (pioneer migrants) are categorized as migrants who come directly from the area of origin to the receiving community and who have not migrated to other areas before. They may have a connection to the receiving community, through their family or neighbor who has lived in those areas, or they just decided to move, driven by the motivation to obtain land and increase their incomes.

Recurrent migrants or multiple movers are those who have already moved into, and out of, the region in surrounding receiving communities more than once. They may have some experience abroad in Malaysia or in Kalimantan, Indonesia. Recurring migrants in Southeast Sulawesi were generally migrants moving from an area near to their current receiving communities but decided to move to other areas in the same province to find better economic opportunities. A summarized excerpt from an interview that illustrates this pattern follows.

> *P, 68 years-old, a cocoa farmer from Kampala Village in Sinjai District, South Sulawesi sailed to Konaweha, Kolaka District, Southeast Sulawesi in the 1970s. He crossed the Gulf of Bone with his youngest child, aged 6 years, bringing five sacks of rice, two cows, and money amounting to IDR 35,000 (≈USD 3.50). After sailing for three days and two nights, he arrived at Kolaka and immediately visited his uncle who had already moved to Konaweha Village. He was helped by his uncle to look for flat land to be used to grow rice. After two days, he found land owned by a native resident who had received the land from the government but was unable to cultivate it. The land was sold cheaply to P. Over time, P's desire for land increased, especially for providing land for his children to equip them for the future. In 1995, P sought land in Lawonua (their current location) assisted by SF. Once the land was obtained, P did not move to Lawonua but still lived in Kolaka. However, in 2000, P finally decided to move to Lawonua with his wife and child. [30]*

**Table 2.** Migrant types and characteristics in Southeast Sulawesi [30].

| Migrant Types | % | Origin Areas | Educational Background (% of Population) | | | | Migrant Age Group (%) | | | | Gender | |
|---|---|---|---|---|---|---|---|---|---|---|---|---|
| | | | Primary | Lower Secondary | Upper Secondary | Tertiary | 16–24 | 35–29 | 40–54 | >54 | M | F |
| **New migrants** | 45.83 | Bone, Bulukumba, Sinjai, Soppeng, Wajo in South Sulawesi | 44.4 | 44.4 | 11.1 | 0 | 1.39 | 22.22 | 18.06 | 5.56 | 14 | 20 |
| **Recurrent migrants** | 38.89 | Bone, Bulukumba, Soppeng, Sinjai, Wajo, Jeneponto, Maros in South Sulawesi | 67.9 | 14.3 | 10.7 | 7 | - | 11.11 | 18.06 | 5.56 | 13 | 14 |
| **Follow-up migrants** | 12.5 | Bone, Bulukumba, Soppeng, Sinjai, Wajo in South Sulawesi | 69.7 | 15.2 | 12.1 | 3 | 2.78 | 6.94 | 2.78 | - | 6 | 3 |

These recurring migrants migrated with the motivation to increase the amount of land they owned. Some of them had financial difficulties and intended to sell land in Pinanggo or Konaweha to overcome their problems. Land in the area was more expensive than land in Lawonua. Selling the land in their area and looking for land in a cheaper place could solve their problems. Some other migrants who were able to accumulate capital deliberately looked for land in new areas to increase the number of their plantations. Recurring migrants usually owned land. The acquired land was later shared as inheritance or dowry for the marriages of their offspring.

Follow-up migrants are descendants of migrants who have lived for a long time in receiving communities. Most of them were born in their origin areas and later moved to the receiving communities as toddlers or teens after their parents had already moved there. They are follow-up migrants who have not received a share of the family land nor been able to obtain their own.

### *3.4. Gender Relations in Communities of Origin*

In describing the gender and age specificity of migration patterns, we used distinctions based on the status of who had migrated, as shown in Table 3. From the two case studies, we could see that for land-based livelihood options, youth or single men and adult or married men migrated by themselves and left their families in the origin area, who might join later. However, the discussion has shown that no married women left their villages for land-based options but perhaps to work in urban areas.

**Table 3.** Livelihood options based on the type of migration, age, and gender.

| Gender of the Migrant | Age Classification | Livelihood Option | | Type of Migration | Condition in Origin Areas |
|---|---|---|---|---|---|
| | | Off-Farm Based | Land-Based | | |
| Only men migrated | Youth (un-married) | Industrial work in urban and rural areas abroad<br>Business (sales) | Plantation labor | SeasonalPermanent | - Abandoned land<br>- Spouse or an extended family member manages the land |
| | Adults (married) | | Sharecropping<br>Land-rent | Seasonal | |
| Only women migrated | Majority youth and unmarried | Industrial and domestic workers (urban and abroad) | No | Seasonal | - Spouse or an extended family member manages the land |
| Men migrated first, women following | Adults (married) | Starting a new business–urban areas | Sharecropping<br>Land-rent<br>Land-purchase | Permanent (combination with seasonal) | - Spouse or an extended family manage the land<br>- Abandoned land |
| Whole family migrated | All ages (family) | Industrial<br>Starting new business | Sharecropping<br>Land-rent<br>Land-purchase<br>Plantation labor | Permanent | - Abandoned land |

Source: Household survey in West Java, Southeast and South Sulawesi, and focus group discussions in West Java and South Sulawesi.

From the case study in West Java, of the family members left behind in the origin communities, some were maintaining their agricultural systems and there were some changes in the roles of family members—when men migrate, the women or other family members maintain their agricultural plots. Statistical analysis of the difference in women's roles showed that women's decision making was significantly increased in paddy rice cultivation and regular maintenance; in particular, for fertilizer and pesticide purchasing, as shown in Table 4. Decision making on other agricultural practices was slightly increased although not significantly. Although women's roles increased in maintaining plots, decision making regarding timber cultivation was still considered to be the man's domain. An excerpt from an interview illustrates this pattern.

> *"When father migrated, I had the responsibility to cultivate our timber garden. Usually, I weed the plot. To me, cultivating a timber garden is not so difficult, just weeding. If we need to harvest it, we need to wait for my husband, who usually knows how to calculate the timber prices and who can negotiate the price. I don't know anything about selling timber and how to calculate the price of timber"* (in-depth interview, Rajadesa Village, West Java).

When we compare the situation of land-based out-migrant and off-farm out-migrant situations, the role of women in decision making is greater in the off-farm, in which the men move to urban areas and have less involvement in agriculture. For land-based out-migrant communities, men and women still worked together on agricultural practices and most decisions were made jointly, or by the man alone, as shown in Table 5.

**Table 4.** Women's workload and power related to decision making based on different periods, West Java.

| Category | Activity and Decision Making | When Male Stayed | When Male Out-Migrated | $p$-Value | Significance |
|---|---|---|---|---|---|
| All | All decisions | 4.48 | 5.01 | 0.00 | √ |
| Farming | Paddy rice | 4.11 | 6.79 | 0.00 | √ |
| | Annual crops | 2.78 | 3.11 | 0.44 | |
| | Trees | 2.78 | 3.11 | 0.44 | |
| | Planting | 3.64 | 4.05 | 0.25 | |
| | Farm labor | 4.24 | 4.78 | 0.18 | |
| | Fertilizer/pesticide application | 3.20 | 3.44 | 0.54 | |
| | Purchasing fertilizer/pesticide | 3.01 | 3.93 | 0.05 | √ |
| | Marketing products | 4.03 | 4.49 | 0.27 | |
| Household | Daily consumption | 8.52 | 9.12 | 0.04 | √ |
| | Schooling | 5.09 | 5.48 | 0.08 | |
| | Purchase house | 4.41 | 3.64 | 0.02 | √ |
| | Purchase land | 4.53 | 3.94 | 0.03 | √ |
| | Purchase vehicle | 4.33 | 3.63 | 0.02 | √ |
| | Sell land | 4.67 | 4.00 | 0.08 | |
| | Financial management | 6.17 | 6.53 | 0.37 | |
| | Home and childcare | 7.76 | 8.71 | 0.00 | √ |

Source: Household survey, West Java (2016).

**Table 5.** Women's workload and power related to decision making based on the type of migration, West Java.

| Category | Activity and Decision Making | Off-Farm Based Out-Migration | Land-Based Out-Migration | $p$-Value | Significance |
|---|---|---|---|---|---|
| All | All decisions | 4.95 | 4.43 | 0.00 | √ |
| Farming | Paddy rice | 5.08 | 5.87 | 0.19 | |
| | Market products | 4.88 | 3.60 | 0.02 | √ |
| | Farm labor | 4.79 | 3.90 | 0.08 | |
| | Annual crops | 4.27 | 2.91 | 0.00 | √ |
| | Purchasing fertilizer/pesticide | 4.19 | 2.43 | 0.00 | √ |
| | Planting time | 4.02 | 3.21 | 0.07 | |
| | Fertilizer/pesticide application | 4.02 | 2.50 | 0.00 | √ |
| | Trees | 3.59 | 2.04 | 0.00 | √ |
| Household | Daily consumption | 8.76 | 8.93 | 0.62 | |
| | Schooling | 5.28 | 5.23 | 0.87 | |
| | Purchase house | 4.00 | 4.05 | 0.91 | |
| | Purchase land | 4.21 | 4.17 | 0.89 | |
| | Purchase vehicle | 4.12 | 4.04 | 0.82 | |
| | Sell land | 4.83 | 4.07 | 0.08 | |
| | Financial management | 6.58 | 6.04 | 0.28 | |
| | House and childcare | 8.55 | 7.83 | 0.07 | |

Source: Household survey, West Java (2016).

We might see that this migration owing to land-based activities does not improve women's role in decision making. However, we could see that this means less burden for women. When the male migrates and women are responsible for the land, the extended family may be involved, or the woman hires labor. However, for a poor family, women's work in agriculture will be increased. We see that in land-based migration, women can still make decisions in certain cases but for more strategic decisions they needed to discuss it with the man, as the land manager. Additionally, in land-based migration, at certain times, i.e., in seasons when source areas need more labor, the male might return to the village to help while for off-farm migrants, opportunities to return to the village were less.

### *3.5. Returnee Migrants*

Some of the migrants who had accumulated considerable capital moved back to their place of origin, cleared land, and started to cultivate. The remittances and investment of the returnee migrant could be in various forms, such as building a mosque or road for the villages of origin. In the West Java study, some of the migrants once back in their village, in addition to managing their agricultural land there, kept their land in Lampung or other destination areas, employing landless farmers from their village of origin to cultivate it. Hence, inducing follow-up migrants.

In West Java, returnee migrants from Lampung used their collected capital and skills they built during the migration experience to manage the coffee system in the Gunung Sawal protected area under a community-based management scheme. De Royer [37] stated that while there seems to be no clear correlation between the beginning of the community forestry program in 2008 and the mass return of migrants, the program has been perceived by migrants as a great opportunity to make money from the skills they obtained in Lampung while re-establishing themselves in West Java. The Gunung Sawal region has since developed into the coffee hub of West Java with a very well-developed market. Here we can see a clear link between the return of migrants and the rapid expansion of land under coffee cultivation during the last two decades. People converted their mixed tree gardens into coffee gardens; those who were not migrants learned from those who were.

In Sulawesi, returnee migrants did not move back to cultivate their land. Some of them tried to find better luck elsewhere with cheaper, or any available, land, thereby creating multiple-time migrants. A few of them used their network and their knowledge of land information to connect people from the origin areas to available land in destination areas, acting as intermediaries or brokers. However, the legality of such land in the market is often "grey". Some land "ownership" does not have a strong legal basis supported by a letter or certificate from the National Land Agency.

## 4. Discussion

These various observations can be interpreted as pieces of a larger puzzle in which migration decisions become linked at human life-cycle scales, as shown in Figure 1, based on the five questions raised in the introduction. We split the first question ("What are the gendered patterns of movement concerning age and life histories in both source and target areas? Do gender norms of behavior influence land-use patterns differentially in source and target areas?") into two parts.

### *4.1. Q1A. Which Conditions within the Community of Origin Have Been the Main Trigger for People to Try Their Luck Elsewhere (Overseas, Urban, or Rural), at Least Temporarily?*

In the two study cases, economic needs were the main trigger for migration. The lack of arable land and lack of capital to intensify farming systems pushed the pioneer farmers in West Java and South Sulawesi to search for land elsewhere. The choice of target areas was usually decided through existing kinship or neighborhood networks. The broker or intermediary mechanism was formed through informal relationships and occurred at almost all stages of migration. In South Sulawesi, during the era of crop development implemented by government programs (the "Green Revolution", agricultural intensification, and agricultural extension), migration occurred spontaneously into those communities that wanted to improve their livelihoods by attracting more labor [38]. Kinship networks were so strongly binding that the functions of intermediaries were not visible. After the boom in cocoa commodity development in Sulawesi, the rate of migration rapidly increased, the primary purpose of which was to improve incomes through the expansion of cocoa plantations. Migration increased massively and spread to other areas; migrants from different villages began to arrive. The roles of intermediaries became stronger during this phase, using networks of kinship, neighborhood, and friendship.

What occurred in the two study areas was similar to that described in a study of motivations to migrate conducted by Amacher and Hyde [39] in remote areas in the Philippines. They indicated that the direction of migration to remote areas was largely determined by the availability of land that was accessible for migrants, resulting in a preference for areas with low population density. Accessible in this context meant that the land had no clear ownership status so that they could easily start cultivating it.

### 4.2. Q1B. What Effects Did Migration Have on Gender-Specific Land Use and Livelihoods in the Areas of Origin?

Feminization of agriculture as described elsewhere [40] under dominantly male migration patterns, appeared to be less common in the two study areas. What occurred was often a pattern where elderly people or other extended family members taking care of grandchildren stayed behind and struggled to maintain their agricultural practices. In the case of the West Java study, male farmers still had the decision-making power on tree-based farming in the origin area. Male farmers would return during harvesting or for any other activity that required their presence. Vice versa, females could undertake temporary migration to help harvesting or other laborious activities. However, in the day-to-day decision making in managing households and annual crops, there were significant differences in decision making compared to when the male was not on migration. Despite the changing roles in the households of a migrant community, balancing the needs of agricultural labor, work seasons, and family needs might be achieved through mutual adjustments of marriage partners [41]. This could be the key to the success of livelihood actions in the areas of origin adapting to change.

Hecht (2015) [42] indicated that off-farm labor migration could result in changes in land uses and forest dependence, following shifts in gender and generational relations. Our findings indicate, though not significant, that women's burden was heavier in the off-farm-based migration situation and abandonment of land could occur more readily. Further exploration is needed to see patterns of work and the success of working relationships in non-land-based migration.

### 4.3. Q2. How Do (Temporary) Migrants Compare the Positive and Negative Aspects of Home and Temporary Abodes?

The influx of migrants into new areas can change the type of farming systems. For example, the Bugis migrants entering Southeast Sulawesi in the era of the rice-focused Green Revolution began to switch their interest to cocoa cultivation. It became easier for them to find new areas that were suitable for planting cocoa because of their migration experience, under the schemes inspired by the Green Revolution, which turned out to be compatible with cocoa cultivation. New technology (such as herbicides and hand tractors) introduced in the Green Revolution for rice fields reduced the length of time on the farm, and therefore, the opportunities to manage more land became greater. Their mobility was increased as well, owing to less time being consumed on-farm and were motivated even more to look for land and manage cocoa plantations in other places. In some of the cases, experience with more intensified land use in the source area (e.g., West Java) enriched agriculture and agroforestry in the new environment (e.g., South Sumatra).

The influx of migrants also triggered a new type of job. The farmers in Southeast Sulawesi who were no longer interested in farming and planting began to turn to other sources of income (off-farm). Those who understood land matters later became intermediaries in land markets, drawing on the pool of people they knew, be they relatives or neighbors, to buy land. Absorption of new labor can start as paid labor and a patron–client relationships or share-cropping but also involve land renting and buying within customary ownership rules (rarely involving formal land certification).

In both Southeast Sulawesi and Lampung, migrants from more densely populated South Sulawesi and Java, respectively, did not arrive on an empty slate but in landscapes where local farming communities of Semendo (a southward expansion of a group in the neighboring province of South Sumatra) and Tolaki ethnic identity had converted forests for extensive land use. The migrants brought traditions of more intensive agroforestry management of coffee and cocoa, leading to a synergy where the group with longer presence opened forest and the newcomers bought existing gardens and intensified them before they returned to fallow vegetation. In the surveys of local ecological knowledge in Sumberjaya in Lampung, the two groups were distinguishable in knowledge and concepts [43].

Similarly, experience in farming cultivation can have an impact on the farming systems in the origin areas. Such is the case with the West Java migrants who, upon returning from Lampung and gaining access to forest land through a social forestry scheme, were able to manage coffee systems that were considered to be more environmentally friendly than timber systems, which were the traditional farming systems in the area.

### *4.4. Q3. How Do Returnees Reintegrate and Modify Land Use, Gender Norms, and Culture of the Areas from Which They Originated?*

There is a tendency that migrants who have failed may find other land uses or urban labor elsewhere, rather than returning home. For those who succeed, these land-based migrants may return home if arable land still exists, as in the case of the West Java study. The migrants, returning or not, will continue supporting the home villages by building a mosque or supporting other social infrastructure development, such as building schools or daycare centers.

In the case of South Sulawesi, the migrant farmers continued exploring other regions to gain more land. In current circumstances, there is very little forest area left that is accessible to migrants. Most of the accessible land is already owned by local communities or is in areas in which conflict between companies, governments, and communities has already emerged. Acquiring land requires contracts with (and payments to) the native or migrant communities who have access (de facto and/or de jure). The actions of land brokers and the migration process grew rapidly, spreading to areas where ample land was available with less population density and less strict local institutions. Simbune Village in Kolaka District firmly rejected migrants. The village head stated that a ban on selling land to new settlers was strictly implemented. It was proven to be the case by the lead researcher of this study, who visited the village in 2014. All land was still controlled by residents.

Ruf [44] analyzed the effects of cocoa development on migration. The model of migration for the development of cocoa plantations was mostly adopted by Bugis migrants and several other groups. Various methods were used to facilitate migrants in obtaining land for cocoa plantations, especially for those who had limited capital. The systems included informal "forest rent" payments to forest authorities, share-cropping, and patron–client relationships. This situation strengthened migrant desire to gain more land in other areas, when they felt to have failed in migrating a previous time or when they needed to expand their livelihood.

### *4.5. Interactions with Development Policies?*

Overall, our analysis suggests that positive aspects in both areas of origin and receiving areas may prevail, with the exchange of knowledge between areas of different land-use intensities spreading agroforestry practices. The latter may well be more effective than the routes through formal knowledge and extension and in some cases is combined with tree germplasm exchange. Feminization of agriculture through preferentially male-based migration is not common in Indonesia but age-based consequences are common, in both urban (or overseas) migration and dispersal to areas of lower population density.

Costs and benefits of social capital in the context of rural–urban migration are organized by gender [45]. Opportunities for urban jobs tend to be gender-specific, with higher perceived success for unmarried women to work in the export factories of West Java so as to remit money to their natal households and, at the same time, gather experience of modernity, has had consequences for the young men that stayed behind [46].

Spontaneous and organized population movements have long been used as a means of promoting a country's goals of development and national integration. At the local level, on the other hand, these movements have frequently done the opposite, fueling local grievances, sharpening group distinctions, and at times creating "sons-of-the-soil" conflicts [47]. In different policy phases associated with environmental governance in Lampung in Indonesia, migrants were defined initially as pioneer entrepreneurs, bringing progress to Indonesia's hinterland but, subsequently, as forest squatters, threatening the cultural and ecological integrity of the province [41]. Rural migrants attempted to resolve their problematic positioning through multi-local livelihoods, which combine access to non-local income through temporary migration with the maintenance of a foothold that signals belonging and legitimate entitlement to state resources.

Migrants coming from more densely populated areas with more intensive agriculture bring know-how and customs that can be new to the areas where they settle but also differ from government regulations. In an irrigation-based Balinese migrant society in Sulawesi, the traditional ("subak") institutions for water management linked to synchronized cropping cycles clashed with the state-regulated water users' association that separated technical water management from the wider scope of subak. Adjustment of perceived property rights to land, water, and irrigation infrastructure was feasible within a transmigration setting but conflicts persist where farmers of different ethnic and religious backgrounds farm close together [48].

Bilocality—in which an individual will spend part of the year in a rural area and the other in an urban area—is increasingly common in Central Java, addressing lack of income in rural areas. It generates remittances but also contributes in limited ways to the commonly anticipated rural development outcomes [49].

A review of the literature on remittances [50] concluded that land use as a driver of migration, livelihood strategies, and the use of remittances in investment in land use need to be studied more coherently than has commonly been the case. Financial remittances are only part of the way relationships are maintained, as may be clear from our case studies.

## 5. Conclusions

Synthesizing across two case studies in Indonesia, we focused on five aspects of (circular) migration.

(1) The conditions within the community of origin that encourage people to migrate are diverse. Most of the decision making to migrate is linked to natural resource and land competition and emergencies, such as natural disasters or increased human conflict. However, decisions are facilitated by networks that take over existing social obligations within the community.

(2) The effect of migration on land use and livelihoods in the areas of origin mean extensification or lowering of the labor/land ratio. Feminization of agriculture, as reported elsewhere, appears to be less common in the study areas. There are differences in decision making between women with migrating and non-migrating spouses but not in managing tree-based systems.

(3) The changes in the receiving community and its environment generally imply intensification. A new way of managing more efficient farming systems was commonly found in both study areas. Absorption of new labor as follow-on migration through expansion of the agricultural area encourages the existence of new jobs as a land broker owing to increases in land renting and buying.

(4) Migrants return with different levels of external success. People coming back with success may help to rebuild the village and its agricultural system and could invest in social capital (mosques, healthcare, schools). Some who have failed may find other land uses or urban labor options elsewhere.

(5) The interaction of migrants with the government and other stakeholders concerning development policies is largely implicit. Government programs on rural and agricultural development often support migrants in the initial phases of migration with the technical know-how of new commodities or farming systems that migrants practice in the origin areas. The development of good

transportation infrastructure allows successful farmers to easily manage agricultural land in both the origin and target areas. Overall, circular migration facilitates the exchange of know-how within the broader Indonesian agroforestry traditions of rural land use on the forest margins, with substantial reliance on economically important tree crops.

**Author Contributions:** Conceptualization, E.M., B.L., M.v.N.; methodology, E.M., B.L.; formal analysis, E.M., B.L.; investigation, E.M.; writing—original draft preparation, E.M., B.L., M.v.N.; writing—review and editing, E.M., B.L., M.v.N. All authors have read and agreed to the published version of the manuscript.

**Acknowledgments:** We acknowledge the willingness of all interviewed respondents in sharing their experience, context, and considerations.

## References

1. De Royer, S.; Visser, L.E.; Galudra, G.; Pradhan, U.; Van Noordwijk, M. Self-identification of indigenous people in post-independence Indonesia: A historical analysis in the context of REDD+. *Int. For. Rev.* **2015**, *17*, 282–297. [CrossRef]
2. Rahmati, S.H.; Tularam, G.A. A critical review of human migration models. *Clim. Chang.* **2017**, *3*, 924–952.
3. Manning, P.; Trimmer, T. *Migration in World History*; Routledge: New York, NY, USA, 2020.
4. Baker, J.L.; Rotimi, C.N.; Shriner, D. Human ancestry correlates with language and reveals that race is not an objective genomic classifier. *Sci. Rep.* **2017**, *7*, 1572. [CrossRef] [PubMed]
5. McColl, H.; Racimo, F.; Vinner, L.; Demeter, F.; Gakuhari, T.; Moreno-Mayar, J.V.; Van Driem, G.; Wilken, U.G.; Seguin-Orlando, A.; De la Fuente Castro, C.; et al. The prehistoric peopling of Southeast Asia. *Science* **2018**, *361*, 88–92. [CrossRef]
6. Iskandar, J.; Iskandar, B.S.; Partasasmita, R. The impact of social and economic change on domesticated plant diversity with special reference to wet rice field and home-garden farming of West Java, Indonesia. *Biodivers. J. Biol. Divers.* **2018**, *19*, 515–527. [CrossRef]
7. Mulyoutami, E.; Rismawan, R.; Joshi, L. Local knowledge and management of simpukng (forest gardens) among the Dayak people in East Kalimantan, Indonesia. *For. Ecol. Manag.* **2009**, *257*, 2054–2061. [CrossRef]
8. De Foresta, H.; Kusworo, A.; Michon, G.; Djatmiko, W.A. *Ketika Kebun Berupa Hutan: Agroforest khas Indonesia Sebuah Sumbangan Masyarakat*; ICRAF: Bogor, Indonesia, 2000.
9. Duguma, L.A.; van Noordwijk, M.; Minang, P.A. COVID-19 and social-ecological system resilience at the forest-agriculture interface. *LAND* **2020**. in review.
10. van Noordwijk, M.; Duguma, L.A.; Dewi, S.; Leimona, B.; Catacutan, D.C.; Lusiana, B.; Öborn, I.; Hairiah, K.; Minang, P.A. SDG synergy between agriculture and forestry in the food, energy, water and income nexus: Reinventing agroforestry? *Curr. Opin. Environ. Sustain.* **2018**, *34*, 33–42. [CrossRef]
11. Black, R.; Adger, W.N.; Arnell, N.W.; Dercon, S.; Geddes, A.; Thomas, D. The effect of environmental change on human migration. *Glob. Environ. Chang.* **2011**, *21*, S3–S11. [CrossRef]
12. McLeman, R.; Smit, B. Migration as an adaptation to climate change. *Clim. Chang.* **2006**, *76*, 31–53. [CrossRef]
13. Fan, C.C. *China on the Move: Migration, the State, and the Household*; Routledge: New York, NY, USA, 2007.
14. Farbenblum, B.; Taylor-Nicholson, E.; Paoletti, S. *Migrant Workers' Access to Justice at Home: Indonesia*. Public Faculty Scholarship at Penn Law. 495. Available online: https://scholarship.law.upenn.edu/faculty_scholarship/495 (accessed on 20 September 2020).
15. Available online: https://www.worldbank.org/en/country/indonesia/overview (accessed on 20 September 2020).
16. Benček, D.; Schneiderheinze, C. *Higher Economic Growth in Poor Countries, Lower Migration Flows to the OECD: Revisiting the Migration Hump with Panel Data*; Working Paper 2145; Kiel Institute for the World Economy: Kiel, Germany, 2020; ISSN 2195–7525.
17. Clemens, M. *The Emigration Life Cycle: How Development Shapes Emigration from Poor Countries*; Center for Global Development: Bonn, Germany, 2020; Available online: https://www.cgdev.org/sites/default/files/emigration-life-cycle-how-development-shapes-emigration-poor-countries.pdf (accessed on 20 September 2020).
18. Bazzi, S. Wealth heterogeneity and the income elasticity of migration. *Am. Econ. J. Appl. Econ.* **2017**, *9*, 219–255. [CrossRef]
19. Syafitri, W. Determinants of labour migration decisions: The case of East Java, Indonesia. *Bull. Indones. Econ. Stud.* **2013**, *49*, 385–386. [CrossRef]

20. Bryan, G.; Morten, M. The aggregate productivity effects of internal migration: Evidence from Indonesia. *J. Polit. Econ.* **2019**, *127*, 2229–2268. [CrossRef]
21. Bazzi, S.; Gaduh, A.; Rothenberg, A.D.; Wong, M. Skill transferability, migration, and development: Evidence from population resettlement in Indonesia. *Am. Econ. Rev.* **2016**, *106*, 2658–2698. [CrossRef]
22. Bazzi, S.; Gaduh, A.; Rothenberg, A.D.; Wong, M. *Unity in Diversity?: Ethnicity, Migration, and Nation Building in Indonesia*; Centre for Economic Policy Research: Jakarta, Indonesia, 2017.
23. Deshingkar, P. Environmental risk, resilience and migration: Implications for natural resource management and agriculture. *Environ. Res. Lett.* **2012**, *7*, 015603. [CrossRef]
24. Pattnaik, I.; Lahiri-Dutt, K.; Lockie, S.; Pritchard, B. The feminization of agriculture or the feminization of agrarian distress? Tracking the trajectory of women in agriculture in India. *J. Asia Pac. Econ.* **2018**, *23*, 138–155. [CrossRef]
25. Lam, T.; Yeoh, B.S. Migrant mothers, left-behind fathers: The negotiation of gender subjectivities in Indonesia and the Philippines. *Gend. Place Cult.* **2018**, *25*, 104–117. [CrossRef] [PubMed]
26. Mueller, V.; Doss, C.; Quisumbing, A. Youth migration and labour constraints in African agrarian households. *J. Dev. Stud.* **2018**, *54*, 875–894. [CrossRef]
27. Burgers, P.P. Livelihood dynamics, the economic crisis, and coping mechanism in Kerinci district, Sumatera. In *Rural Livelihoods, Resources and Coping with Crisis in Indonesia. A Comparative Study*; University of Utrecht: Utrecht, The Netherlands, 2008; pp. 71–90.
28. Colfer, C.J.P.; Achdiawan, R.; Roshetko, J.M.; Mulyoutami, E.; Yuliani, E.L.; Mulyana, A.; Moeliono, M.; Adnan, H. The balance of power in household decision-making: Encouraging news on gender in southern Sulawesi. *World Dev.* **2015**, *76*, 147–164. [CrossRef]
29. Mulyoutami, E.; Roshetko, J.M.; Martini, E.; Awalina, D. Gender roles and knowledge in plant species selection and domestication: A case study in South and Southeast Sulawesi. *Int. For. Rev.* **2015**, *17*, 99–111.
30. Mulyoutami, E.; Wahyuni, E.S.; Kolopaking, L.M. *Agroforestry and Forestry in Sulawesi Series: Unravelling Rural Migration Networks: Land-Tenure Arrangements among Bugis Migrant Communities in Southeast Sulawesi*; Working Paper 225; World Agroforestry Centre (ICRAF): Bogor, Indonesia, 2016; p. 58. [CrossRef]
31. Sub Direktorat Statistik Ketahanan Wilayah. *Statistik Potensi Desa Indonesia 2014*; Badan Pusat Statistik: Jakarta, Indonesia, 2014.
32. BPS. Diseminasi Hasil Sensus 2010. Available online: https://sp2010.bps.go.id/index.php/metadata/index (accessed on 20 September 2020).
33. van Noordwijk, M.; Mulyoutami, E.; Sakuntaladewi, N.; Agus, F. *Swiddens in Transition: Shifted Perceptions on Shifting Cultivators in Indonesia*; ICRAF: Bogor, Indonesia, 2008.
34. Li, T.M. Local histories, global markets: Cocoa and class in upland Sulawesi. *Dev. Chang.* **2002**, *33*, 415–437.
35. Pasya, G.; Mudikdjo, K.; Tjondronegoro, S.M.P.; Kusmana, C.; Nurbaya, S. Analysis of Causes dan Styles of Environmental Conflict in Protection Forest Zone Management (A Case Study in Register 45B Bukit Rigis Protection Forest Zone, Lampung Province). *Forum Pascasarj.* **2011**, *34*, 215–229.
36. Verbist, B.; Pasya, G. Perspektif sejarah status kawasan hutan, konflik dan negosiasi di Sumberjaya, Lampung Barat, Propinsi Lampung. *AGRIVITA* **2004**, *26*, 20–28.
37. De Royer, S.; Pradhan, U.; Fauziyah, E.; Sulistyati, T. *Community-Based Forest Management: Who Is Benefiting? Community-Based Forest Management, Coffee and Migration Patterns in Ciamis Regency, West Java, Indonesia*; Brief 44; World Agroforestry Centre (ICRAF): Bogor, Indonesia, 2014.
38. Ruf, F. Cacao migrants from boom to bust. In *Agriculture in Crisis: People, Commodities, and Natural Resources in Indonesia, 1996–2000*; Gerard, F., Ruf, F., Eds.; Routledge: London, UK, 2001.
39. Amacher, G.S.; Cruz, W.; Grebner, D.; Hyde, W.F. Environmental motivations for migration: Population pressure, poverty, and deforestation in the Philippines. *Land Econ.* **1998**, 92–101. Available online: https://www.jstor.org/stable/3147215 (accessed on 20 September 2020). [CrossRef]
40. Tamang, S.; Paudel, K.P.; Shrestha, K.K. Feminization of Agriculture and its Implications for Food Security in Rural Nepal. *J. For. Livelihood.* **2014**, *12*, 20–32.
41. Elmhirst, R. Displacement, resettlement, and multi-local livelihoods: Positioning migrant legitimacy in Lampung, Indonesia. *Crit. Asian Stud.* **2012**, *44*, 131–152. [CrossRef]
42. Hecht, S.; Yang, A.L.; Basnett, B.S.; Padoch, C.; Peluso, N.L. *People in Motion, Forests in traNsition: Trends in Migration, Urbanization, and Remittances and Their Effects on Tropical Forests*; CIFOR: Bogor, Indonesia, 2015; Volume 142.

43. Joshi, L.; Schalenbourg, W.; Johansson, L.; Khasanah, N.M.; Stefanus, E.; Fagerström, M.H.; van Noordwijk, M. Soil and water movement: Combining local ecological knowledge with that of modellers when scaling up from plot to landscape level. In *Belowground Interactions in Tropical Agroecosystems*; CAB International: Wallingford, UK, 2004; pp. 349–364.
44. Ruf, F. Tree crops as deforestation and reforestation agents: The case of cacao in Cote d'Ivoire and Sulawesi. In *Agricultural Technologies and Tropical Deforestation*; Angelsen, A., Kaimowitz, D., Eds.; CABI Publishing: Wallingford, UK; Center for International Forestry Research: Bogor, Indonesia, 2001; pp. 291–316.
45. Silvey, R.; Elmhirst, R. Engendering social capital: Women workers and rural–urban networks in Indonesia's crisis. *World Dev.* **2003**, *31*, 865–879. [CrossRef]
46. Elmhirst, R. Tigers and gangsters: Masculinities and feminised migration in Indonesia. *Popul. Space Place* **2007**, *13*, 225–238. [CrossRef]
47. Côté, I. Internal migration and the politics of place: A comparative analysis of China and Indonesia. *Asian Ethn.* **2014**, *15*, 111–129. [CrossRef]
48. Roth, D. Property and authority in a migrant society: Balinese irrigators in Sulawesi, Indonesia. *Dev. Chang.* **2009**, *40*, 195–217. [CrossRef]
49. Rudiarto, I.; Hidayani, R.; Fisher, M. The bilocal migrant: Economic drivers of mobility across the rural-urban interface in Central Java, Indonesia. *J. Rural Stud.* **2020**, *74*, 96–110. [CrossRef]
50. Cole, R.; Wong, G.; Brockhaus, M. *Reworking the Land: A Review of Literature on the Role of Migration and Remittances in the Rural Livelihoods of Southeast Asia*; CIFOR: Bogor, Indonesia, 2015.

# 4

# People-Centric Nature-Based Land Restoration through Agroforestry: A Typology

**Meine van Noordwijk [1,2,3,*], Vincent Gitz [4], Peter A. Minang [5], Sonya Dewi [1], Beria Leimona [1], Lalisa Duguma [5], Nathanaël Pingault [4] and Alexandre Meybeck [4]**

1. World Agroforestry (ICRAF), Bogor 16155, Indonesia; S.Dewi@cgir.org (S.D.); L.Beria@cgiar.org (B.L.)
2. Plant Production Systems, Wageningen University & Research, 6700 AK Wageningen, The Netherlands
3. Agroforestry Research Group, Brawijaya University, Malang 65145, Indonesia
4. Centre for International Forestry Research (CIFOR), Bogor 16115, Indonesia; V.Gitz@cgiar.org (V.G.); nathanael.pingault@gmail.com (N.P.); A.Meybeck@cgiar.org (A.M.)
5. World Agroforestry (ICRAF), Nairobi 00100, Kenya; A.Minang@cgiar.org (P.A.M.); L.Duguma@cgiar.org (L.D.)

* Correspondence: m.vannoordwijk@cgiar.org

**Abstract:** Restoration depends on purpose and context. At the core it entails innovation to halt ongoing and reverse past degradation. It aims for increased functionality, not necessarily recovering past system states. Location-specific interventions in social-ecological systems reducing proximate pressures, need to synergize with transforming generic drivers of unsustainable land use. After reviewing pantropical international research on forests, trees, and agroforestry, we developed an options-by-context typology. Four intensities of land restoration interact: R.I. Ecological intensification within a land use system, R.II. Recovery/regeneration, within a local social-ecological system, R.III. Reparation/recuperation, requiring a national policy context, R.IV. Remediation, requiring international support and investment. Relevant interventions start from core values of human identity while addressing five potential bottlenecks: Rights, Know-how, Markets (inputs, outputs, credit), Local Ecosystem Services (including water, agrobiodiversity, micro/mesoclimate) and Teleconnections (global climate change, biodiversity). Six stages of forest transition (from closed old-growth forest to open-field agriculture and re-treed (peri)urban landscapes) can contextualize interventions, with six special places: water towers, riparian zone and wetlands, peat landscapes, small islands and mangroves, transport infrastructure, and mining scars. The typology can help to link knowledge with action in people-centric restoration in which external stakeholders coinvest, reflecting shared responsibility for historical degradation and benefits from environmental stewardship.

**Keywords:** assisted natural regeneration (ANR); co-investment; ecosystem services; environmental stewardship; equity; forest and landscape restoration (FLR); landscape approach; rights-based approach; tree planting; water

---

## 1. Introduction

With the Bonn Challenge (2011) [1], the New York Declaration on Forests (2014) [2] and the UN Decade on ecosystem restoration [3] launched in March 2019, forest and landscape restoration (FLR) is gaining traction on the global political agenda. Within the goal of reversing centuries of damage to forests, wetlands, and other ecosystems, getting it right will be key to putting the planet back on a sustainable course. However, despite the high level of political engagement; despite the number and diversity of actors and institutions involved, from public and private sectors, civil society and local communities, research, and academia, at all levels, from local to global; and beyond some success stories, restoration is not happening at scale. As agriculture is a major driver of degradation but remains a primary source of rural livelihoods, international agricultural research can be part of the solution, contributing to the design of successful restoration approaches to be implemented at scale in the coming years.

As part of the international agricultural research consortium (CGIAR) as food security-oriented international agricultural research body [4], the partners in the Forests, Trees and Agroforestry (FTA) program, building on many decades of research, identified the need to uncover the diverse understandings and perspectives about 'restoration' and to construct a typology that can help to clarify contrasts, similarities and possible synergies across the many starting points for targeted interventions that are currently propagated and implemented under the common heading of 'restoration' of forests, landscapes and/or land. The aim of such typology is to better describe links between evolving knowledge, stakeholder-driven action, and achievement of Sustainable Development Goals.

As a first step toward such a typology of restoration as part of international agricultural research, three Common Research Programs (CRPs) of the CGIAR—Forests, Trees and Agroforestry (FTA), Policies, Institutions and Markets (PIM) and Water, Land and Ecosystems (WLE)—conducted a joint stocktaking of CGIAR work on forest and landscape restoration [5]. They covered a wide range of field projects and case studies, decision making supporting tools, modeling, mapping, conceptual approaches, and frameworks across the geographical area of interest of the CGIAR, i.e., the tropics and sub-tropics in Africa, Asia, and Latin America. They covered a broad range of issues directly or indirectly related to restoration, including sustainable land and water management; seed supply systems and genetic diversity; climate change adaptation and mitigation: land tenure security and land governance reform. They showed how restoration efforts could contribute to SDG2 by supporting smallholder farmers' ability to increase food production, while also addressing SDG15 "protect, restore and promote sustainable use of terrestrial ecosystems" and "halt and reverse land degradation", contingent on SDG16 "peace, justice and strong institutions", assist with SDG6 "availability and sustainable management of water and sanitation for al", and most, if not all the 17 sustainable development goals [6]. They unveiled the many drivers of land degradation, not only biophysical, but also socio-economic, political, and institutional and the variety of actors involved. They suggest different ways to categorize restoration interventions.

Building on the results of this survey, the main objective of this paper is to elaborate a typology of restoration options by context applicable in a wide range of situations across the tropics and sub-tropics. We adopt here a people-centric nature-based perspective focusing on land restoration through agroforestry. For that purpose, Section 2 discusses the underlying concepts and definitions and presents our approach. Section 3 focuses on land degradation, its symptoms, drivers, and indicators. Section 4 suggests and discusses a possible typology of restoration options by context.

## 2. Underlying Concepts, Definitions and Approach

### *2.1. Beyond Tree Planting, the Various Aspects of Restoration*

Tree planting as a way to restore local ecosystems appeals to many farmers, local communities, national policymakers, and private companies, for a variety of reasons. Tree planting ceremonies as a symbol of peace and commitment to stability and prosperity have an important place in the diplomatic world and national policy agendas for post-disaster contexts [7]. The 'tree planter hero' portrayal has strong emotional appeal [8–10]. Claims of the millions, billions [11,12] or trillions [13] of trees planted capture the public imagination and get news headlines. More than seven billion humans share the planet with approximately three trillion trees [14] 46% less trees than at the start of human history. Approximately 1.36 trillion of these trees exist in tropical and subtropical regions 0.84 trillion in temperate regions and 0.84 trillion in the boreal region; overall nearly one-third are outside forests [15]. Tree diversity in agroforestry landscapes varies over three orders of magnitude, from 1–1000 [16].

Enthusiasm for tree planting only partially aligns with the focus of thousands of experts around the world who have dedicated their professional lives to the protection and restoration of ecosystems [17]. Guidelines for restoration include recommendations to first consider and deal with root causes of degradation, to work with nature, rather than against it with technical means, and to work with people. The guidelines suggest that experts can provide advice on where and how tree planting can be helpful. The idea, however, that planting trees is at the core of restoration ignores the expertise of millions of agroforesters (practitioners) around the world who learned that people-centric and nature-based land restoration through agroforestry can be as simple as 'assisted natural regeneration', with selective retention of the right trees growing in the right places [18,19]. While the practice may be as old as agriculture, agroforestry as a branch of applied science started four decades ago. 'Restoration' was and still is one of its main motivators [20–23]. Where it is ecologically and socially feasible [24–27], approaches such as assisted natural regeneration need to be upfront part of a 'restoration typology' message and list of options to be considered for any given context. In a more comprehensive 'options by context' typology for restoration a wider spectrum of activities beyond 'tree planting' is needed, as we will explore in this contribution.

The growing consensus on the relevance of a social-ecological systems perspective has seen restoration ecology [28–30] and forestry-oriented implementation guidelines [31–33] evolving to restoration science [34–37]. This meant a stronger social orientation [38–40], with attention to success factors for community forestry [41,42]. Yet agendas on food security [43] and public health [44] remain poorly connected to the dominant restoration discourse. A gap still exists between Land Restoration from an agricultural perspective, as perceived in CGIAR efforts [9], and the main ideas in Forest and Landscape Restoration. The wording of principles of Forest and Landscape Restoration [45], such as "Engage stakeholders and support participatory governance", "Taylor to the local context using a variety of approaches", "Manage adaptively for long-term resilience" suggest a genuine attempt to connect with bottom-up farmer perspectives, but also serious challenges to actually achieve that. Principles such as "Restore multiple functions for multiple benefits" may be redundant if local stakeholders have a real say in what happens on the ground. A list of common governance challenges for Forest and Landscape Restoration in a recent review [46] included (1) Poor alignment across levels and sectors of government, (2) Environmental and social heterogeneity, (3) Lack of enabling conditions and implementation capacity. A non-involved reader might wonder whose agenda such type of restoration actually is. Thus, a recent review of FLR practice concluded that "Existing guidelines and best practices documents do not satisfy, at present, the need for guiding the implementation of Forest and Landscape Restoration (FLR) based on core principles. Given the wide range of FLR practices and the varied spectrum of actors involved, a single working framework is unlikely to be effective, but tailored working frameworks can be co-created based on a common conceptual framework" [47]. While FLR is supposed to support sustainable agricultural production, there is very little discussion on how this can be achieved. When farmers are asked about their own decision making with regards to landscape restoration, responses may be surprising. Recent efforts to obtain a deeper understanding of farmer decision making for landscape restoration in Malawi revealed that the expectations that 'planting more trees will attract reliable rains' figured prominently in local perspectives, before expected benefits for soil fertility or beekeeping [48]. One of the milestones of restoration science is the "reference ecosystem", specifying the desired successional stage of recovery, the species (or group of species) that are the target of rehabilitation and the expected time of recovery of the degraded ecosystems after initial treatment (change in management). Follow-up questions are whether the time involved is socially accepted, economically feasible and ecologically reasonable.

One of the key propositions of the paper is to advance restoration science to a more complex set of objectives and functions to be restored and to put local land users at the center of the process. The confrontation, managing synergies, and trade-offs between such functions, and at nested scales, is fundamental to the challenge, and in line with decades of analysis of landscape multifunctionality and integrated natural resource management in international agricultural research [49,50]. Rather than a static reference ecosystem continued change and agility will be needed as a vision for multi-functional restoration [51]. Therefore, we propose to extend the concept of "reference ecosystem" to the "reference social-ecological system" and resulting functions.

Innovation, development, intensification, adaptation, rejuvenation, and restoration appear to refer to different actions, with emphasis on the new, the existing or the past. In practice, however, similarities exceed the differences as all need to deal with motivation, rights, know-how, markets, environment from local environmental effects to global effects through bio-geochemical and hydrological cycles and their global teleconnections [52–54]. This means that all efforts need to match options for interventions to context at the nested scales of farms, landscapes, nation-states, and the changing global context. This means involving a multiplicity of actors: farmers, communities, private sector, public sector, and global investors [55–59]. Interventions also range from projects and programs to wider policies at national or even international (regional) level [60]. Unless these are people-centric, however, their chance of sustainable success is small [61,62]. Land restoration, as we present it and contrarily to what the word often means in other contexts, is not backward-looking but forward-looking. Innovative restoration (or restorative innovation) reconciles historical path-dependency of the degraded status quo with forward-looking theories of induced change that are empirically grounded, rather than wish-lists of over-optimistic planners. It requires science-based and across-scales diagnosis of the underlying causes ('driving forces') that shaped current context, mobilizing a wide range of conceptual frameworks to understand social-institutional constraints, drivers of change and sustainers of long-term action. It then relates that context to options for interventions. Common interventions in land restoration focus on modifying land cover or structural land surface properties but are aimed at improvement of land use and functionality in support of multiple goals.

Land cover change can be achieved through natural regeneration (with various degrees of human assistance and farmer management), tree and grass planting (with its dependence on seed supply and nursery value chains), or remediating management of soil and water. Land-use change depends on who is allowed to be a user, for what purpose and for what use, their motivation, preferences, restrictions, the know-how of managing land in local contexts, market opportunities, concerns about local environmental impacts and external co-investment in land stewardship. Across all modes of regeneration there are tree genetic resources issues [63–65], agronomic options in context [66,67],value chains [68–70], hydrology [71], global teleconnections through climate [72,73], and biodiversity [74–76], as well as policy reform of rights [77], cross-scale incentive systems for stewardship [78–80], and distributional and process concerns over equity and inclusiveness [81]. Key constituencies (including policy-makers, funders, local stakeholders, scientists) need commonly understood metrics to achieve progress [82,83]. Restoration is commonly differentiated by geographic contexts, such as China [84], Southeast Asian fallow to forest transitions [85], East and South African drylands [86], Horn of Africa [87], or Brazil [88]. Still, a more incisive way of describing similarities and differences is needed.

### 2.2. *Definitions*

The Society for Ecological Restoration (SER) [89], in line with the Intergovernmental Science-Policy Platform on Biodiversity and Ecosystem Services (IPBES) [90] defines ecological restoration as "any activity with the goal of achieving substantial ecosystem recovery relative to an appropriate reference model, regardless of the time required to achieve recovery. Reference models used for ecological restoration projects are informed by native ecosystems, including many traditional cultural ecosystems". Both the SER and the IPBES distinguish ecological restoration from rehabilitation. The latter refers to restoration activities that aim at "reinstating a level of ecosystem functioning for renewed and ongoing provision of ecosystem services potentially derived from nonnative ecosystems as well" [79] but may "fall short of fully restoring the biotic community to its pre-degradation state" [80]. Both see restoration activities as a continuum, from rehabilitation to ecological restoration, aiming at initiating or accelerating the recovery of an ecosystem from a degraded state. These various acceptances of the term 'restoration' reflect the various perspectives, motivations and behaviors of the different actors involved. While ecologists, scientists, and activists might focus more on ecological restoration as the return to a pristine state of natural ecosystems, governments and economists may be more interested in a discourse focusing on regaining ecological functionality in degraded landscapes in order to enhance food security and livelihoods, reduce poverty and contribute to sustainable development. In line with this latter perspective, in the context of the Bonn Challenge, the International Union for Conservation of Nature (IUCN) and other partners adopted the following definition [91]: "Forest landscape restoration (FLR) is the ongoing process of regaining ecological functionality and enhancing human well-being across deforested or degraded forest landscapes. FLR is more than just planting trees–it is restoring a whole landscape to meet present and future needs and to offer multiple benefits and land uses over time".

For the purpose of this paper, the term 'restoration' encompasses the whole continuum of restoration activities, from rehabilitation to ecological restoration, covers any kind of ecosystems (natural forest, agroforest or agricultural landscapes) and gives a central place to the concept of 'ecological functions' (Figure 1), i.e., the functions that allow ecosystems to generate various regulating, supporting, provisioning and cultural benefits to people or 'services' [92], including those generating direct economic value.

Building on these considerations, this paper adopts the following definitions:

- Degradation: Loss of functionality of e.g., land or forests, usually from a specific human perspective, linked to a change in land cover with consequences for (at least one category of) ecosystem services,
- Degraded lands: Lands that have lost functionality beyond what can be recovered by natural processes and existing land use practices in a defined, policy-relevant time frame,
- Syndrome: Set of concurrent diagnostical indicators or symptoms that can be the result of different and often interacting causes or drivers,
- Restoration: Efforts to halt ongoing and reverse past degradation, by aiming for increased functionality of ecosystems supporting land use (not necessarily recovering past system states)

Restoration covers a broad range of changes (innovations) relative to the current system state—from land use practices and land cover changes to physical infrastructure and institutional changes. As highlighted above, the objective here is to regain ecological functionality and enhance human well-being, not necessarily to go back to the initial ecosystem state or function. That may simply be impossible in some places because of the change in local demographic conditions. Living with the current 7 billion people on the planet would not allow that. Moreover, the final 'restored' state of the ecosystem shall be self-sustaining. This means that in a particular context, given the set of ecological functions to 'restore', restoration interventions need to lead to social, economic, and ecological benefits lasting in the long-term. While in other aspects of human life binary classifications have been recognized as being problematic, past distinctions between 'Nature' (ecosystems, wilderness) and human endeavors (agro-ecosystems, plantations, (peri)urban systems) remain evident in the way 'ecological restoration' is distinguished from 'forest and landscape restoration'. Under the heading 'land restoration' we aim to bridge this divide. Therefore, we propose here a broader umbrella to clarify the full perspective of 'stopping degradation *plus* recovering damage' and addressing in a sustainable way the underlying drivers, including those related to the production function of the lands and to the needs of people living in it and from it. Reconciling these perspectives is critical to the sustained, long-term success of land restoration and especially relevant in the pan-tropical domain. Dealing with the generic drivers of degradation, rather than area-specific pressures, builds on decades of policy development, although progress on Aichi 2020 target of the Convention of Biological Diversity on pollution control is less than that on target 11 achieving an increase in protected areas. [93].

### *2.3. The Social-Ecological Cascade Framework*

The distinctions between land cover (as structure) and land use (as a set of services derived) are compatible with a Social-Ecological System (SES) cascade (Figure 1A) [94]. Mainstream 'forest' definitions combine aspects of scale (minimum area), structure (tree cover, tree size), function (primary designated purpose) and institutional control by forestry agencies, segregating trees into 'forestry' and 'agricultural' ones [95,96]. Important parts of the structure are also the condition of the topsoil (including its soil organic matter [97], protective litter layer, macroporosity [98] and soil biota), while the related functions include processes such as rainfall infiltration and (absence of) surface runoff and erosion [99].

Feedback options from 'stakeholders' based on the values at stake for them to 'actors' in the landscape make the cascade flow into a social-ecological system with self-adjustment or learning ability. These feedbacks link 'bio-geo-physical' units, social actors and institutions, across scales. Many changes in the landscape increase some and decrease other functions and values, and as such they are either degradation or restoration depending on the weight given to various functions by stakeholders, or the strength of their voice in public discourse. Post-logging forest management to increase the growth of the most desirable species, may imply degradation from an ecological perspective, for example. Draining swamps for improved public health, implies ecological degradation, as a second example. The feedback loops that aim to shift ongoing degradation toward restoration require specific ways of linking knowledge (on options in context) with action (getting societal traction on issues and agreement on goals) (Figure 1B). Five issue cycle steps depend on and strengthen the knowledge-action linkages: A. Agenda setting, B. Better and shared understanding of what is at stake, C. Commitment to common principles, often based on coalitions, D. Devolved details of design and delivery, dealing with trade-offs, and E. Efforts to evaluate, and where necessary restart [50,100].

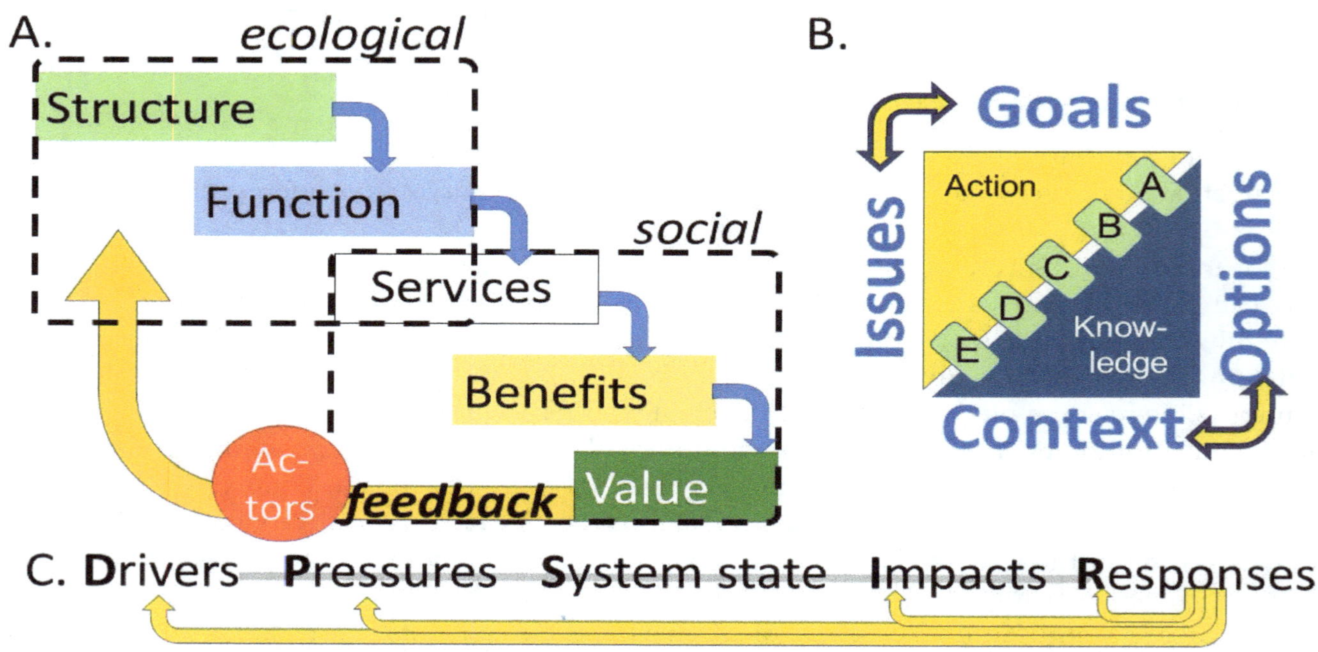


**Figure 1.** (**A**) Social-ecological system (SES) as a cascade with feedbacks via actors (modified from [101]), (**B**) Multiple links between knowledge and action along the intervention cycle, with two-way interactions between issues and goals, as well as between options and context (modified from [50]), (**C**). Responses can target drivers, pressures, impacts or the emergence of responses itself.

The SES cascade framework and the above-mentioned knowledge-action linkages are conceptually close to a third commonly used framework, which depicts how Drivers, Pressures, State change, Impact and Responses (DPSIR) interact in a feedback loop, with responses addressing the immediate pressures and/or the underlying causes (drivers) (Figure 1C). Often Drivers, Pressures and Impacts operate in nested scales, necessitating Responses (including 'restoration') to do the same to be effective.

### 2.4. Our Approach

Interventions to support the restoration of the functionality of lands, forests and landscapes can be many folds, and operate at different scales (local, national, global), involve various combinations of private and public initiatives and investment, relate to the rights and authority of local actors and stakeholders in multiple ways (ranging from eviction to full consultation and respect of Free and Prior Informed Consent (FPIC) or public and private support for local initiatives). These interventions seek various entry points into Social-Ecological Systems other than directly addressing land cover (e.g., by tree planting), such as modified rights, enhanced know-how, supported markets, or incentive systems. These are always related to specific contexts that are in turn very diverse depending on the social-ecological system in place and its historical path-dependency.

The dual purpose of our proposed typology is:

1. to allow a more effective exchange of knowledge and experience between settings where restoration (of any kind) is initiated, planned or in process, based on recognized similarities and contextualized differences, enabling a better confrontation and/or integration of the corresponding evidence-base, and
2. to assist planning and priority setting, especially where scarce public resources are involved.

The first purpose is to reflect an actor-centric (bottom-up) perspective, the second a planner (top-down) one.

As typologies of goals and issues can be derived from other frameworks, such as the set of Sustainable Development Goals (SDG) targets and indicators (or any other national or local level framework), we focus here on the knowledge side of the interaction and thus on a combined typology of intervention options and contexts, within the main steps of issue cycles, as illustrated in Figure 1B [102].

Restoration interventions are chosen out of a wide array of options, supposedly finetuned to the local context. Therefore, to elaborate a typology of restoration options by context, we cross two different typologies:

- a typology of contexts/situations, focused on types and levels of ecological degradation in their social contexts, at the system state level of DPSIR, but with attention to the pressures, drivers and impacts as levels of analysis;
- a typology of options for restoration interventions that address symptoms of ecological degradation and/or its drivers, and/or support the social conditions that sustain further improvement.

The resulting 'restoration' typology positions (combinations of) interventions in a specific context, expecting considerable 'endogeneity' in what is attempted and has success where. Endogeneity is one of the main obstacles to interpreting observable patterns of associations (e.g., forest cover and human well-being) in terms of replicable mechanisms and generic theories of change. Real learning from track records of any intervention elsewhere requires contextualization of its initiation and operationalization.

A technical perspective on restoration takes 'ecosystem structure' as the direct target for interventions, triggering the cascade (Figure 1A) to function, services and human benefits, but a social-ecological perspective starts from the Response part of the DPSIR cycle (Figure 1C) and identifies leverage points (preferably at Driver level), leading to land use change that leads to changes in ecosystem structure (Figure 2).

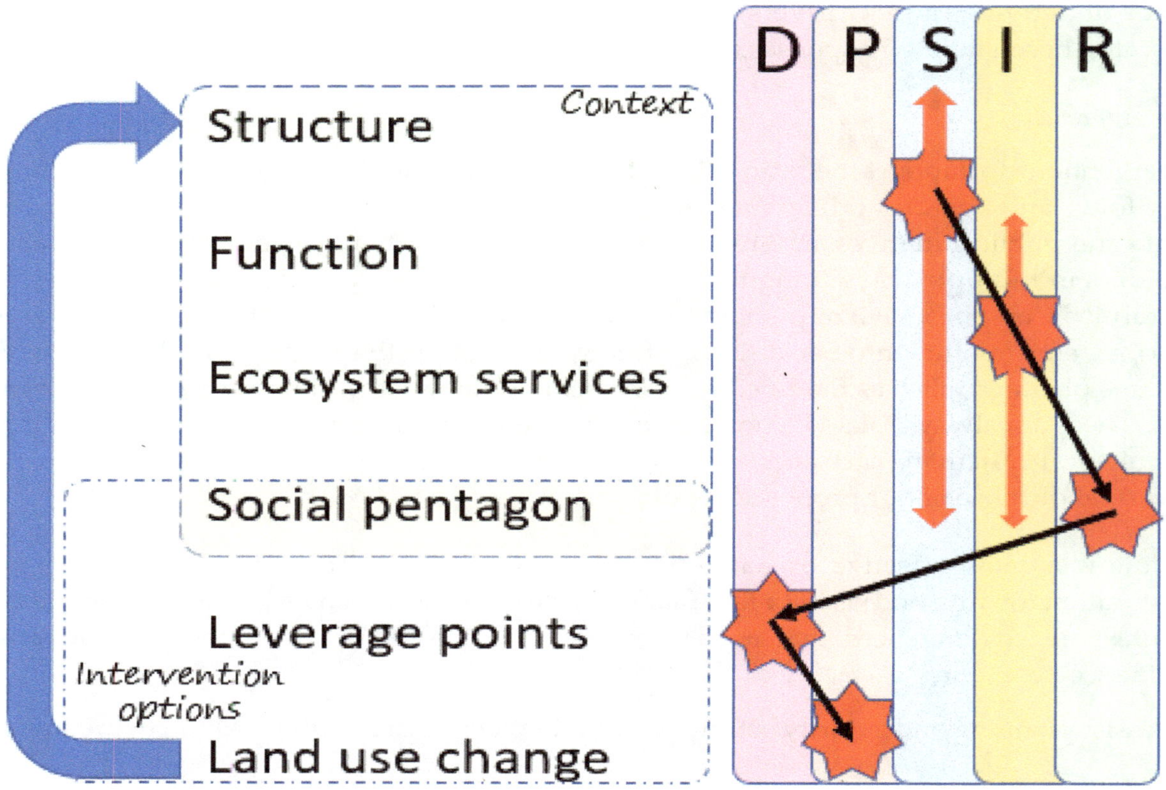


**Figure 2.** Relating the cascade and DPSIR concepts of Figures 1A and 1C, respectively, in social-ecological systems that shift from degradation to restoration; the social pentagon is shorthand for rights, know-how, markets, local ES issues and teleconnections.

## 3. Land Degradation: Symptoms, Drivers, and Indicators

### *3.1. Symptoms, Syndromes and Diagnostics*

In a social-ecological system interpretation, the importance of breaking current trends of degradation (loss of functionality) emerges in many different contexts, with a wide range of 'entry points' that get issues inscribed into an agenda for action. These starting points can be compared with 'symptoms' in the medical tradition, signs that something is amiss with system health, but requiring diagnosis. As in the medical tradition a determined set of symptoms can appear concurrently, a syndrome, giving further indications for diagnosis. Diagnosis aims to identify the location-specific pressures (Table 1) and their underlying generic drivers as targets for interventions beyond the symptom level.

**Table 1.** Examples of diagnostic links between symptoms along the SES cascade and the underlying pressures (that themselves respond to generic 'drivers' outside the targeted landscape).

| Starting Point | Symptom | Example of Contributing Pressures |
|---|---|---|
| *Structure* | Loss of land cover | Uncontrolled conversion |
| | Loss of perennials | Overgrazing |
| | Loss of tree cover | Overlogging, overharvest |
| | Loss of plant diversity | Specialization, markets |
| | Skewed tree age distribution | Lack of rejuvenation investment |
| | Loss of soil structure, carbon | Over-cropping |
| *Function* | Reduced primary production | Local climate change, soil fertility loss |
| | Disturbed hydrology: quantity, quality, timing of river flow; freshwater retention | Local climate change, loss of filter and buffer functions by vegetation |
| | Loss of soil retention, downstream sedimentation, landslides | Loss of effective land cover, increased rainfall intensity |
| | Spread of uncontrolled fire | Loss of functional fire-breaks |
| | Increase of pest, disease, weed pressures | Disturbed ecology and food-webs |
| *Services* | Loss of harvestable and marketable production ('provisioning services') | Loss of productivity, consumer trust in responsible production |
| | Decline in usable water, increased floods | Disturbed hydrology |
| | Loss of net greenhouse gas sequestration | Loss of soil functions, vegetation |
| | Loss of human health | Loss of healthy ecosystems |
| | Loss of cultural and spiritual value | Loss of respect and recognition |
| *Benefits* | Loss of local livelihoods | Loss of harvestable/marketable production |
| | Loss of secure and trusted value chains | Production costs, market prices, loss of trust |
| | Loss of existence value of global biodiversity | Awareness of existence and threats |
| *Value* | Increased resource use conflicts | Lack of rights, lack of law implementation |
| | Increased inequity and gender inequity | Lack of voice in decisions |
| | Loss of options for young people | Land shortage, lack of livelihood options |
| | Loss of local ecological knowledge, rules | Lack of attention, respect and rejuvenation |

Behind the location-specific pressures that lead to degradation in Table 1, there are generic underlying causes (such as population growth, economic growth ambitions, globalization, urbanization and changes in diets and lifestyles) that require generic responses [103]. Such ultimate causes of land degradation cannot be ignored by restoration scientists and practitioners. However, a long history of claims that intensifying agriculture, especially in the tropical forest margins, would by itself reduce environmental impact (known as the 'Borlaug' hypothesis) [104,105] was rejected based on evidence from the field. Agricultural land abandonment, providing space for restoration, may be expected in less-favorable conditions where agricultural labor moves to cities [106], rather than as a direct consequence of intensified agriculture elsewhere. Population policies, both the migration and birth rate side of them are closely linked to national identity issues, and hardly modifiable by directly environmental concerns or agricultural research [4]. SDG 12 "responsible production and consumption" may be the closest approximation among the SDGs to deal with market-related drivers of degradation and restoration, but the goal is largely aspirational, with weak operationalization [107].

### *3.2. Structural Indicators: Tree Cover Linked to Ecological Functions*

Restoration issues can start with any of the five elements of the SES cascade of Figure 1: structure, function, services, benefits, or value, and need to identify where 'lack of' function is caused by 'loss of' function. At the highest level of this cascade is the category Land or Land Health, as vegetation and surface soil are parts of the structure—which also is most readily observed. Remote sensing has so far had a dominating role in informing top-down priority setting for restoration interventions, but it does not always allow separating anthropogenic change from natural patterns of variation. Anthropogenic changes (over time) in land cover and surface soil structure such as tree cover transitions (also known as forest transition (FT)) interact with two other dimensions of tree cover variation: latitude and topography, jointly shaping climate and soils. The natural spatial distribution of vegetation types responds to climate and soil, with temperature and rainfall varying by both latitude (from tropics to boreal) and topography and water availability in any given climate by topographical variation in water acquisition and drainage (Figure 3).

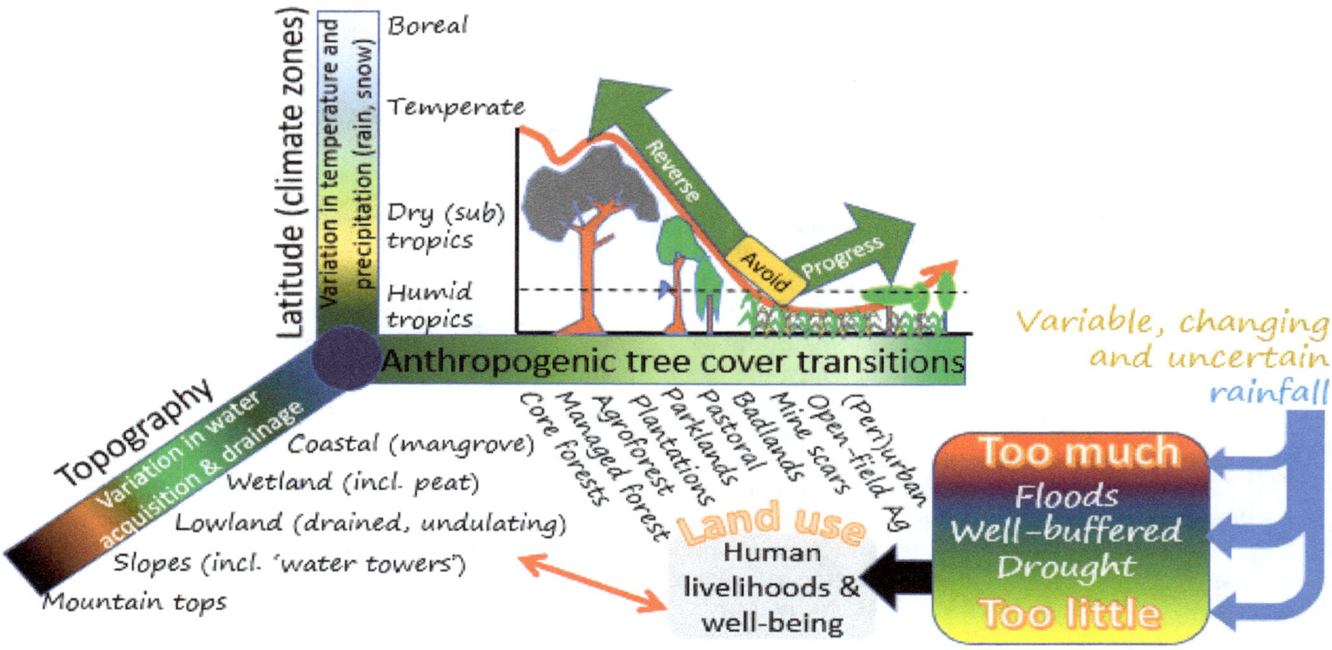


**Figure 3.** Three major sources of spatial variation in tree cover: latitude, topography and place-based anthropogenic tree cover transition in interaction with rainfall regimes as the cause of water excess, shortage or well-buffered conditions (modified from [54]).

Tree cover in itself cannot be a universal proxy for land degradation or restoration, or for assessing the human influence in the SES cascade. For example, vegetation with 20% tree cover could be the natural vegetation in semi-arid conditions or on mountain slopes but indicates human influence elsewhere (e.g., increased fire frequency in anthropogenic savanna conditions). With several of the worst effects of 'degradation' related to disturbed hydrology (e.g., floods and droughts), one needs to unpack anthropogenic effects from natural background variation.

The forest transitions (FT) typology [108], illustrated in Figure 3 and presented in more detail in Section 4.4, clarified that the highest human population densities (FT6) in a pantropical study were associated with around 30% tree cover at sub-watershed scale (Figure 4A), while areas with less tree cover (e.g., open-field agriculture) are associated with lower population densities (Figure 4B). Within the humid and (per)humid tropics there is a strong negative relationship of the 'more people, less forest' type (Figure 4D), but in arid and semiarid an opposite trend (more people, more trees) can be noted (Figure 4C).

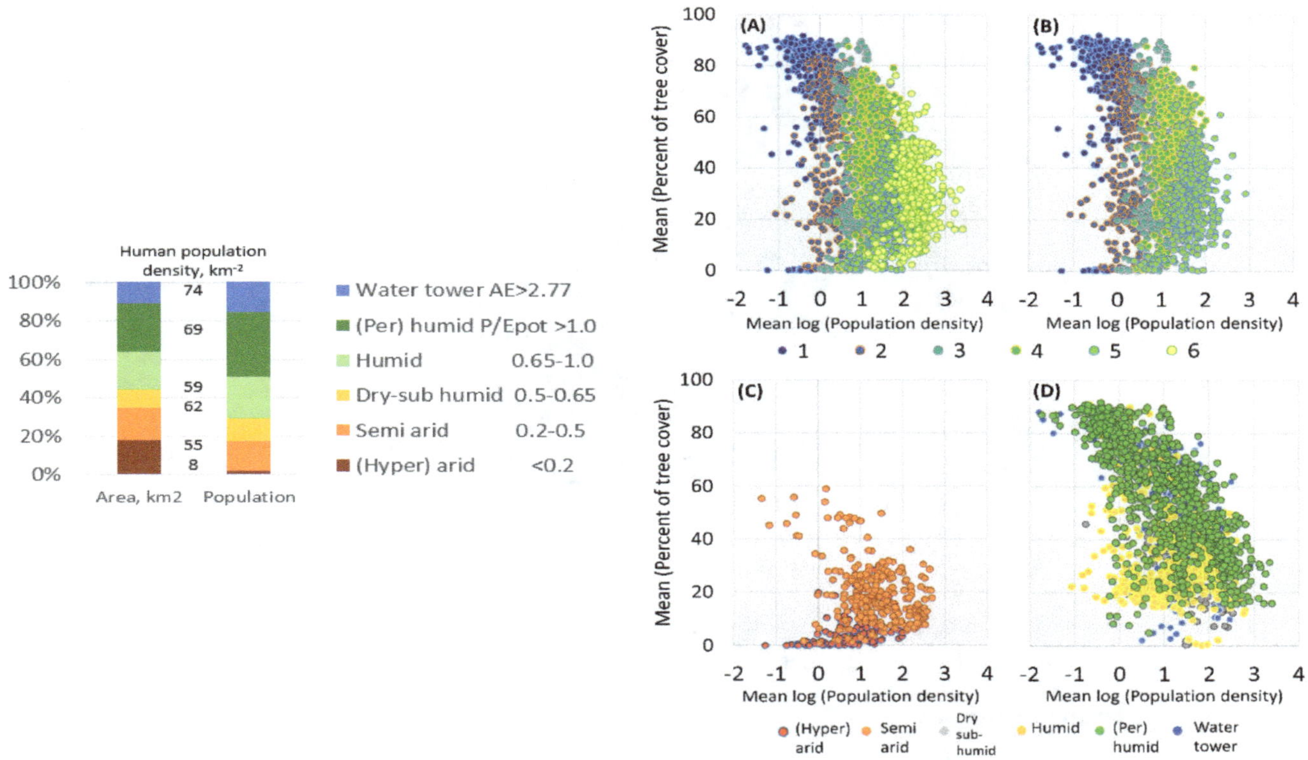


**Figure 4.** (**Left panel**): area and human population size in the tropics in five climatic zones (defined on the basis of the ratio of precipitation P and potential evapotranspiration $E_{pot}$ and the water tower configuration as defined in [108]); (**Right panel**): pantropical relationships at sub-watershed level between human population density and tree cover, for six identified stages of forest transition FT and the six climate zones as defined in [108]; (**A**) for all sub-watersheds differentiated by forest-transition stages; (**B**) only for sub-watersheds of stages 1–5; (**C**) for sub-watersheds within (hyper-) arid and semi-arid zones; (**D**) for sub-watersheds in the remaining agro-ecological zones.

Reductions, rather than the expected increases, in streamflow are a common result of tree planting and forest restoration [109] that do not have to come as a surprise. Depending on the environmental conditions an intermediate tree density can be optimal from a groundwater recharge perspective [110], while desirable effects on flood-causing peak flows can be stronger than undesirable reductions in annual water yield [111,112]. Flood control may depend on using trees as part of river restoration [113] rather than blanket reforestation of landscapes. Increasing attention to the rainfall-generating effects of high evapotranspiration [114,115] justifies renewed attention to the savanna-forest hysteresis theory of local climate influences [116,117].

### *3.3. Social Indicators: Services, Benefits And Values*

At the lower part of Table 1 we see closely connected, primarily socially defined, indicators of services, benefits, and value. While there are many ways to unpack the complexity further (e.g., the 'five capital' framing), we found five aspects that operate at the landscape scale and capture important dimensions of human well-being. These five aspects illustrated by the 'social pentagon' (Figure 5), are discussed in more detail in the next section.

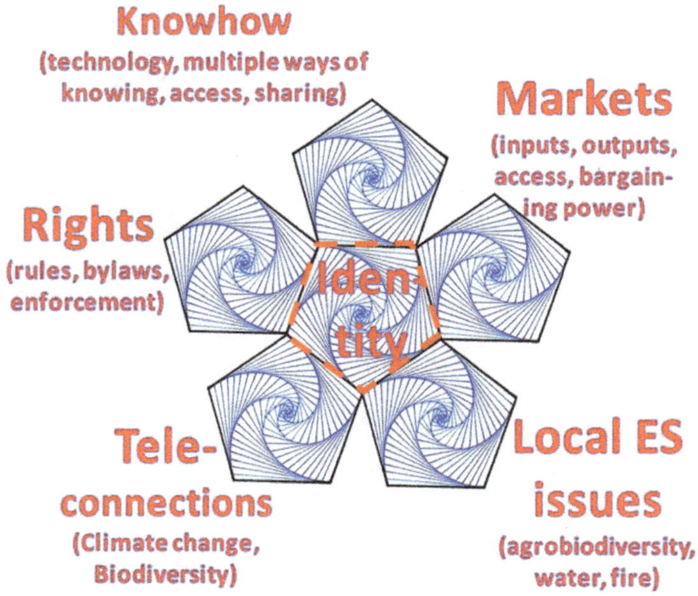


**Figure 5.** Representation of six key attractors of attention in Social-Ecological Systems that relate to both degradation and restoration phases.

## 4. Typology of Interventions

### *4.1. The Social Pentagon as a Starting Point for a Typology of Restoration Options*

At the heart of the social pentagon (Figure 5) a lost or modified sense of identity may be the worst impact of degradation and has to be on the basis of any success in restoration. Human identity (self-image) relates to local institutions (that include education, religion, cultural values, collective action), motivation, usually stratified by age- and gender, as well as to other dimensions such as social stratifiers of wealth and influence (e.g., caste, social class, patronage, entrepreneurship, or other aspects of political ecology).

Identity and the human capacity to adapt to gradual degradation processes can be the main obstacle to transformative change or to tackling the underlying degradation drivers, often requiring an external event such as a disaster to trigger or catalyze the needed transformative changes. As noted in a review for Southeast Asia, many examples of locally led restoration have a dearly paid locally learned lesson of a disaster (including landslides, floods, drought, fire) as their turning point in local history, generating the energy needed to overcome vested interests and status quo degradation [118].

Identity interacts with rights (as defined at national scale in laws governing forests, tenure and inheritance rules, and land-use planning, or in local bylaws clarifying stewardship and collective action), know-how (accumulated in local knowledge and interacting with externally supported ways of knowing), markets (for both inputs such as planting materials and products), local ecosystem service issues (often with water, microclimate, agrobiodiversity and fire as focal points of concern) and global teleconnections, interactions and feedback loops (especially those regarding climate change and biodiversity conservation).

A substantial body of case studies and action research engagements in Africa and Asia tested several ecological, economic, social, and governance propositions on the way 'Payments for Ecosystem Services (PES)' may have to be renamed 'Co-investment in Environmental Stewardship' to understand sensitivities and misunderstandings that arise from the most commonly used terminology [119]. The risk of 'crowding' social motivation for pro-environmental behavior by a focus on financial transactions was demonstrated to exist experimentally [120] but depends on the communication of programs and opportunities for local re-interpretation of terms [121].

The social pentagon interacting with identity defines not only the ultimate effect of degradation, it also forms the starting point of interventions for change (Figure 6), challenging the neat two-way options by context typology by emphasizing that it needs to be considered along an issues-cycle of awareness, motivation for change and steps toward a break with the past.

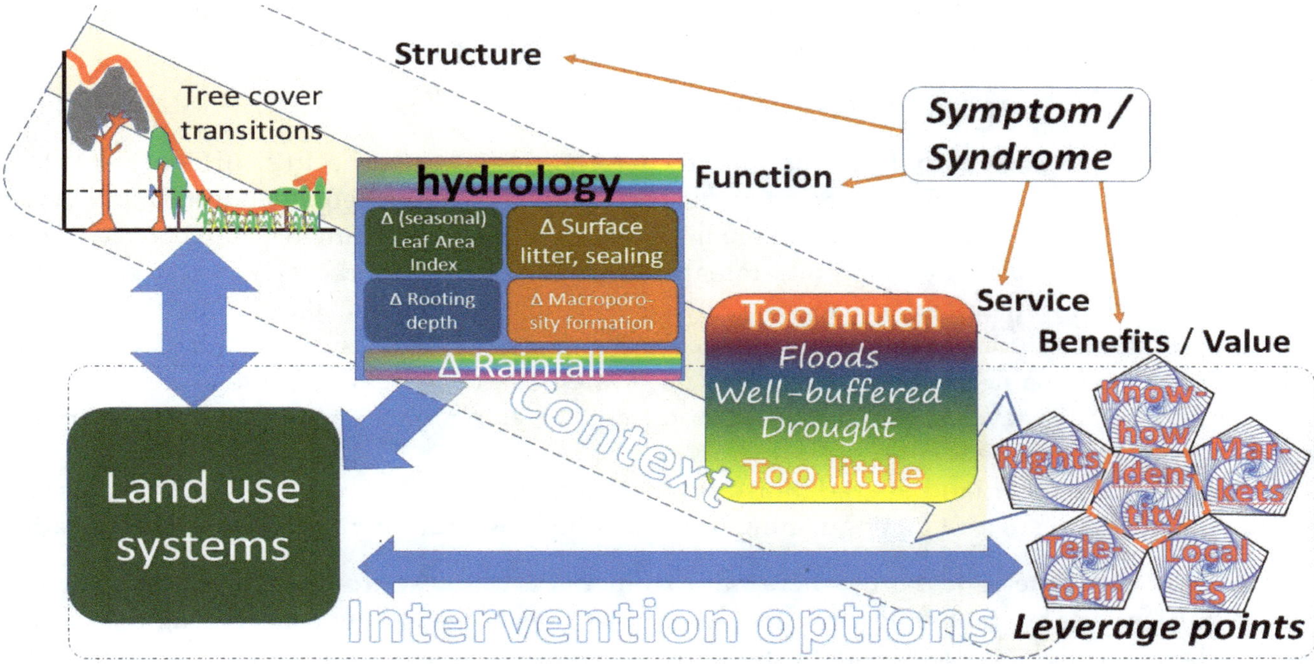


**Figure 6.** Connecting typologies of degradation/restoration context with intervention options through leverage points in the social pentagon and core identities.

While specific 'development' organizations take rights-, knowledge-, or market-based approaches as their starting point for restoration interventions in local social-ecological systems, trying to nudge current degradation into a restoration and innovation trajectory, 'conservation' organizations have focused on globally important teleconnections (climate, biodiversity, footprints of commodity trade) and/or local ecosystem services and associated local livelihoods. Regardless of the starting point, however, the relevance of an integrative livelihood orientation that relates to all five aspects has become clear to all actors. For example, lessons learned in the past 'Integrated Conservation and Development' projects were (partially) learned in designs for subsequent 'Reducing Emissions from Deforestation and forest Degradation' (REDD+) efforts [122], and lessons learned in REDD+ pilots can inform current Forest Landscape Restoration efforts [123]. Policy reform in community forestry was more difficult to achieve then was expected, as the balance between visionary 'prophets', a practical profit orientation for the main stakeholders and the transparency requirements of 'prove it' agencies (within and outside of formal government, at local, national and international scales) is hard to achieve [124]. Long-term efforts to get operational programs to reduce emissions from deforestation and forest degradation (REDD+) have had much less success than expected, revealing complexities in implementing ideas that at first had large appeal [125].

Engagement in the Sumberjaya landscape in Indonesia, a hot spot of degradation and conflict around 2000 and an inspiration for watershed restoration and resolution of similar conflicts elsewhere went through three phases: first addressing tenure conflicts in the forest margins, then providing incentives for 'river-care' efforts to reduce sediment loads by engagement with the hydro-power company, and thirdly support for marketing environment-friendly products (especially coffee) through more rewarding channels [126].

A case study on restoring traditional water harvesting structures in India showed that groundwater recharge could indeed be enhanced, facilitating an extra crop and fruit tree production, while reducing the need for seasonal migration to a nearby urban center, but negative impacts on water capture by a downstream dam suggest that tradeoffs across scales are complex [127].

Some of the appearances of integrated approaches, however, may be informed by opportunities to tap into financial incentive streams that focus on specific entry points that represent current donor/investor priorities. Although many projects claim to be 'people-centric', their trust in specific 'theories of change' and publicly declared targets in terms of area or number of trees planted can be at odds with adaptive management and local control over process and speed.

*4.2. Land-Use Change as the Target of Interventions*

While land cover refers to the (bio)physical cover observed on the Earth's surface (FAO, 2005), land use is characterized by all "the arrangements, activities and inputs people undertake in a certain land cover type to produce, change or maintain it" [128–130]. As an important interface between 'actors' and 'land cover', the concept of 'land use' integrates social, economic and ecological aspects. It thus forms the target of restoration interventions. Only where land use is functional from a local perspective, restoration efforts, including land cover changes and changes in the ecosystem structure, will have a chance to be sustainable. Typologies of land use systems have to deal with life-cycle accounting, e.g., in swidden/fallow cycles, or rotational plantation or grazing management, and environmental impacts that are time-dependent.

Based on the level of land degradation, and the intended impact of restoration on land use and land-use changes, we suggest distinguishing here the following four intensities/levels of restoration:

- R.I. Ecological intensification: where improvements to the resource base are possible within existing land use by combining provisioning, regulating and regenerative aspects of agro-ecosystem functioning, within a context of supportive input and output markets. It may include a re-integration of livestock on farms that specialized into arable-only types of farming, as 'leys' as part of a rotation can be both productive and support the recovery of desirable soil properties.
- R.II Recovery/regeneration: where forms of fallow, resting land, exclosures from grazing, fire control and assisted natural regeneration can bring back conditions within which ecological intensification is possible. This level often entails a change in land use, at least temporarily.
- R.III Reparation/recuperation: where more intense action than recovery/regeneration is performed (e.g., tree planting), with additional external support, e.g., by creating access to nurseries for diversified germplasm, knowledge not locally available, inputs (including soil amendments) not currently used, supporting local institutions (and bridging social capital with institutions outside the landscape) not currently effective and/or changing tenurial relations with the state or private sector.
- R IV. Remediation: where past activities such as long-term unsustainable land use, mining, soil pollution or deep drainage have substantially or completely destroyed the ecosystem, preventing its natural functioning or its sustainable exploitation for forestry or agricultural production. This level requires intense specific, typically externally supported, and financed efforts and economic reparation of past damage, e.g., by those who benefited from the unsustainable resource exploitation.

The overarching goal of restoration is to progress across this restoration intensity scale down to the first level. In other words, the aim is to disrupt existing degradation spirals and transform lives and landscapes to bring them progressively back into the domain were 'ecological intensification' (R.I) becomes possible again.

The degree of needed intervention/support is likely to increase across the four levels, and so does the perimeter of the system and the reach of main institutions to be mobilized. Ecological intensification (R I) is generally applicable within the current local land use system at the farm or landscape scale. Recovery/regeneration (R II) within a local social-ecological system; reparation/recuperation (R III) within a broader national policy context; while remediation (R IV) usually requires a stronger external, or even international support and investment (Figure 7).

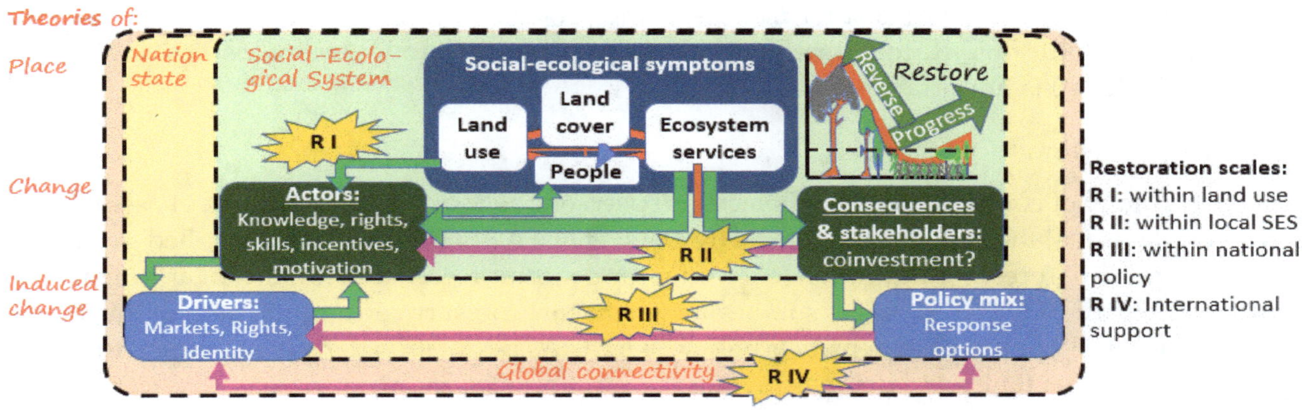


**Figure 7.** Four scales of restoration interventions and enabling actions in a nested social-ecological system perspective.

Enabling actions can be identified across these four levels that relate land use to current farm-gate profitability, additional local efforts to internalize externalities, changes in the national policy mix that influence the profitability of alternative land uses and allow sustainable land use planning and integrated landscape management, and/or global co-investment and efforts to address global challenges and strengthen global value chains. Responses at one scale (for example local SES) may help to 'scale up' to another scale, via pressures to national policies or international support, or by the cumulative effect to global drivers.

*4.3. Reconciling Bottom-Up and Top-Down Restoration Initiatives*

The strength of local motivation for change is probably the switch for any 'restoration' success, but often external support is needed to make change sustainable. Where 'restoration' is to be managed as a program or project, it requires 'metrics' as markers of progress and clarity on targets. To do so it needs to link bottom-up local initiatives and drives for return/increase of ecological functionality with the understanding of local social-ecological systems that are needed to define the elements of project designs that can attract funding and investment.

Some of the interventions targeting the surface soil structure (such as stone-rows, Zai pits and terracing in Sahelian agriculture [131]) are aimed at modifying lateral flows of soil and water in the landscape. Their effectiveness is likely to depend on scale and position on a topo-sequence, as they combine water harvesting source and sink zones. Physical interventions in surface soil structure tend to be labor demanding, with limited opportunities for mechanization, triggering a search for low-labor alternatives such as the naturally vegetated strips (started by not-plowing contour strips in intensively used slopes in the Philippines) as an alternative to tree planting and hedgerow pruning.

Local preferences are also an important aspect of obtaining the right mix between three primary ways of obtaining a change in land cover:

- Leave alone: rely on seed banks and seed rains (depending on the dispersal mode of (tree) seeds, distances to 'mother' trees, and presence of flying or walking animals as dispersal agents) as the basis of a diverse, locally adapted vegetation,
- Assisted/managed natural regeneration of vegetation or of land infrastructure: dealing with fire, grazing and other pressures and selective retention of desirable trees that can still derive from seedbanks and seed rain,
- Tree and grass planting, land infrastructure building: taking full control of the vegetation and infrastructures that will form the next land cover.

Especially a choice for the latter has important implications for local input markets, in terms of tree seeds and/or local nurseries that can provide diverse and good quality planting materials at an affordable cost. The widespread tendency to provide cost-free externally produced planting materials is now seen as a way to achieve short-term success (and project deliverables) at the cost of the long-term sustainability of solutions. Reconciling such with a gradual shift to lower intensities of restoration (toward the RI level) is important.

*4.4. Typology of Contexts*

Our typology of contexts ('theory of place') is based on the forest transition concept as applied to a pantropical dataset at the subwatershed level by Dewi et al. [108]. The classification relied primarily on (i) forest fraction, (ii) forest configuration (core, edge, and mosaic forest) and (iii) the logarithm of human population density. Stage 1 represents sub-basins where core forest covers around 80% of the total area of the sub-basins and population density is below 1 $km^{-2}$. As population density increased, the fraction of core natural forest tended to decrease while the fraction of non-forest increased, and the ratio of planted to natural tree cover (both classified as forest in national statistics) increased. While some may envision a 'transition' to have distinct stages, in a large data set these are markers along a continuum of gradual change. In a parallel paper [132] the typology is applied to thirteen pantropical landscapes (from up-river Suriname in FT1 to densely polluted East Java in Indonesia in FT6, with key issues in the forest-water-people nexus identified for each landscape but changing in character. Generic aspects of degradation that occur across a range of forest transition stages are discussed elsewhere for a range of landscapes in Southeast Asia [118]: Forest classification conflicts in FT2-4, Over-intensified monocropping in FT3-5, Degraded hillslopes in FT3-6 and Fire-climax coarse-grass lands in FT3-6.

Beyond this general typology of contexts, six 'special places' have so far been identified that deserve specific attention in the analysis of pressures, drivers, and restoration options, because of their specific importance in the interactions between ecosystem functions and human activities:

Water towers—areas that generate river flow for neighboring landscapes but tend to have an above-average human population density and opportunities to supply local markets with vegetables and other commodities, as well as providing the highest quality types of coffee to global markets [108]. These are prominent in East and West Africa and various parts of Asia, often with substantial downstream areas depending on the rivers that originate in such water towers. They can include 'cloud forests' that capture moisture beyond what precipitation gauges register.

Riparian zone & wetlands—As riparian zones often have fertile soils, sedimented from uphill erosion over long periods, offer easy ways of transport and access to water for irrigation and human use, as well as fishing as a complementary source of food security and nutrition, they have been the parts of the landscape with the longest settlement history in many parts of the world. Exceptions, with a ridge-based settlement pattern can be related to either prevalence of human disease vectors, or invading human enemies using the same river. Controlling disease vectors has often been a primary reason for draining wetlands, further allowing for increases in human population density. Many of the problem-solving interventions, however, displace pressures, such as the increase in downriver flood frequency if local flooding risk is reduced by increased drainage or removal of riparian vegetation that slowed down river flow. Restoring upstream water storage capacity and flow buffering is one of the primary targets of watershed restoration, but typically requires new upstream-downstream coordination of land use patterns and redistribution of economic benefits ('payments for ecosystem services').

Peat landscapes—Limited in area but now recognized in Congo and Amazon basin beyond their better-studied examples in Southeast Asia, they are disproportionately important in terrestrial carbon storage. Peat domes and lowland peat areas developed where drainage was restricted, and a year-round level of water saturation reduced organic matter decomposition to rates below the annual above- and belowground inputs [133]. Current understanding is that restoration focus should be on the peatland hydrological units (from dome to the river) essential for the continued function of peat domes including riparian zones that are not classified as peat soils themselves, spanning all land from river to river across the dome [134,135]. Beyond a 'zoning' perspective, focused on conservation of the dome, restoration should target the landscape, as human livelihoods in the riparian zone are both part of the problem (as the starting point for exploitation of the peat) and at the core of any livelihoods-focused solution.

Small islands, and mangroves in coastal zones—Small islands are miniature universes, where sectors of society are not as divergent as in larger on main-lands, even when national structures promote segregation. Small islands often have limited sources and supplies of freshwater, which is particularly challenging when tourism discovers the attraction of coastal zones and possibly adjacent coral reefs [136]. Developing tourism increases pressure on natural resources but may also provide a financial basis for restorative innovation, as seen in pioneering mangrove restoration on the tourist island Bali in Indonesia. Similar issues and opportunities to link terrestrial and marine ecosystems exist in the mangroves and other coastal vegetation of larger islands and continents. Attractive economic returns on mangrove destruction, e.g., for shrimp farming, as drivers of degradation affect the disproportionately high concentrations of people in coastal areas, exposing them to sea level rise, and surges due to typhoons and tsunami's [137–139].

Mining scars—areas where economic interests sparked particularly destructive change, especially where open-cast mining is used, leaving scars in the landscape where all vegetation and soil is disturbed to such a degree that natural recovery tends to be very slow. As there are frequently high metal concentrations in mine spoils affecting downstream water quality, remediating action is urgent [140]. While large-scale mining permits, and international pressure on transnational mining companies have 'internalized' such externalities by obligations to leave landscapes behind in a multifunctional condition, past mining and small and medium scale enterprises are not effectively bound by the same rules. External involvement in cleaning up the mess left by rogue mining companies can have a 'moral hazard' aspect, as the costs should morally be borne by those who benefited from the destruction.

Transport infrastructure—High investments in roads, canals and power lines make such, typically linear, landscape elements specifically vulnerable to floods, landslides, and similar disasters—to which they often contributed by disturbing hydrology and cutting into mountain slopes. As for the mines, engineers and constructors have a specific responsibility to avoid and mitigate such effects, but some of the past damage may require external 'restoration' support.

### *4.5. A Typology of Restoration Intervention Options by Context*

Table 2 combines the two typologies developed in the previous sections (typology of contexts in Section 4.4; and typology of restoration options in Section 4.1), giving further specification at cell-level (possible options for a specific context), and identifying some among the special "hotspots".

Further descriptors of context that are likely to be relevant for restoration are the climatic zones (Figure 4A) and soil properties. Soil quality is a key context parameter for restoration: it is determined by soil type, texture and depth (as it relates to hydrology and erosion), topography (slope angle and length) and indicators of current soil condition relative to what could be expected for soil in a given location under undisturbed conditions, such as the ratios of current soil carbon and bulk density, to their reference values [141–143]. Such characteristics can be complemented by a characterization of the social, institutional, and economic context that often conditions both the drivers of degradation and the possibility to overcome them, often linking to goals. As the relations between tree cover, soil quality, and human population density within a given climatic zone are fairly strong, we can similarly identify areas where tree cover is above or below what would be expected for the same demographic condition (e.g., with a 10% bandwidth around the expected value). Further classification within the six columns for options are needed and can build on existing assessment methods and typologies for (gender-specific) rights, knowledge, and market access, and for ecosystem services and associated co-investment prototypes.

**Table 2.** Proposed typology of restoration options by social-ecological context (with further distinctions by climate zone, human development index) and options for inducing change, with examples of intervention targets.

| Context / Options | Interventions in Support of Any Combination of | | | | | | | | | |
|---|---|---|---|---|---|---|---|---|---|---|
| | **Motivation, Local Insti-Tutions** | **Know-how Tree Soil Management** | | **Rights** | **Markets Inputs Outputs** | | **Local ES, Human Water Health** | | **Teleconnections: Climate Biodi-versity** | |
| **SDG-links** | **2,3,4,5,10,16** | **5,10,16** | | **4,5,8,17** | **1,2,8,9,12** | | **3, 6,7,11** | | **13,17** | **14,15,17** |
| FT-stage1: Core forests, low population density, swiddening | | FGR | | | | EcoT | | | REDD+ | PA |
| FT-stage2: e.g., Logged over forests, swiddening | | FGR/RIL /CBFM | | | | Logs NTFP | | | REDD+ | PA |
| FT-stage3: e.g., Mosaic of agriculture, secondary forests, agroforests | Local identity and initiatives, empowered and supported by participation and co-investment by external stakeholders | CBFM/ AF/TGR | | National law & local bylaw reform, forest classification and land use rights planning | Nurseries, Tree seed sources | AFP, Plant | Disease vector control, clean water supply, sanitation | Water quantity, regularity of flow (floods/droughts), quality | REDD+/ NDC | PA |
| FT-stage4: Mosaic of agriculture, secondary forests and plantations | | AF/TGR | ISFM | | | AFP, Plant | | | NDC | PA |
| FT-stage5: Open-field agriculture | | AF/TGR | ISFM | | | AFP | | | NDC | PA |
| FT-stage6: Peri-urban, tree cover higher than FT5 | | AF/TGR | | | | AFP | | | NDC | PA |
| *Special places:* Water towers | | | | Water rights | | | | | | |
| Riparian zone & wetlands | | | | RC | | | | | | |
| Peat | | Palud | Rewet | Drainage rights RC | | | | | RC | |
| Small islands, Mangroves | | | | Hotel permits | | | | | | |
| Transport infrastructure | | | Land-slides | Development contracts | | | | | | |
| Mining scars | | | | Permits, RC | | | | | | |

***Acronyms used:*** AFP (agroforestry product), CBFM (Community-based forest management), EcoT (Ecotourism), ES (Ecosystem services), FGR (Forest genetic resources), ISFM (Integrated spoil fertility management), Log (Logging), NDC (Nationally determined contribution), NTFP (non-timber forest product), PA (Protected area), Palud (Paludiculture), RC (Restoration concession), REDD$^+$ (Reducing emissions from deforestation and forest degradation), RIL (Reduced impact logging), TGR (Tree genetic resources).

As the option-by-context cells are mostly of a target 'land use' nature (see Figure 6 and Section 4.2), existing efforts to achieve generic classifications of land-use systems and intensity of land use, such as used and further developed in the ASB matrix studies can help [144,145]. Participatory land use planning methods and the LUMENS (land use for multiple environmental services) procedures are relevant here [83,146].

### *4.6. Discussion: Linking the Options in Context Typology to Issues and Goals Across Scales*

With a basic option-by-context typology (Table 2), we can revisit the links with an issue-and-goals level typology (as indicated in Figure 1B). Figure 8 suggests how the six critical enabling or resulting dimensions/aspects of degradation and restoration (earlier indicated as a social pentagon) relate across scales to the 17 sustainable development goals (SDGs). Most, if not all, goals relate to conditioning factors for restoration success (e.g., SDG16 on governance and rights, SDG4 on education or SDG5 and 10 on gender and generic equity), but also as policy domains that can benefit from the successful restoration of the land and livelihoods base of national economies. The challenge of coordinated approaches, however, exists within international organizations, as much as it does within national governments [147].

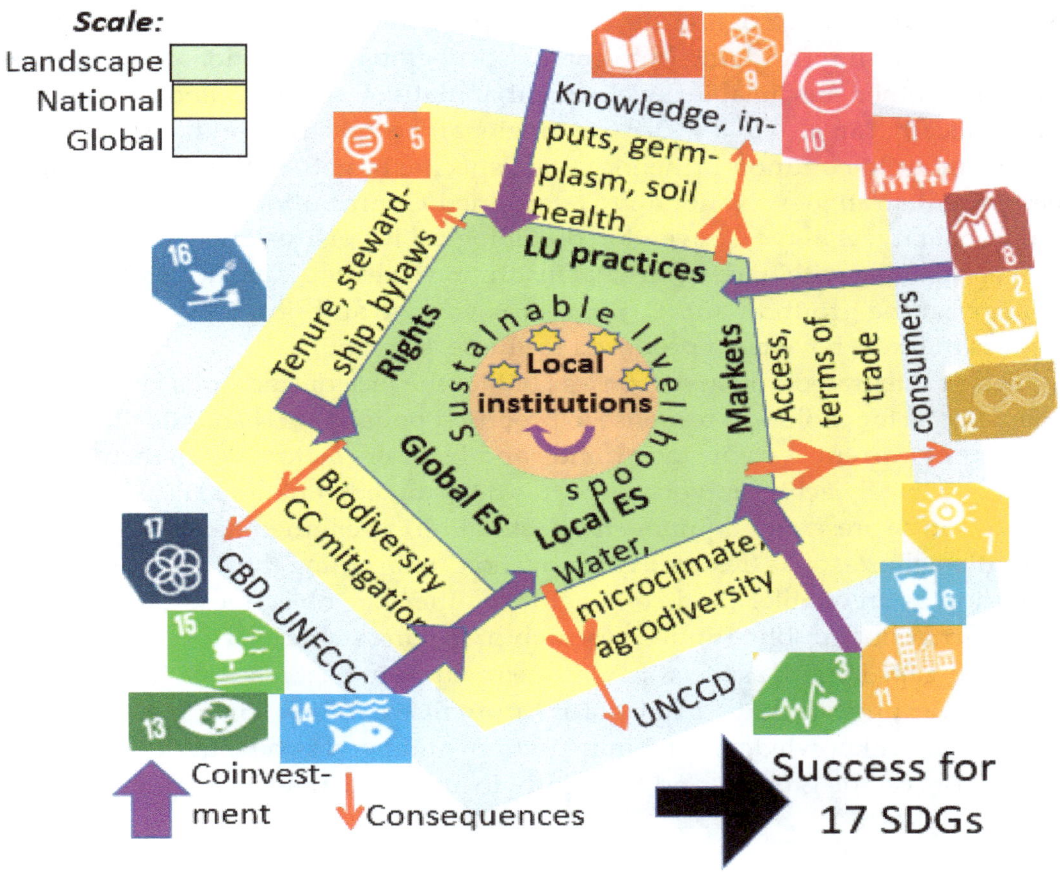


**Figure 8.** Cross-scale (local, national, global) linkage of the five determinants of local livelihoods, and the local institutional core needed to achieve the 17 sustainable development goals (SDGs).

Boundary work to relate the spheres of knowledge to the arenas of action [148] is needed to take steps from the 'symptoms' (Table 1) to potentially effective restoration actions (Table 2). Figure 9 suggests key researchable questions that can support a shared understanding among stakeholders of existing land users and use (who?, where?, how?), its consequences (so what?), stakeholders (who cares?) and the underlying drivers and pressures that need to be tackled (Why?), across scales (compare Figure 8).

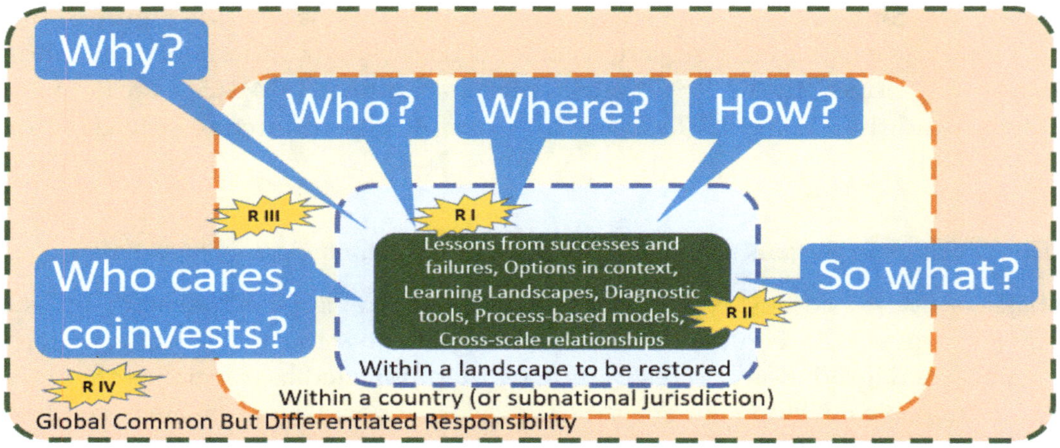


**Figure 9.** Questions that can drive efforts to understand theories of place and change for local social -ecological systems and their connections with higher-level pressures and drivers.

The boundary work to link existing and emerging knowledge to desirable action to make a difference on the ground has to recognize the different pace and dynamic of the five steps indicated in Figure 1B: A. Agenda setting (bridging between existing global and national 'restoration', 'climate change adaptation' or other 'sustainable development' initiatives), B. Better and shared understanding of what is at stake locally, but also in a wider (e.g., regional) perspective and how it interconnects with others, C. Commitment to common principles (e.g., specific targets within the SDG agenda and/or national development strategies, Nationally Determined Contributions to the Paris agreement on climate change, Aichi targets in the Convention on Biological Diversity (CBD)), D. Devolved details of design and delivery, dealing with trade-offs (what institutional change is needed to provide the essential context for change on the ground), and E. Efforts to evaluate and provide a basis for adjustments. Ten-point progress markers have been proposed for these steps [50].

Development of integrated policy responses [149] in the face of trade-offs [150],has to effectively deal with anticipating actor choices in response to proposed policies [151] and the diversity of opinions and interpretations of current system state, trends and leverage factors [152], including the balance between what can be locally achieved versus what is determined at national scale [153].

Finally, efforts can be presented as 'innovative restoration', strengthening the livelihoods dimension of existing restoration efforts, or as 'restorative innovation', supporting the integrative aspects of SDGs, pursuing restoration as an enabler (and co-benefit) of other key objectives. These are two sides of the same coin. International public-funded research programs such as FTA can support knowledge and implementation gaps in both aspects. Lessons learned from past research for development work in the restoration area suggests that a priority for future research investments is to co-develop with stakeholders system-level knowledge, combining pattern and process type understanding, to provide real solutions to actors on the ground, and the agility to answer to new, emerging issues.

## 5. Conclusions

Restoration is increasingly an object of interest for a multitude of institutional actors and groups of interests with specific objectives and perspectives as well as of diverse scientific disciplines and approaches. Each category of actors, each scientific approach has its own definition of restoration modeled by its specific perspective. This multiplicity of definitions, of ideal visions of what restoration should be is a significant impediment to collective, long-term engagement.

The typology presented in its paper can support a broader and more precise understanding of the very notion of restoration, in its diversity, generated by the diversity of contexts, objectives and solutions. Moreover, because it is precisely grounded on the interests, objectives and perspectives of the actors engaged in restoration it can help the diverse categories of actors to be involved understand the diversity of their objectives and find common ground uniquely adapted to the specificity of the situation.

The following are some of the key points arising from the assessment and analysis presented in the study.

- There is no single generalizable approach to restoration despite the general notion underlying all restoration interventions is improving or rehabilitating the functionality of the system. Restoration involves multiple entry points most of which arise from either the state of the social-ecological systems or the choices of the people who depend on it.

- Despite the goal similarity (see point 1 above), the entry points to restoration also vary widely. The entry points are defined based on the specific contexts (both the place and its social-ecological aspects) of the system to be restored. It is thus important to understand how the identified entry point could lead to the desired state of the landscape or location and what benefits (e.g., ecosystem services) should be generated from the process of restoration for the effort to be characterized as being impactful.

- Progress along the recovery trajectory in a given landscape or location needs to embrace structural, functional, and social attributes all of which are required to form an agile system with its own improved structural, functional, and social 'identity' or characteristics.

- The typology proposed in this paper is a useful one in view of the lack of such structured interpretations of the typologies. It can support further studies on the effectiveness of interventions in different contexts.

- The framework can facilitate extrapolations from a single case, combining an intervention in a context, to other potential combinations. Such a framework can facilitate the establishment of sound research findings, public goods applicable in a diversity of cases, from a reduced number of well- studied cases in the long-term. This multiplication factor is particularly important given the length of restoration projects that increase both the time to observe results and the risks of non or bad intervention.

- Such a typology, grounded on the two axes of contexts and intervention options, can also orient comparative studies in order to support decision making. In particular, it could facilitate estimations ex-ante of costs and benefits, whether marketable or not, of a specific intervention in a specific context by comparison with past comparable interventions in comparable contexts. Such evidence-based estimations are necessary for actors to engage in long-term actions that are often disrupting established interests and to attract long-term investments of external actors, public and private.

- It is also important to move the restoration beyond the forest and agricultural systems and include the 'special' places of high social and ecological values. Such places may include water towers, riparian zone and wetlands, peat landscapes, small islands, and mangroves in coastal zones, mining scars and transport infrastructure. The global emphasis on greenhouse gases and food security has given generic forest and agricultural land issues high visibility in claims of tens or hundreds of Mha of restoration, while other critical ecosystems are not getting sufficient attention for their specific needs, which may imply higher costs per unit of land for greater societal benefits.

**Author Contributions:** Conceptualization, M.v.N., V.G. and P.A.M.; Investigation, M.v.N., S.D., B.L., L.D., N.P. and A.M.; Writing—original draft, M.v.N.; Writing—review & editing, V.G., P.A.M., S.D., B.L., L.D., N.P. and A.M. All authors have read and agreed to the published version of the manuscript.

**Acknowledgments:** We acknowledge the ideas, reflections and inputs of many participants in Forests, Trees and Agroforestry (FTA) workshops held in August 2018 in Nairobi and January 2019 in Rome.

## References

1. Available online: http://www.bonnchallenge.org (accessed on 18 July 2020).
2. Available online: http://forestdeclaration.org (accessed on 18 July 2020).

3. Available online: https://www.decadeonrestoration.org/ (accessed on 18 July 2020).
4. Tomich, T.P.; Lidder, P.; Coley, M.; Gollin, D.; Meinzen-Dick, R.; Webb, P.; Carberry, P. Food and agricultural innovation pathways for prosperity. *Agric. Syst.* **2019**, *172*, 1–15. [CrossRef]
5. Gitz, V.; Place, F.; Koziell, I.; Pingault, N.; van Noordwijk, M.; Meybeck, A.; Minang, P. *A Joint Stocktaking of CGIAR Work on Forest and Landscape Restoration*; Working Paper 4; The CGIAR Research Program on Forests, Trees and Agroforestry (FTA): Bogor, Indonesia, 2020.
6. Katila, P.; Colfer, C.J.P.; de Jong, W.; Galloway, G.; Pacheco, P.; Winkel, G. (Eds.) *Sustainable Development Goals: Their Impacts on Forests and People*; Cambridge University Press: Cambridge, UK, 2019; p. 653.
7. Tidball, K.G. Seeing the forest for the trees: Hybridity and social-ecological symbols, rituals and resilience in postdisaster contexts. *Ecol. Soc.* **2014**, *19*, 25. [CrossRef]
8. Giono, J. *The Man Who Planted Trees*; Chelsea Green Publishing Co.: Chelsea, VT, USA, 1985.
9. Robbins, J. *The Man Who Planted Trees: Lost Groves, Champion Trees, and an Urgent Plan to Save the Planet*; Spiegel & Grau Random House Group: New York, NY, USA, 2012.
10. McLauchlan, L. A multispecies collective planting trees: Tending to life and making meaning outside of the conservation heroic. *Cult. Stud. Rev.* **2019**, *25*, 135–153. [CrossRef]
11. Available online: https://www.nature.org/en-us/get-involved/how-to-help/plant-a-billion/ (accessed on 15 July 2020).
12. Available online: https://www.bbc.com/news/world-africa-50813726 (accessed on 15 July 2020).
13. Available online: https://en.wikipedia.org/wiki/Trillion_Tree_Campaign (accessed on 15 July 2020).
14. Crowther, T.W.; Glick, H.B.; Covey, K.R.; Bettigole, C.; Maynard, D.; Thomas, S.M.; Smith, J.R.; Hintler, G.; Duguid, M.C.; Amatulli, G.; et al. Mapping tree density at a global scale. *Nature* **2015**, *525*, 201. [CrossRef]
15. Creed. I.F.; van Noordwijk, M. Forests, trees and water on a changing planet: A contemporary scientific perspective. In *Forest and Water on a Changing Planet: Vulnerability, Adaptation and Governance Opportunities: A Global Assessment Report (No. 38)*; Creed, I.F., van Noordwijk, M., Eds.; International Union of Forest Research Organizations (IUFRO): Vienna, Austria, 2018; pp. 13–24.
16. van Noordwijk, M.; Rahayu, S.; Gebrekirstos, A.; Kindt, R.; Tata, H.L.; Muchugi, A.; Ordonez, J.C.; Xu, J. Tree diversity as basis of agroforestry. In *Sustainable Development Through Trees on Farms: Agroforestry in its Fifth Decade*; van Noordwijk, M., Ed.; World Agroforestry (ICRAF): Bogor, Indonesia, 2019; pp. 17–44.
17. Available online: https://www.decadeonrestoration.org/Interactive/tree-planting-and-ecosystem-restoration-crash-course (accessed on 29 July 2020).
18. Garrity, D.P.; Bayala, J. Zinder: Farmer-managed natural regeneration of Sahelian parklands in Niger. In *Sustainable Development Through Trees on Farms: Agroforestry in its Fifth Decade*; van Noordwijk, M., Ed.; World Agroforestry (ICRAF): Bogor, Indonesia, 2019; pp. 151–172.
19. Duguma, L.A.; Minang, P.A.; Kimaro, A.A.; Otsyina, R.; Mpanda, M. Shinyanga: Blending old and new agroforestry to integrate development, climate change mitigation and adaptation in Tanzania. In *Sustainable Development Through Trees on Farms: Agroforestry in its Fifth Decade*; van Noordwijk, M., Ed.; World Agroforestry (ICRAF): Bogor, Indonesia, 2019; pp. 121–131.
20. van Noordwijk, M. (Ed.) *Sustainable Development Through Trees on Farms: Agroforestry in its Fifth Decade*; World Agroforestry (ICRAF): Bogor, Indonesia, 2019.
21. Miccolis, A.; Peneireiro, F.M.; Marques, H.R.; Vieira, D.L.M.; Arcoverde, M.F.; Hoffmann, M.R.; Rehder, T.; Pereira, A.V.B. *Agroforestry Systems for Ecological Restoration: How to Reconcile Conservation and Production. Options for Brazil's Cerrado and Caatinga Biomes*; Instituto Sociedade, População e Natureza—ISPN/World Agroforestry Centre (ICRAF): Brasília, Brasil, 2016; Available online: https://www.worldagroforestry.org/publication/agroforestry-systems-ecological-restoration-how-reconcile-conservation-and-production (accessed on 29 July 2020).
22. Cornelius, J.P.; Miccolis, A. Can market-based agroforestry germplasm supply systems meet the needs of forest landscape restoration? *New For.* **2018**, *49*, 457–469. [CrossRef]
23. HLPE. Agroecological and other innovative approaches for sustainable agriculture and food systems that enhance food security and nutrition. In *A Report by the High Level Panel of Experts on Food Security and Nutrition of the Committee on World Food Security*; FAO: Rome, Italy, 2019.
24. Shono, K.; Cadaweng, E.A.; Durst, P.B. Application of assisted natural regeneration to restore degraded tropical forestlands. *Restor. Ecol.* **2007**, *15*, 620–626. [CrossRef]

25. Friday, K.S. *Imperata Grassland Rehabilitation Using Agroforestry and Assisted Natural Regeneration*; World Agroforestry Centre: Bogor, Indonesia, 1999.
26. Reij, C.; Garrity, D. Scaling up farmer-managed natural regeneration in Africa to restore degraded landscapes. *Biotropica* **2016**, *48*, 834–843. [CrossRef]
27. Chazdon, R.L.; Guariguata, M.R. Natural regeneration as a tool for large-scale forest restoration in the tropics: Prospects and challenges. *Biotropica* **2016**, *48*, 716–730. [CrossRef]
28. Perrow, M.R.; Davy, A.J. (Eds.) *Handbook of Ecological Restoration*; Cambridge University Press: Cambridge, UK, 2002; Volume 2.
29. Van Andel, J.; Aronson, J. *Restoration Ecology*; Blackwell Publishing: Malden, MA, USA, 2006.
30. Vaughn, K.J.; Porensky, L.M.; Wilkerson, M.L.; Balachowski, J.; Peffer, E.; Riginos, C.; Young, T.P. Restoration ecology. *Nat. Educ. Knowl.* **2010**, *3*, 66.
31. Clewell, A.; Rieger, J.; Munro, J. *Guidelines for Developing and Managing Ecological Restoration Projects*, 2nd ed.; Society for Ecological Restoration: Tucson, AZ, USA, 2005; Available online: http://c.ymcdn.com/sites/www.ser.org/resource/resmgr/custompages/publications/ser_publications/Dev_and_Mng_Eco_Rest_Proj.pdf (accessed on 29 July 2020).
32. Buckingham, K.; Weber, S. *Assessing the ITTO Guidelines for the Restoration, Management, and Rehabilitation of Degraded Secondary Tropical Forests—Case Studies of Ghana, Indonesia, and Mexico*; International Tropical Tree Organization (ITTO) Consultancy with the World Resources Institute (WRI): Yokahama, Japan, 2015; Available online: http://www.itto.int/direct/topics/topics_pdf_download/topics_id=4632&no=1 (accessed on 29 July 2020).
33. Stanturf, J.; Mansourian, S.; Kleine, M. (Eds.) *Implementing Forest Landscape Restoration, A Practitioner's Guide*; International Union of Forest Research Organizations, Special Programme for Development of Capacities (IUFRO-SPDC): Vienna, Austria, 2017; p. 128.
34. Chazdon, R.L.; Brancalion, P.H.; Lamb, D.; Laestadius, L.; Calmon, M.; Kumar, C. A policy-driven knowledge agenda for global forest and landscape restoration. *Conserv. Lett.* **2017**, *10*, 125–132. [CrossRef]
35. Clewell, A.F.; Aronson, J. *Ecological Restoration: Principles, Values, and Structure of an Emerging Profession*; Island Press: Washington, DC, USA, 2013.
36. Duguma, L.A.; Minang, P.A.; Mpanda, M.; Kimaro, A.; Alemagi, D. Landscape restoration from a social-ecological system perspective? In *Climate-Smart Landscapes: Multifunctionality in Practice*; World Agroforestry Centre (ICRAF): Nairobi, Kenya, 2015; pp. 63–72.
37. Erbaugh, J.T.; Oldekop, J.A. Forest landscape restoration for livelihoods and well-being. *Curr. Opin. Environ. Sustain.* **2018**, *32*, 76–83. [CrossRef]
38. Kerr, J.; Verbist, B.; Suyanto; Pender, J. Placement of a payment for watershed services program in indonesia: Social and ecological factors. In *Co-Investment in Ecosystem Services: Global Lessons from Payment and Incentive Schemes*; Namirembe, S., Leimona, B., van Noordwijk, M., Minang, P., Eds.; World Agroforestry Centre (ICRAF): Nairobi, Kenya, 2017.
39. IUCN/WRI. *A Guide to the Restoration Opportunities Assessment Methodology (ROAM): Assessing Forest Landscape Restoration Opportunities at the National or Sub-National Level*; Working Paper (Road-test edition); IUCN: Gland, Switzerland, 2014; p. 125. Available online: http://www.bonnchallenge.org/content/restoration-opportunities-assessment-methodology-roam (accessed on 29 July 2020).
40. Ghazoul, J.; Chazdon, R. Degradation and recovery in changing forest landscapes: A multiscale conceptual framework. *Annu. Rev. Environ. Resour.* **2017**, *42*, 161–188. [CrossRef]
41. de Souza, S.E.F.; Vidal, E.; Chagas, G.D.F.; Elgar, A.T.; Brancalion, P.H. Ecological outcomes and livelihood benefits of community-managed agroforests and second growth forests in Southeast Brazil. *Biotropica* **2016**, *48*, 868–881. [CrossRef]
42. Baynes, J.; Herbohn, J.; Smith, C.; Fisher, R.; Bray, D. Key factors which influence the success of community forestry in developing countries. *Glob. Environ. Chang.* **2015**, *35*, 226–238. [CrossRef]
43. HLPE. *Sustainable Forestry for Food Security and Nutrition: A Report by the High-Level Panel of Experts on Food Security and Nutrition of the Committee on World Food Security*; FAO: Rome, Italy, 2017; Available online: http://www.fao.org/3/a-i7395e.pdf (accessed on 29 July 2020).
44. IRP. *Land Restoration for Achieving the Sustainable Development Goals: An International Resource Panel Think Piece*; Herrick, J.E., Abrahamse, T., Abhilash, P.C., Ali, S.H., Alvarez-Torres, P., Barau, A.S., Branquinho, C., Chhatre, A., Chotte, J.L., Von Maltitz, G.P., Eds.; United Nations Environment Programme: Nairobi, Kenya, 2019.

45. Besseau, P.; Graham, S.; Christophersen, T. *Restoring Forests and Landscapes: The Key to a Sustainable Future*; IUFRO On Behalf of the Global Partnership on Forest and Landscape Restoration: Vienna, Austria, 2018.
46. Chazdon, R.L.; Wilson, S.J.; Brondizio, E.; Guariguata, M.R.; Herbohn, J. Key challenges for governing forest and landscape restoration across different contexts. *Land Use Policy* **2020**, 104854. [CrossRef]
47. Chazdon, R.L.; Gutierrez, V.; Brancalion, P.H.; Laestadius, L.; Guariguata, M.R. Co-creating conceptual and working frameworks for implementing forest and landscape restoration based on core principles. *Forests* **2020**, *11*, 706. [CrossRef]
48. Djenontin, I.N.S.; Zulu, L.C.; Ligmann-Zielinska, A. Improving representation of decision rules in LUCC-ABM: An example with an elicitation of farmers' decision making for landscape restoration in central malawi. *Sustainability* **2020**, *12*, 5380. [CrossRef]
49. Lee, D.R.; Barrett, C.B. (Eds.) *Tradeoffs or Synergies? Agricultural Intensification, Economic Development, and the Environment*; CABI: Wallingford, UK, 2000; p. 538.
50. van Noordwijk, M. Integrated natural resource management as pathway to poverty reduction: Innovating practices, institutions and policies. *Agric. Syst.* **2019**, *172*, 60–71. [CrossRef]
51. Jackson, L.; van Noordwijk, M.; Bengtsson, J.; Foster, W.; Lipper, L.; Pulleman, M.; Said, M.; Snaddon, J.; Vodouhe, R. Biodiversity and agricultural sustainagility: From assessment to adaptive management. *Curr. Opin. Environ. Sustain.* **2010**, *2*, 80–87. [CrossRef]
52. Rockström, J.; Steffen, W.; Noone, K.; Persson, Å.; Chapin, F.S.; Lambin, E.; Lenton, T.M.; Scheffer, M.; Folke, C.; Schellnhuber, H.J. Planetary boundaries: Exploring the safe operating space for humanity. *Ecol. Soc.* **2009**, *14*, 1–33. [CrossRef]
53. Rockström, J.; Falkenmark, M.; Allan, T.; Folke, C.; Gordon, L.; Jägerskog, A.; Kummu, M.; Lannerstad, M. The unfolding water drama in the Anthropocene: Towards a resilience-based perspective on water for global sustainability. *Ecohydrology* **2014**, *7*, 1249–1261. [CrossRef]
54. van Noordwijk, M.; Bargues-Tobella, A.; Muthuri, C.W.; Gebrekirstos, A.; Maimbo, M.; Leimona, B.; Bayala, J.; Ma, X.; Lasco, R.; Xu, J.; et al. Agroforestry as part of nature-based water management. In *Sustainable Development through Trees on Farms: Agroforestry in its Fifth Decade*; van Noordwijk, M., Ed.; World Agroforestry (ICRAF): Bogor, Indonesia, 2019; pp. 305–334.
55. Sayer, J.; Sunderland, T.; Ghazoul, J.; Pfund, J.-L.; Sheil, D.; Meijaard, E.; Venter, M.; Boedhihartono, A.K.; Day, M.; Garcia, C.; et al. Ten principles for a landscape approach to reconciling agriculture, conservation, and other competing land uses. *Proc. Natl. Acad. Sci. USA* **2013**, *110*, 8349–8356. [CrossRef] [PubMed]
56. Freeman, O.E.; Duguma, L.A.; Minang, P.A. Operationalizing the integrated landscape approach in practice. *Ecol. Soc.* **2015**, *20*, 24. Available online: http://www.ecologyandsociety.org/vol20/iss1/art24/ (accessed on 29 July 2020). [CrossRef]
57. Minang, P.A.; van Noordwijk, M.; Freeman, O.E.; Mbow, C.; de Leeuw, J.; Catacutan, D. (Eds.) *Climate-Smart Landscapes: Multifunctionality In Practice*; World Agroforestry Centre (ICRAF): Nairobi, Kenya, 2015; p. 444. ISBN 978-92-9059-375-1. Available online: http://asb.cgiar.org/climate-smart-landscapes/ (accessed on 29 July 2020).
58. Reed, J.; van Vianen, J.; Deakin, E.L.; Barlow, J.; Sunderland, T. Integrated landscape approaches to managing social and environmental issues in the tropics: Learning from the past to guide the future. *Glob. Chang. Biol.* **2016**, *22*, 2540–2554. [CrossRef]
59. Sayer, J.; Margules, C.; Boedhihartono, A.K.; Sunderland, T.; Langston, J.D.; Reed, J.; Riggs, R.A.; Buck, L.E.; Campbell, B.M.; Kusters, K.; et al. Measuring the effectiveness of landscape approaches to conservation and development. *Sustain. Sci.* **2016**, *12*, 465–476. [CrossRef]
60. Wilson, S.J.; Cagalanan, D. Governing restoration: Strategies, adaptations and innovations for tomorrow's forest landscapes. *World Dev. Perspect.* **2016**, *4*, 11–15. [CrossRef]
61. Evans, K.A.; Guariguata, M.R. Success from the ground up: Participatory monitoring and forest restoration. In *Occasional Paper 159*; Center for International Forestry Research (CIFOR): Bogor, Indonesia, 2016. Available online: http://www.cifor.org/publications/pdf_fies/OccPapers/OP-159.pdf (accessed on 20 July 2020).
62. FAO. *Global Guidelines for the Restoration of Degraded Forests and Landscapes in Drylands: Building Resilience and Benefiting Livelihoods*; FAO Forestry Paper 175; Food and Agricultural Organization of the United Nations: Rome, Italy, 2015; Available online: http://www.fao.org/3/a-i5036e.pdf (accessed on 20 July 2020).
63. Bozzano, M.; Jalone, R.; Thomas, E.; Boschier, D.; Gallo, L.; Cavers, S.; Bordacs, S.; Smith, P.; Loo, J. (Eds.) *Genetic Considerations in Ecosystem Restoration Using Native Tree Species*; FAO and Bioversity International: Rome, Italy, 2014; Available online: http://www.fao.org/3/a-i3938e.pdf (accessed on 29 July 2020).

64. Dawson, I.K.; Leakey, R.; Clement, C.R.; Weber, J.C.; Cornelius, J.P.; Roshetko, J.M.; Vinceti, B.; Kalinganire, A.; Tchoundjeu, Z.; Masters, E.; et al. The management of tree genetic resources and the livelihoods of rural communities in the tropics: Non-timber forest products, smallholder agroforestry practices and tree commodity crops. *For. Ecol. Manag.* **2014**, *333*, 9–21. [CrossRef]
65. Jalonen, R.; Valette, M.; Boshier, D.; Duminil, J.; Thomas, E. Forest and landscape restoration severely constrained by a lack of attention to the quantity and quality of tree seed: Insights from a global survey. *Conserv. Lett.* **2018**, *11*, e12424. [CrossRef]
66. Sinclair, F.L.; Coe, R. The options by context approach: A paradigm shift in agronomy. *Exp. Agric.* **2019**, *55*, 1–13. [CrossRef]
67. van Noordwijk, M.; Coe, R. Methods in agroforestry research across its three paradigms. In *Sustainable Development Through Trees on Farms: Agroforestry in Its Fifth Decade*; van Noordwijk, M., Ed.; World Agroforestry (ICRAF): Bogor, Indonesia, 2019; pp. 379–402.
68. Mithöfer, D.; van Noordwijk, M.; Leimona, B.; Cerutti, P.O. Certify and shift blame, or resolve issues? Environmentally and socially responsible global trade and production of timber and tree crops. *Int. J. Biodivers. Sci. Ecosyst. Serv. Manag.* **2017**, *13*, 72–85. [CrossRef]
69. Leimona, B.; van Noordwijk, M.; Mithöfer, D.; Cerutti, P.O. Environmentally and socially responsible global production and trade of timber and tree crop commodities: Certification as a transient issue-attention cycle response to ecological and social issues. *Int. J. Biodivers. Sci. Ecosyst. Serv. Manag.* **2018**, *13*, 497–502. [CrossRef]
70. van Noordwijk, M.; Dewi, S.; Minang, P.A.; Simons, A.J. Deforestation-free claims: Scams or substance? In *Zero Deforestation: A Commitment to Change*; Pasiecznik, N., Savenije, H., Eds.; Tropenbos International: Wageningen, The Netherlands, 2017; pp. 11–16.
71. Creed, I.F.; van Noordwijk, M. (Eds.) *Forest and Water on a Changing Planet: Vulnerability, Adaptation and Governance Opportunities: A Global Assessment Report (No. 38)*; International Union of Forest Research Organizations (IUFRO): Vienna, Austria, 2018.
72. Harris, J.A.; Hobbs, R.J.; Higgs, E.; Aronson, J. Ecological restoration and global climate change. *Restor. Ecol.* **2006**, *14*, 170–176. [CrossRef]
73. Meli, P.; Holl, K.D.; Benayas, J.M.R.; Jones, H.P.; Jones, P.C.; Montoya, D.; Mateos, D.M. A global review of past land use, climate, and active vs. passive restoration effects on forest recovery. *PLoS ONE* **2017**, *12*, e0171368. [CrossRef] [PubMed]
74. Mansourian, S.; Dudley, N.; Vallauri, D. Forest landscape restoration: Progress in the last decade and remaining challenges. *Ecol. Restor.* **2017**, *35*, 281–288. [CrossRef]
75. Swinfield, T.; Afriandi, R.; Antoni, F.; Harrison, R.D. Accelerating tropical forest restoration through the selective removal of pioneer species. *For. Ecol. Manag.* **2016**, *381*, 209–216. [CrossRef]
76. Harrison, R.D.; Tan, S.; Plotkin, J.B.; Slik, F.; Detto, M.; Brenes, T.; Itoh, A.; Davies, S.J. Consequences of defaunation for a tropical tree community. *Ecol. Lett.* **2013**, *16*, 687–694. [CrossRef]
77. de Royer, S.; van Noordwijk, M.; Roshetko, J. Does community-based forest management in Indonesia devolve social justice or social costs? *Int. For. Rev.* **2018**, *20*, 167–180. [CrossRef]
78. van Noordwijk, M.; Leimona, B. Principles for fairness and efficiency in enhancing environmental services in Asia: Payments, compensation, or co-investment? *Ecol. Soc.* **2010**, *15*, 17. [CrossRef]
79. Leimona, B.; van Noordwijk, M.; de Groot, R.; Leemans, R. Fairly efficient, efficiently fair: Lessons from designing and testing payment schemes for ecosystem services in Asia. *Ecosyst. Serv.* **2015**, *12*, 16–28. [CrossRef]
80. Namirembe, S.; Leimona, B.; van Noordwijk, M.; Minang, P.A. *Co-Investment in Ecosystem Services: Global Lessons from Payment and Incentive Schemes*; World Agroforestry (ICRAF): Nairobi, Kenya, 2017; Chapter 1; Available online: https://www.worldagroforestry.org/sites/default/files/u884/Ch1_IntroCoinvest_ebook.pdf (accessed on 20 July 2020).
81. Meinzen-Dick, R.; Quisumbing, A.; Doss, C.; Theis, S. Women's land rights as a pathway to poverty reduction: Framework and review of available evidence. *Agric. Syst.* **2017**, *172*, 72–82. [CrossRef]
82. Kusters, K.; Buck, L.; de Graaf, M.; Minang, P.A.; van Oosten, C.; Zagt, R. Participatory planning, monitoring and evaluation of multi-stakeholder platforms in integrated landscape initiatives. *Environ. Manag.* **2017**, *2*, 170–181. [CrossRef] [PubMed]
83. van Noordwijk, M.; Kim, Y.S.; Leimona, B.; Hairiah, K.; Fisher, L.A. Metrics of water security, adaptive capacity and agroforestry in Indonesia. *Curr. Opin. Environ. Sustain.* **2016**, *21*, 1–8. [CrossRef]
84. Ahrends, A.; Hollingsworth, P.M.; Beckschäfer, P.; Chen, H.; Zomer, R.J.; Zhang, L.; Wang, M.; Xu, J. China's fight to halt tree cover loss. *Proc. R. Soc. B* **2017**, *284*. [CrossRef]

85. Wangpakapattanawong, P.; Kavinchan, N.; Vaidhayakarn, C.; Schmidt-Vogt, D.; Elliott, S. Fallow to forest: Applying indigenous and scientific knowledge of swidden cultivation to tropical forest restoration. *For. Ecol. Manag.* **2010**, *260*, 1399–1406. [CrossRef]
86. Chirwa, P.W.; Mahamane, L. Overview of restoration and management practices in the degraded landscapes of the Sahelian and dryland forests and woodlands of East and southern Africa. *South. For. J. For. Sci.* **2017**, *79*, 87–94. [CrossRef]
87. Mokria, M.; Tolera, M.; Sterck, F.J.; Gebrekirstos, A.; Bongers, F.; Decuyper, M.; Sass-Klaassen, U. The frankincense tree Boswellia neglecta reveals high potential for restoration of woodlands in the Horn of Africa. *For. Ecol. Manag.* **2017**, *385*, 16–24. [CrossRef]
88. Miccolis, A. Restoration through agroforestry: Options for reconciling livelihoods with conservation in the cerrado and caatinga biomes in Brazil. *Exp. Agric.* **2017**, *55*, 208–225. [CrossRef]
89. Gann, G.D.; McDonald, T.; Walder, B.; Aronson, J.; Nelson, C.R.; Jonson, J.; Hallett, J.G.; Eisenberg, C.; Guariguata, M.R.; Liu, J.; et al. International principles and standards for the practice of ecological restoration. Second edition. *Restor. Ecol.* **2019**, *27*, S1–S46. [CrossRef]
90. IPBES. *The IPBES Assessment Report on Land Degradation And Restoration*; Montanarella, L., Scholes, R., Brainich, A., Eds.; Secretariat of the Intergovernmental Science-Policy Platform on Biodiversity and Ecosystem Services: Bonn, Germany, 2018; p. 44. Available online: https://ipbes.net/sites/default/files/2018_ldr_full_report_book_v4_pages.pdf (accessed on 29 July 2020).
91. Available online: https://www.iucn.org/theme/forests/our-work/forest-landscape-restoration (accessed on 29 July 2020).
92. Millennium Ecosystem Assessment. *Ecosystems and Human Well-Being: Current State and Trends*; Island Press: Washington, DC, USA, 2005; Volume 5.
93. Available online: https://www.iucn.org/news/protected-areas/201905/enhancing-progress-towards-aichi-target-11#:~{}:text=As%20reported%20in%20the%20latest,and%2010%25%20of%20the%20sea.&text=In%20an%20effort%20to%20ramp,Partnership%20on%20Aichi%20Target%2011 (accessed on 18 July 2020).
94. Potschin-Young, M.; Haines-Young, R.; Görg, C.; Heink, U.; Jax, K.; Schleyer, C. Understanding the role of conceptual frameworks: Reading the ecosystem service cascade. *Ecosyst. Serv.* **2018**, *29*, 428–440. [CrossRef]
95. Van Noordwijk, M.; Minang, P.A. *If We Cannot Define It, We Cannot Save It: Forest Definitions and REDD*; Tropenbos International: Wageningen, The Netherlands, 2009.
96. Chazdon, R.L.; Brancalion, P.H.; Laestadius, L.; Bennett-Curry, A.; Buckingham, K.; Kumar, C.; Wilson, S.J. When is a forest a forest? Forest concepts and definitions in the era of forest and landscape restoration. *Ambio* **2016**, *45*, 538–550. [CrossRef]
97. Dignac, M.F.; Derrien, D.; Barré, P.; Barot, S.; Cécillon, L.; Chenu, C.; Chevallier, T.; Freschet, G.T.; Garnier, P.; Guenet, B.; et al. Increasing soil carbon storage: Mechanisms, effects of agricultural practices and proxies. A review. *Agron. Sustain. Dev.* **2017**, *37*, 14. [CrossRef]
98. Suprayogo, D.; van Noordwijk, M.; Hairiah, K.; Meilasari, N.; Rabbani, A.L.; Ishaq, R.M.; Widianto, W. Infiltration-Friendly Agroforestry Land Uses on Volcanic Slopes in the Rejoso Watershed, East Java, Indonesia. *Land* **2020**, *9*, 240. [CrossRef]
99. Shepherd, K.D.; Shepherd, G.; Walsh, M.G. Land health surveillance and response: A framework for evidence-informed land management. *Agric. Syst.* **2015**, *132*, 93–106. [CrossRef]
100. Tomich, T.P.; Chomitz, K.; Francisco, H.; Izac, A.M.N.; Murdiyarso, D.; Ratner, B.D.; Thomas, D.E.; van Noordwijk, M. Policy analysis and environmental problems at different scales: Asking the right questions. *Agric. Ecosyst. Environ.* **2004**, *104*, 5–18. [CrossRef]
101. Namirembe, S.; Leimona, B.; van Noordwijk, M.; Minang, P.A. *Co-Investment in Ecosystem Services: Global Lessons from Payment and Incentive Schemes*; World Agroforestry (ICRAF): Nairobi, Kenya, 2019.
102. van Noordwijk, M.; Duguma, L.A.; Dewi, S.; Leimona, B.; Catacutan, D.M.; Lusiana, B.; Öborn, I.; Hairiah, K.; Minang, P.A. SDG synergy between agriculture and forestry in the food, energy, water and income nexus: Reinventing agroforestry? *Curr. Opin. Environ. Sustain.* **2018**, *34*, 33–42. [CrossRef]
103. Minang, P.A.; van Noordwijk, M.; Duguma, L.A. Policies for ecosystem services enhancement. In *Sustainable Development Through Trees on Farms: Agroforestry in its Fifth Decade*; van Noordwijk, M., Ed.; World Agroforestry (ICRAF): Bogor, Indonesia, 2019; pp. 361–376.
104. Rudel, T.K. Did a green revolution restore the forests of the American South. In *Agricultural Technologies and Tropical Deforestation*; Angelsen, A., Kaimowitz, D., Eds.; CABi: Wallingford, UK, 2001; pp. 53–68.

105. Minang, P.A.; van Noordwijk, M.; Kahurani, E. *Partnership in the Tropical Forest Margins: A 20-Year Journey in Search of Alternatives to Slash-and-Burn*; World Agroforestry: Nairobi, Kenya, 2014; p. 241.
106. Li, S.; Li, X. Global understanding of farmland abandonment: A review and prospects. *J. Geogr. Sci.* **2017**, *27*, 1123–1150. [CrossRef]
107. Schröder, P.; Antonarakis, A.S.; Brauer, J.; Conteh, A.; Kohsaka, R.; Uchiyama, Y.; Pacheco, P. SDG 12: Responsible consumption and production–Potential Benefits and impacts on forests and livelihoods. In *Sustainable Development Goals: Their Impacts on Forests and People*; Katila, P., Colfer, C.J.P., de Jong, W., Galloway, G., Pacheco, P., Winkel, G., Eds.; Cambridge University Press: Cambridge, UK, 2019; pp. 386–418.
108. Dewi, S.; van Noordwijk, M.; Zulkarnain, M.T.; Dwiputra, A.; Prabhu, R.; Hyman, G.; Gitz, V.; Nasi, R. Tropical forest-transition landscapes: A portfolio for studying people, tree crops and agro-ecological change in context. *Int. J. Biodivers. Sci. Ecosyst. Serv. Manag.* **2017**, *13*, 312–329. [CrossRef]
109. Filoso, S.; Bezerra, M.O.; Weiss, K.C.; Palmer, M.A. Impacts of forest restoration on water yield: A systematic review. *PLoS ONE* **2017**, *12*, e0183210. [CrossRef]
110. Ilstedt, U.; Tobella, A.B.; Bazié, H.R.; Bayala, J.; Verbeeten, E.; Nyberg, G.; Sanou, J.; Benegas, L.; Murdiyarso, D.; Laudon, H.; et al. Intermediate tree cover can maximize groundwater recharge in the seasonally dry tropics. *Sci. Rep.* **2016**, *6*. [CrossRef]
111. van Noordwijk, M.; Tanika, L.; Lusiana, B. Flood risk reduction and flow buffering as ecosystem services—Part I: Theory on a flow persistence indicator. *Hydrol. Earth Syst. Sci.* **2017**, *21*, 2321–2340. [CrossRef]
112. van Noordwijk, M.; Tanika, L.; Lusiana, B. Flood risk reduction and flow buffering as ecosystem services—Part II: Land use and rainfall intensity effects in Southeast Asia. *Hydrol. Earth Syst. Sci.* **2017**, *21*, 2341–2360. [CrossRef]
113. Dufour, S.; Piégay, H. From the myth of a lost paradise to targeted river restoration: Forget natural references and focus on human benefits. *River Res. Appl.* **2009**, *25*, 568–581. [CrossRef]
114. Ellison, D.; Morris, C.; Locatelli, B.; Sheil, D.; Cohen, J.D.M.; Murdiyarso, D.; Gutierrez, V.; Van Noordwijk, M.; Creed, I.F.; Pokorny, J.; et al. Trees, forests and water: Cool insights for a hot world. *Glob. Environ. Chang.* **2017**, *43*, 51–61. [CrossRef]
115. Bonnesoeur, V.; Locatelli, B.; Guariguata, M.R.; Ochoa-Tocachi, B.F.; Vanacker, V.; Mao, Z.; Stokes, A.; Mathez-Stiefel, S.L. Impacts of forests and forestation on hydrological services in the Andes: A systematic review. *For. Ecol. Manag.* **2019**, *433*, 569–584. [CrossRef]
116. Sternberg, L.D. Savanna-forest hysteresis in the tropics. *Glob. Ecol. Biogeogr.* **2001**, *10*, 369–378. [CrossRef]
117. Hirota, M.; Holmgren, M.; Van Nes, E.H.; Scheffer, M. Global resilience of tropical forest and savanna to critical transitions. *Science* **2011**, *334*, 232–235. [CrossRef]
118. van Noordwijk, M.; Ekadinata, A.; Leimona, B.; Catacutan, D.C.; Martini, E.; Tata, H.L.; Öborn, I.; Hairiah, K.; Wangpakapattanawong, P.; Mulia, R.; et al. *Agroforestry Options for Degraded Landscapes in Southeast Asia*; Dagar, J.C., Gupta, S.R., Teketay, D., Eds.; Agroforestry for Degraded Landscapes Springer: Singapore, 2020; in press.
119. Leimona, B.; van Noordwijk, M.; Kennedy, S.; Namirembe, S.; Minang, P.A. Synthesis and lessons on ecological, economic, social and governance propositions. In *Co -Investment in Ecosystem Services: Global Lessons from Payment and Incentive Schemes*; Namirembe, S., Leimona, B., van Noordwijk, M., Minang, P.A., Eds.; World Agroforestry Centre (ICRAF): Nairobi, Kenya, 2019; Chapter 38.
120. Van Noordwijk, M.; Leimona, B.; Jindal, R.; Villamor, G.B.; Vardhan, M.; Namirembe, S.; Catacutan, D.; Kerr, J.; Minang, P.A.; Tomich, T.P. Payments for Environmental Services: Evolution toward efficient and fair incentives for multifunctional landscapes. *Annu. Rev. Environ. Resour.* **2012**, *37*, 389–420. [CrossRef]
121. Chapman, M.; Satterfield, T.; Wittman, H.; Chan, K.M. A payment by any other name: Is Costa Rica's PES a payment for services or a support for stewards? *World Dev.* **2020**, *129*, 104900. [CrossRef]
122. Minang, P.A.; van Noordwijk, M. Design challenges for achieving reduced emissions from deforestation and forest degradation through conservation: Leveraging multiple paradigms at the tropical forest margins. *Land Use Policy* **2013**, *31*, 61–70. [CrossRef]
123. Barr, C.M.; Sayer, J.A. The political economy of reforestation and forest restoration in Asia–Pacific: Critical issues for REDD+. *Biol. Cons.* **2012**, *154*, 9–19. [CrossRef]
124. van Noordwijk, M. Prophets, profits, prove it: Social forestry under pressure. *One Earth* **2020**, *2*, 394–397. [CrossRef]
125. Minang, P.A.; Van Noordwijk, M.; Duguma, L.; Alemagi, D.; Do, T.H.; Bernard, F.; Agung, P.; Robiglio, V.; Catacutan, D.; Suyanto, S.; et al. REDD+ Readiness progress across countries: Time for reconsideration. *Clim. Policy* **2014**, *14*, 685–708. [CrossRef]

126. van Noordwijk, M.; Leimona, B.; Amaruzaman, S. Sumber Jaya from conflict to source of wealth in Indonesia: Reconciling coffee agroforestry and watershed functions. In *Sustainable Development Through Trees on Farms: Agroforestry in its Fifth Decade*; van Noordwijk, M., Ed.; World Agroforestry (ICRAF): Bogor, Indonesia, 2019; pp. 177–192.
127. Singh, R.; van Noordwijk, M.; Chaturvedi, O.P.; Garg, K.K.; Dev, I.; Wani, S.P.; Rizvi, J. Bundelkhand: Public co-investment in groundwater recharge in India. In *Sustainable Development Through Trees on Farms: Agroforestry in its Fifth Decade*; van Noordwijk, M., Ed.; World Agroforestry (ICRAF): Bogor, Indonesia, 2019; pp. 201–208.
128. FAO/UNEP. *Terminology for Integrated Resources Planning and Management*; United Nations Environmental Programme: Kenya, Nairobi; FAO: Rome, Italy, 1999.
129. FAO. *Land Cover Classification System Classification Concepts and User Manual Software Version (2)*; Di Gregorio, A., Jansen, L.J.M., Eds.; FAO: Rome, Italy, 2005; Available online: http://www.fao.org/3/y7220e/y7220e00.htm (accessed on 20 July 2020).
130. FAO/ITPS. *Status of the World's Soil Resources (SWSR)—Main Report*; Food and Agriculture Organization of the United Nations and Intergovernmental Technical Panel on Soils: Rome, Italy, 2015; p. 650. Available online: http://www.fao.org/3/a-i5199e.pdf (accessed on 20 July 2020).
131. Bayala, J.; Sanou, J.; Teklehaimanot, Z.; Ouedraogo, S.J.; Kalinganire, A.; Coe, R.; Van Noordwijk, M. Advances in knowledge of processes in soil–tree–crop interactions in parkland systems in the West African Sahel: A review. *Agric. Ecosyst. Environ.* **2015**, *205*, 25–35. [CrossRef]
132. van Noordwijk, M.; Speelman, E.; Hofstede, G.J.; Farida, A.; Abdurrahim, A.Y.; Miccolis, A.; Hakim, A.L.; Wamucii, C.N.; Lagneaux, E.; Andreotti, F.; et al. Sustainable agroforestry management: Changing the game. *Land* **2020**, *9*, 243. [CrossRef]
133. Saragi-Sasmito, M.F.; Murdiyarso, D.; June, T.; Sasmito, S.D. Carbon stocks, emissions, and aboveground productivity in restored secondary tropical peat swamp forests. *Mitig. Adapt. Strat. Glob. Chang.* **2018**, *24*, 521–533. [CrossRef]
134. Widayati, A.; Tata, H.L.; van Noordwijk, M. Agroforestry in peatlands: Combining productive and protective functions as part of restoration. In *Agroforestry Options for ASEAN Series No. 4*; World Agroforestry Centre (ICRAF): Bogor, Indonesia, 2016.
135. van Noordwijk, M.; Matthews, R.B.; Agus, F.; Farmer, J.; Verchot, L.; Hergoualc'h, K.; Persch, S.; Tata, H.L.; Lusiana, B.; Widayati, A.; et al. Mud, muddle and models in the knowledge value- chain to action on tropical peatland issues. *Mitig. Adapt. Strat. Glod. Chang.* **2014**, *19*, 863–885.
136. van Noordwijk, M. Small-island agroforestry in an era of climate change and sustainable development goals. In *Sustainable Development Through Trees on Farms: Agroforestry in its Fifth Decade*; van Noordwijk, M., Ed.; World Agroforestry (ICRAF): Bogor, Indonesia, 2019; pp. 233–248.
137. Chan, H.T.; Baba, S. *Manual on Guidelines for Rehabilitation of Coastal Forests Damaged by Natural Hazards in the Asia-Pacific Region*; International Society for Mangrove Ecosystems and International Tropical Timber Organization (ITTO): Yokahama, Japan, 2009; Available online: http://www.preventionweb.net/fies/13225_ISMEManualoncoastalforestrehabilita.pdf (accessed on 20 July 2020).
138. Global Nature Fund. *Mangrove Restoration Guide, Best Practices and Lessons Learned from a Community-Based Conservation Project*; Global Nature Fund: Radolfzell, Germany, 2015; Available online: https://www.globalnature.org/bausteine.net/f/8281/GNF_Mangrove_Handbook_2015.pdf?fd=0 (accessed on 20 July 2020).
139. Sasmito, S.D.; Murdiyarso, D.; Friess, D.A.; Kurnianto, S. Can mangroves keep pace with contemporary sea level rise? A global data review. *Wetl. Ecol. Manag.* **2016**, *24*, 263–278. [CrossRef]
140. Wong, M.H. Ecological restoration of mine degraded soils, with emphasis on metal contaminated soils. *Chemosphere* **2003**, *50*, 775–780. [CrossRef]
141. Hairiah, K.; van Noordwijk, M.; Sari, R.R.; Saputra, D.D.; Widianto; Suprayogo, D.; Kurniawan, S.; Prayogo, C.; Gusli, S. Soil carbon stocks in Indonesian (agro)forest transitions: Compaction conceals lower carbon concentrations in standard accounting. *Agric. Ecosyst. Environ.* **2020**, *294*, 106879. [CrossRef]
142. van Noordwijk, M.; Goverse, T.; Ballabio, C.; Banwart, S.; Bhattacharyya, T.; Goldhaber, M.; Nikolaidis, N.; Noellemeyer, E.; Zhao, Y.; Mline, E. Soil carbon transition curves: Reversal of land degradation through management of soil organic matter for multiple benefits. In *Soil Carbon: Science, Management and Policy for Multiple Benefits*; Banwart, S.A., Noellemeyer, E., Milne, E., Eds.; CAB International: Harpenden, UK, 2015; pp. 26–46.

143. Minasny, B.; Malone, B.; McBratney, A.B.; Angers, D.; Arrouays, D.; Chambers, A.; Chaplot, V.; Chen, Z.-S.; Cheng, K.; Das, B.S.; et al. Soil carbon 4 per mille. *Geoderma* **2017**, *292*, 59–86. [CrossRef]
144. Tomich, T.P.; van Noordwijk, M.; Budidarsono, S.; Gillison, A.N.; Kusumanto, Y.; Murdiyarso, D.; Stolle, F.; Fagi, A.M. Agricultural intensification, deforestation, and the environment: Assessing tradeoffs in Sumatra, Indonesia. In *Tradeoffs or Synergies? Agricultural Intensification, Economic Development and the Environment*; Lee, D.R., Barrett, C.B., Eds.; CAB-International: Wallingford, UK, 2001; pp. 221–244.
145. Tomich, T.P.; van Noordwijk, M. Trade-off matrix between private and public benefits of land-use systems (ASB Matrix). In *Negotiation-Support Toolkit for Learning Landscapes*; van Noordwijk, M., Lusiana, B., Leimona, B., Dewi, S., Wulandari, D., Eds.; World Agroforestry Centre (ICRAF): Bogor, Indonesia, 2013; pp. 118–119.
146. Dewi, S.; Ekadinata, A.; Galudra, G.; Agung, P.; Johana, F. LUWES: Land use planning for low emission development strategy. *World Agrofor. Cent. Bogor. Indones* **2011**. 47p. Available online: http://www.worldagroforestry.org/output/land-use-planning-low-emission-development-strategy-luwes (accessed on 20 July 2020).
147. Neely, C.; Bourne, M.; Chesterman, S.; Kouplevatskaya-Buttoud, I.; Bojic, D.; Vallée, D. *Implementing 2030 Agenda for Food and Agriculture: Accelerating Impact through Cross-Sectoral Coordination*; World Agroforestry Centre (ICRAF) and Food and Agricultural Organization of The United Nations (FAO): Nairobi, Kenya; Rome, Italy, 2017; Available online: http://www.fao.org/3/a-i7749e.pdf (accessed on 20 July 2020).
148. Clark, W.C.; Tomich, T.P.; Van Noordwijk, M.; Guston, D.; Catacutan, D.C.; Dickson, N.M.; McNie, E. Boundary work for sustainable development: Natural resource management at the Consultative Group on International Agricultural Research (CGIAR). *Proc. Natl. Acad. Sci. USA* **2016**, *113*, 4615–4622. [CrossRef]
149. Robiglio, V.; Reyes, M. Restoration through formalization? Assessing the potential of Peru's Agroforestry Concessions scheme to contribute to restoration in agricultural frontiers in the Amazon region. *World Dev. Perspect.* **2016**, *3*, 42–46. [CrossRef]
150. Villamor, G.B.; van Noordwijk, M.; Leimona, B.; Duguma, L. Tradeoffs. In *Co-Investment in Ecosystem Services: Global Lessons from Payment and Incentive Schemes*; Namirembe, S., Leimona, B., van Noordwijk, M., Minang, P.A., Eds.; ICRAF: Nairobi, Kenya, 2017; Chapter 3; Available online: http://www.worldagroforestry.org/sites/default/files/chapters/Ch3%20Trade-offs_ebookB-DONE2.pdf (accessed on 20 July 2020).
151. Suwarno, A.; van Noordwijk, M.; Weikard, H.P.; Suyamto, D. Indonesia's forest conversion moratorium assessed with an agent-based model of Land-Use Change and Ecosystem Services (LUCES). *Mitig. Adapt. Strat. Glob. Chang.* **2018**, *23*, 211–229. [CrossRef]
152. Amaruzaman, S.; Leimona, B.; van Noordwijk, M.; Lusiana, B. Discourses on the performance gap of agriculture in a green economy: A Q-methodology study in Indonesia. *Int. J. Biodivers. Sci. Ecosyst. Serv. Manag.* **2017**, *13*, 233–247. [CrossRef]
153. Langston, J.; McIntyre, R.; Falconer, K.; Sunderland, T.J.C.; van Noordwijk, M.; Boedihartono, A.K. Discourses mapped by Q-method show governance constraints motivate landscape approaches in Indonesia. *PLoS ONE* **2019**, *14*, e0211221. [CrossRef]

# 5

# Gendered Species Preferences Link Tree Diversity and Carbon Stocks in Cacao Agroforest

**Rika Ratna Sari [1,2,*], Danny Dwi Saputra [1,2], Kurniatun Hairiah [1], Danaë M. A. Rozendaal [2,3], James M. Roshetko [4] and Meine van Noordwijk [2,4]**

1 Soil Science Department, Faculty of Agriculture, Universitas Brawijaya, Malang 65145, Indonesia; danny_saputra@ub.ac.id (D.D.S.); kurniatun_h@ub.ac.id (K.H.)

2 Plant Production Systems Group, Wageningen University and Research, 6708 PB Wageningen, The Netherlands; danae.rozendaal@wur.nl (D.M.A.R.); m.vannoordwijk@cgiar.org (M.v.N.)

3 Centre for Crop Systems Analysis, Wageningen University and Research, 6708 PB Wageningen, The Netherlands

4 World Agroforestry Centre, ICRAF Southeast Asia, Bogor 16001, Indonesia; j.roshetko@cgiar.org

* Correspondence: rr.sari@ub.ac.id

**Abstract:** The degree to which the maintenance of carbon (C) stocks and tree diversity can be jointly achieved in production landscapes is debated. C stocks in forests are decreased by logging before tree diversity is affected, while C stocks in monoculture tree plantations increase, but diversity does not. Agroforestry can break this hysteresis pattern, relevant for policies in search of synergy. We compared total C stocks and tree diversity among degraded forest, complex cacao/fruit tree agroforests, simple shade-tree cacao agroforestry, monoculture cacao, and annual crops in the Konawe District, Southeast Sulawesi, Indonesia. We evaluated farmer tree preferences and the utility value of the system for 40 farmers (male and female). The highest tree diversity (Shannon–Wiener H index 2.36) and C stocks (282 Mg C ha$^{-1}$) were found in degraded forest, followed by cacao-based agroforestry systems (H index ranged from 0.58–0.93 with C stocks of 75–89 Mg ha$^{-1}$). Male farmers selected timber and fruit tree species with economic benefits as shade trees, while female farmers preferred production for household needs (fruit trees and vegetables). Carbon stocks and tree diversity were positively related ($R^2 = 0.72$). Adding data from across Indonesia ($n = 102$), agroforestry systems had an intermediate position between forest decline and reforestation responses. Maintaining agroforestry in the landscape allows aboveground C stocks up to 50 Mg ha$^{-1}$ and reduces biodiversity loss. Agroforestry facilitates climate change mitigation and biodiversity goals to be addressed simultaneously in sustainable production landscapes.

**Keywords:** carbon storage; cacao agroforestry; farmer tree preference; utility value

## 1. Introduction

The global relevance of managing landscapes simultaneously for resilient production (climate change adaptation), carbon (C) storage (climate change mitigation), biodiversity, and watershed functions has raised interest in the degree to which such functions tend to correlate [1,2], with some authors claiming causal links beyond correlations [3]. Synergy between combatting climate change (mitigation) and conserving biodiversity in production landscapes is desirable, and may be attained in agroforestry systems [4]. Theoretically, there is a positive feedback between biodiversity and ecosystem functioning [5,6], as trait diversity reduces vulnerability to external shocks, enhances niche differentiation and productivity [7], and thus carbon stocks. However, there are plateaus for each function considered at relatively low diversity levels which can be maintained by a few species in a particular functional group [8].

Biological diversity promotes various ecosystem functions at multiple spatial and temporal scales [9]. Within forest ecosystems, it allows species to access more available resources, and facilitate other species, resulting in enhanced resistance to disturbance and generally enhanced stability [10] at the system level. Various tree species from multiple aboveground strata provide different types of litter input, which contribute to soil surface protection and supply organic matter to support the nutrient cycle. Litter input maintains soil fertility, promotes biomass growth and enhances carbon dioxide sequestration from the atmosphere. The patterns are more complex within intact tropical forests [4,6,11,12]. Several studies concluded that there was a positive relationship between biodiversity and ecosystem services (including C stock) in intact tropical forest [6,12,13], while other studies found weak [14–16] or even a negative [17] relationships. These differences may be caused by differences in forest types and the biodiversity metrics that were used [9]. However, the degree of causality and the relevance of such relationships beyond (modified) natural forests is still uncertain and its relevance in the face of global climate change is debated [18].

In managed tropical forest landscapes, relationships between biodiversity and C storage may be less clear. Intact tropical forests are known for both their high biodiversity (including tree diversity) and high C stocks [12,19]. Converting forests to agricultural use may start with logging and/or land clearing through slash-and-burn agriculture that reduces plot-level aboveground biodiversity as well as C stock to zero, but in subsequent recovery the two characteristics are not necessarily aligned (Figure 1). Logging tropical forests reduces the terrestrial C stock, before it negatively affects biodiversity [20,21] (the "degradation leg"; Figure 1). Murdiyarso et al. [22] assessed the relationship between relative C stocks and relative plot-level biodiversity for a range of forest-derived land uses in the humid tropics, and found that the loss of C stocks was higher than that of plant species richness (data from 3 continents). Restoration can take place through replanting in plantations (low diversity), but high biodiversity as a result of natural seed dispersal will usually be achieved after C stocks increased (the "restoration leg"; Figure 1). Natural regeneration (secondary succession) can simultaneously increase C stocks and diversity over time, but C stocks recover faster than diversity [23]. Agroforestry systems are expected to occupy an intermediate position between the degradation and restoration leg, because they support a higher diversity.

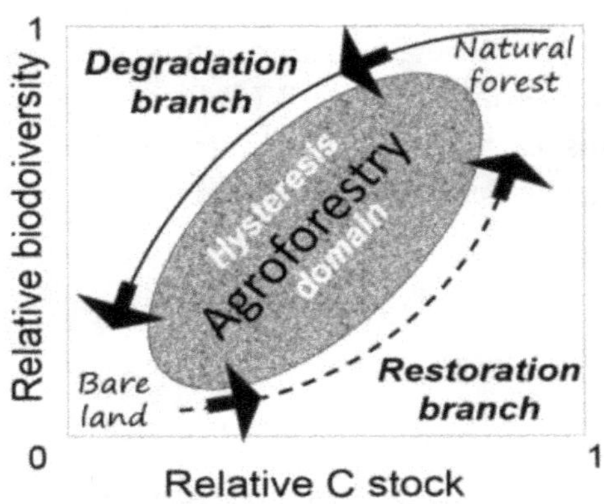


**Figure 1.** Relationship between carbon stocks and biodiversity across forest-derived land uses with agroforestry intermediate to degradation and restoration legs (modified from [22]).

The agroforestry literature for various climatic and biogeographical zones covers three orders of magnitude of tree diversity: 1–10, 10–100 and 100–1000 tree species [24]. Agroforests in Indonesia can harbour hundreds of tree species where a high influx of seeds from surrounding forests provides opportunities for farmers to selectively retain what appears to be of value. Forest conversion to other land-use systems at the forest/agriculture interface in Sulawesi [25] has followed a dual economy track of providing for local needs as well as income [26], with limited outsourcing of local food production when market-based income became reliable and the terms of trade favourable [27]. While becoming the largest cacao producing region of Indonesia, Sulawesi boomed in cacao-based land-use systems after 1980 [28], with further increased from 55,000 ha in 1990 to 230,000 ha in 2010, particularly in South and Southeast Sulawesi [29]. Due to its suitability to a broad range of climatic conditions, cacao may be as great a threat to tropical rainforests [18,30] as the more widely debated oil palm. Subsequent intensification in cacao agroforestry systems leads to shade tree removal, shifting the systems to become monoculture. The loss of diversity influences litter production, which is a major input to soil organic matter and influences the soil quality of the systems, plant growth and biomass production (including C stocks) [31,32]. Therefore, their potential to provide goods and services for local livelihoods, as well as global environmental services, may be reduced [33]. Expected cacao yield benefits may be short-lived. By 2013, cacao production in Indonesia decreased to 420,000 tons per year (around 40% of total production in 1990s) due to pests, diseases, and poor land management [34], and failed cacao intensification programs.

Cacao is traditionally grown in agroforestry systems with limited fertilizer inputs. Most of the cacao in Sulawesi is produced by smallholders in agroforestry systems with various shade trees, including fruit trees, timber and multipurpose tree species (MPTS) such as *Gliricidia sepium ('mother of cacao')*. The diverse mixture of trees in agroforestry systems has great potential for climate change mitigation through increased carbon storage, biodiversity conservation, and economic benefits from a range of products from shade trees [35]. Including shade trees may also maintain other ecosystem services that forests provide, such as temperature and humidity regulation, input of organic matter [36], and water and nutrient cycling [37].

However, management intensity directly affects stand composition and structural complexity [33], which influences the ecosystem services provided by agroforestry systems. The way farmers cultivate and integrate trees and cacao is likely tightly linked to their knowledge [30]. Farmers also determine how long agroforestry systems persist in the landscape. In addition, different knowledge, experience, and strategies of male and female farmers may lead to different shade tree management choices [38].

In this study, we explore the potential of agroforestry systems in Indonesia for supporting biodiversity maintenance and carbon storage, accounting for farmer preference in tree selection. We address the following research questions: (1) To what extent do cacao-based agroforestry, remaining forest and cacao monoculture systems differ in C stocks and species diversity? (2) What are the shade tree preferences based of male and female farmers? (3) Do C stocks in agroforestry systems correlate with tree diversity? We hypothesize that tree diversity and C stocks in agroforestry systems are positively related and that they are distinct from forest degradation and restoration curves due to farmer management actions (Figure 1).

## 2. Materials and Methods

### 2.1. Study Site

Field work was conducted in the Konawe District, Southeast Sulawesi, Indonesia, in the Konaweha watershed (3°15′0″–5°13′0″ S and 121°22′30″–122°31′0″ E), where cacao (*Theobroma cacao*) is commonly cultivated. In the greater study area, forests have been converted to intensive agricultural systems. After a few years of annual crop cultivation (maize, paddy and patchouli), land is often converted to cacao monoculture systems or simple cacao agroforestry systems. In the older plots (complex cacao agroforestry), fruit trees were inter-planted with cacao to increase income and address daily needs. Cacao is spaced at 3m × 3m, and the age of cacao ranges from 9–14 years. The average annual rainfall recorded is generally 1500–1900 mm [39] and the average temperature varies from 24 to 31 °C.

We selected plots using stratified sampling methods. We conducted initial rapid surveys in three villages (Lawonua, Wonuahoa, and Asinua Jaya) to characterize land-use system domination by analysing 45 land use system points. Based on this, we selected five land-use systems which represented the whole watershed: degraded forest (DF), complex cacao agroforestry (CAF) combined with fruit trees, simple shade cacao agroforestry (SAF), monoculture cacao (CM), and annual crops (CR). Per land-use system, three plots of 20 × 100 m were included in the study, based on common criteria: a minimum cacao age of 9 years, slope ranging from 0% to 15%, and similar soil texture (silt loam to silty clay loam). Subplots of 20m × 20m (sub-plot) were used for C stocks measurements for trees with 5–30 cm DBH. If a big tree (with DBH > 30 cm) was found in a subplot, then we extended the plot size to 100 m × 20 m (plot) for measuring the large tree component only (tree or necromass for trees > 30 cm DBH). Tree diversity was measured in the plot of 100 m × 20 m.

### 2.2. Plot Measurements

In each plot, the diameter at breast height (1.3 m above soil surface; DBH) of all trees with DBH > 5 cm was measured, and trees were identified by species. All trees < 5 cm DBH were included as understorey. Tree density was calculated based on the number of trees per ha. Plot basal area, the proportion of the area occupied by trees per ha, was calculated ($m^2$ $ha^{-1}$) as $\frac{(\frac{1}{4}\pi \sum_i DBH^2)}{land\ area}$, where $DBH_i$ is the DBH of tree $i$. Cacao agroforestry systems were selected in which cacao contributed less than 80% to the total basal area of the system [40]. We included both cacao and shade trees when calculating tree density, basal area, tree diversity and C stocks in the plots. Biodiversity was estimated using the Shannon–Wiener diversity index (H) $= -\sum_{i=1}^{s}(p_i)(\ln p_i)$, where $p_i$ is the proportion of the number of individuals of species $i$ divided by the total number of individuals [41].

### 2.3. C stock Estimation

We assessed C stocks using the RaCSA (Rapid Carbon Stock Appraisal) protocol [42], by quantifying five carbon pools including tree biomass, understorey biomass, necromass, litter and soil organic C [43]. Aboveground tree biomass was estimated using the allometric equation for humid tropical forest [44] and species-specific allometric equations for some plant species found in agroforestry systems, such as cacao [45], banana [46], and palms (Table 1). Wood density (WD) values were obtained from the wood density database of the World Agroforestry Centre (http://db.worldagroforestry.org/wd). Ten samples of understory biomass and litter (leaves and branches) were collected within each plot using a 0.5 × 0.5 m frame (Figure 2). Litter samples were oven-dried for 48 hours at 60 °C. Root biomass was estimated using the default shoot:root ratio of 4:1 in the tropics [47]. Biomass was converted into carbon by multiplying by 0.46 [42]. Ten soil samples per plot were collected at the same point beneath the understorey and litter frame at 3 depths (0–10 cm, 10–20 cm, and 20–30 cm). A composite soil sample for each plot was used to determine soil organic C. Soil C content was determined based on the Walkley and Black method [48]. The soil C stock was calculated by multiplying soil C content (%) with the bulk density (g $cm^{-3}$). Total C stocks (Mg $ha^{-1}$) were calculated as the sum of aboveground C (tree biomass, understorey biomass, necromass) and belowground C (roots and soil C stock (Mg $ha^{-1}$)).

**Table 1.** Allometric equations for several plant species ($AGB_{est}$ = estimated aboveground biomass (kg); D = diameter at breast height (cm); WD = wood density (g $cm^{-3}$).

| Plant species | Equation | Source |
|---|---|---|
| Tropical trees (moist forest) | $AGB_{est} = WD*\exp(-1.499 + 2.148\ln(D) + 0.148\ln(D) + 0.207(\ln(D))^2 - 0.0281(\ln(D))^3)$ | Chave et al., 2005 |
| Cacao | $AGB_{est} = 0.1208\ D^{1.98}$ | Yuliasmara, 2008 |
| Banana | $AGB_{est} = 0.030\ D^{2.13}$ | Arifin, 2001 |
| Palms | $AGB_{est} = 0.118\ D^{2.53}$ | Brown, 1997 |

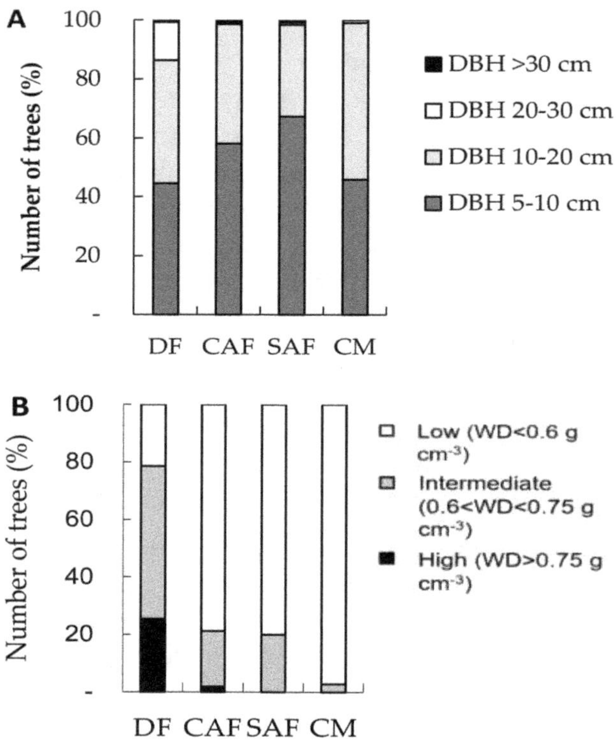


**Figure 2.** (**A**) The distribution of tree diameter for degraded forest (DF), cacao complex agroforestry (CAF), cacao simple agroforestry (SAF), and cacao monoculture (CM); (**B**) Number of trees based on wood density (WD) classes.

### *2.4. Farmer Preferences for Shade Tree Species*

Informal in-depth (semi-structured) interviews of approximately 90 minutes were conducted with each farmer to assess whether farmers have preferences for certain tree species. We selected 20 agroforestry farm families from three villages using purposive sampling technique, which included farmers who managed the included agroforestry systems. We separately interviewed male and female farmers from each household. The preferences of both men and women determine what plant species are planted, as both male and female are actively involved in land management. In total, 40 farmers were interviewed, some of the farmers were Tolaki people (local people) and the others were Bugis people (immigrants from another province). Each interview consisted of two parts, with approximately 50 minutes focused on the question which trees farmers selected to maintain (remnant forest trees), plant, or tolerate in their agroforestry systems. The second part of approximately 40 minutes focused on the utility value of their shade trees, which ranged from 1 (not important) to 4 (very important).

### *2.5. Compilation of C Stocks and biodiversity Data from Previous Research*

We evaluated additional published [15,49] and unpublished [50–54] data on tree diversity and C stocks that were collected using the same methodology, in order to compare the relationship between C stocks and tree diversity across a broader range of land-use systems. Data for 7 plots in primary forest, 23 plots in degraded forest, 42 plots in agroforestry systems, and 18 plots in plantations from Indonesia were included.

### *2.6. Data Analysis*

Tree diversity, basal area and carbon stocks were compared among cacao agroforestry systems, degraded forest, and cacao monocultures (land-use systems) using a one-way ANOVA. Post hoc multiple comparisons between land-use systems were performed using Fisher's protected LSD. Prior to analysis, basal area, tree biomass, understorey biomass, necromass and litter biomass were log (base10) transformed to meet the normality assumption for analysis of variance. Differences were regarded as significant at $\alpha = 0.05$. We used linear and power regressions to evaluate the relationship between C stocks and tree diversity in agroforestry systems (at the local and regional scale). All statistical analyses were performed using Genstat (18th edition).

## 3. Results

### 3.1. Tree Diversity

Tree diversity was significantly different ($P < 0.05$) among land-use systems. The highest tree diversity (number of species and species diversity) was found in DF followed by cacao-based agroforestry systems. The Shannon–Wiener index (H) in DF was low (2.36), ranging from 2.26 to 2.52, but was substantially higher than values in cacao agroforestry and monoculture (Table 2). A total of 28 species were recorded in DF, dominated by *Myrtaceae, Fagaceae, Salicaceae, Clusiaceae, Elaeocarpaceae,* and *Primulaceae*. The five dominant tree species were *Sloanea sp., Castanopsis buruana, Fagraea fragrans, Homalium foetidum,* and *Metrosideros petiolata*. In contrast, only 18 species were found in CAF and SAF, where most species were fruit trees planted by farmers such as *Durio zibethinus, Lansium domesticum, Nephelium lappaceum,* and *Mangifera indica*. The shade tree *Gliricidia sepium* was also common. Other species, such as *Albizia procera* and *Fagraea fragrans*, were remnant forest trees.

**Table 2.** Structure and diversity of cacao systems compared to degrded forest (H = Shannon index; DF = Degraded forest; CAF = Cacao complex agroforestry; SAF = Cacao simple agroforestry; CM = Cacao monoculture).

| Land Use Systems | Tree Density, Trees $ha^{-1}$ | Total BA, $m^2\ ha^{-1}$ | Number of Species | H | Dominant/Codominant Species | Benefit |
|---|---|---|---|---|---|---|
| DF | 1275$^a$ | 18.42$^a$ | 28 | 2.36$^a$ | *Metrosideros petiolata, Homalium foetidum* | Timber |
| CAF | 1317$^a$ | 9.14$^b$ | 18 | 0.93$^b$ | *Theobroma cacao, Durio zibethinus, Lansium domesticum* | Fruits |
| SAF | 1267$^a$ | 7.63$^b$ | 4 | 0.58$^b$ | *Theobroma cacao, Gliricidia sepium* | Fruits, Fodder |
| CM | 900$^b$ | 8.32$^b$ | 2 | 0.24$^c$ | *Theobroma cacao* | Fruits |

Note: Values not followed by the same letter were statistically different at $P < 0.05$.

### 3.2. Vegetation Structure

The highest tree density (including both cacao and shade trees) was found in CAF (1317 trees $ha^{-1}$) followed by DF and SAF (Table 2). Most of the shade trees in CAF were fruit trees, such as *D. zibethinus, L. domesticum, N. lappaceum,* and *M. indica*. Shade trees were planted less frequently in SAF than in CAF. Most of the shade trees planted in SAF were *G. sepium*, which is an important fodder resource for livestock. Total basal area significantly ($P < 0.01$) differed among land-use systems. The highest tree basal area was recorded in DF followed by the cacao agroforestry systems and monoculture (Table 2). Trees from 5 to 20 cm DBH (approximately 85% of all trees) dominated the DF plots (Figure 2A), but some trees were larger (> 20 cm DBH; 14% of all trees). Larger trees belonged to species medium-to-very-heavy WD, such as *H. foetidum* and *M, petiolata*. Cacao-based land-use systems were dominated by trees from 5 to 10 cm DBH (57% of all trees) and 10 to 20 cm DBH (42% of all trees). A few large trees (DBH > 30 cm) were found in CAF and SAF, but these accounted for just 1% of the total number of trees. However, almost 20% of the trees in CAF had high WD, which were *N. lappaceum* (WD > 0.9 g $cm^{-3}$) and *Citrus sinensis* (WD 0.78 g $cm^{-3}$) trees of 10–20 cm DBH (Figure 2B).

### 3.3. *Tree Biomass, Necromass and C Stocks*

The highest total tree biomass found was in DF (Table 3), whereas tree biomass in CM had the lowest value because of its low shade tree density. The total necromass found in CAF was relatively high compared to the other two cacao systems (41.5 Mg ha$^{-1}$). Land management was less intensive, because farmers did not transport the necromass. Ethnicity influenced the way farmers managed their cacao systems. Most of the CAF was managed by Tolaki people who tend to minimize the intensification of their land because of the lack of labour. In contrast, the Bugis people had more intensely managed cacao systems. Average litter biomass in cacao agroforestry systems was 5.6 Mg ha$^{-1}$ and ranged from 4.3–7.0 Mg ha$^{-1}$. Thus, biomass stored in the litter layer contributed little to aboveground biomass across all land-use systems.

**Table 3.** Tree biomass, understorey biomass, necromass, and litter in cacao-based agroforestry systems (DF = Degraded forest; CAF = Cacao complex agroforestry; SAF = Cacao simple agroforestry; CM = Cacao monoculture; CR = Annual crops).

| Land Use Systems | Tree, Mg ha$^{-1}$ | Understorey, Mg ha$^{-1}$ | Woody necromass, Mg ha$^{-1}$ | Litter, Mg ha$^{-1}$ |
|---|---|---|---|---|
| DF | 288.41$^{a}$ | 0.64$^{b}$ | 137.54$^{a}$ | 6.98$^{a}$ |
| CAF | 52.64$^{b}$ | 0.27$^{b}$ | 41.53$^{ab}$ | 7.02$^{a}$ |
| SAF | 54.00$^{b}$ | 0.11$^{b}$ | 5.68$^{b}$ | 5.51$^{a}$ |
| CM | 35.25$^{bc}$ | 0.19$^{b}$ | 3.31$^{b}$ | 4.26$^{a}$ |
| CR | 0$^{c}$ | 0.91$^{a}$ | 0$^{b}$ | 0$^{b}$ |

Note: Values not followed by the same letter were statistically different at $P < 0.05$

The total C stock differed significantly ($P < 0.01$) among land-use systems. Total C stock was highest in DF (282 Mg ha$^{-1}$), followed by CAF (89 Mg ha$^{-1}$), SAF (75 Mg ha$^{-1}$), CM (56 Mg ha$^{-1}$) and CR (40 Mg ha$^{-1}$) (Figure 3). Aboveground C stocks (tree biomass, understorey biomass, necromass, and litter) contributed 80% to the total in DF, more than half of the total C stock (62%) in CAF, 48% in SAF and 43% in CM. Shade trees (fruit trees) in CAF contributed almost 30% of the aboveground C stock, while in SAF, the contribution was higher (almost 40%). Larger trees were found in SAF, but they had lower WD compared to CAF.

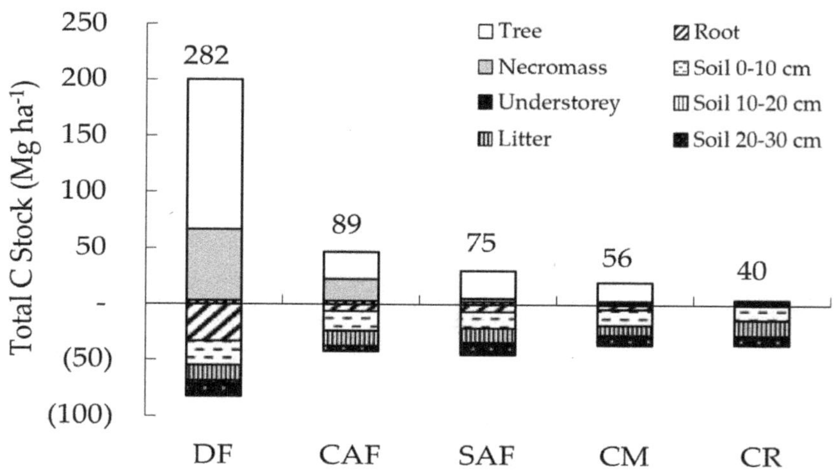


**Figure 3.** Total carbon stock in cacao-based agroforestry systems in 5 C pools compared to degraded forest and annual crop systems (DF = Degraded forest; CAF = Cacao complex agroforestry; SAF = Cacao simple agroforestry; CM = Cacao monoculture; CR=Annual crops).

Belowground C stocks (roots and soil organic matter) accounted for 18% of the total in DF, 41% in CAF, 52% in SAF and 57% in CM. Soil organic carbon was on average 38.4 Mg ha$^{-1}$ and did not significantly differ ($P > 0.05$) among land-use systems.

### 3.4. Relationships between C Stocks and Tree Diversity in Agroforestry Systems

Within the cacao-based agroforestry systems in this study, we found a strong and positive relationship between the total C stock and tree diversity (Figure 4A). However, there was a weak relationship between tree diversity and soil C stock ($R^2 = 0.33$). When comparing the relationship between C stocks and tree diversity across a larger range of agroforestry systems in Indonesia (Figure 4B), variation was larger, but C stocks and tree diversity tended to be positively related.

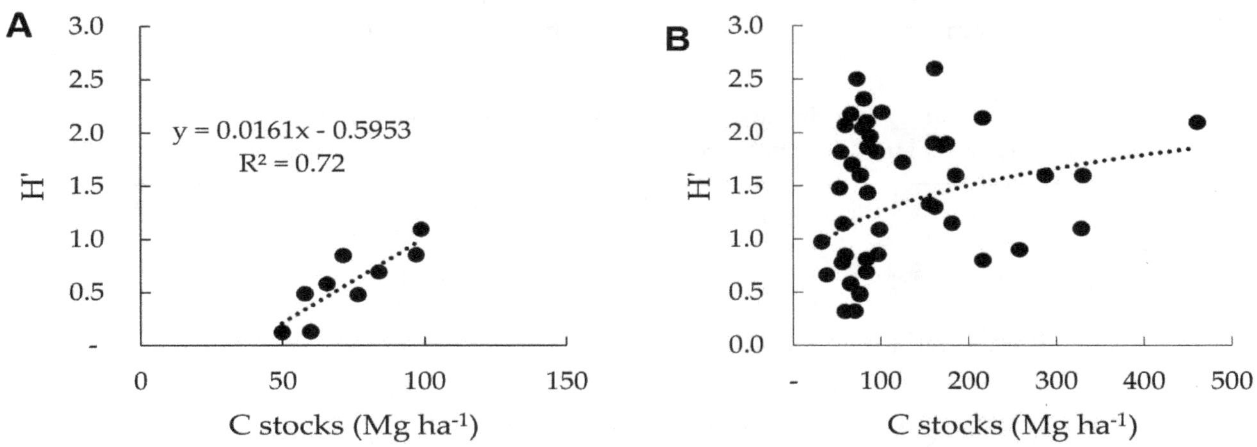


**Figure 4.** (**A**) The relation between total C stock and tree diversity (H index) in the cacao-based agroforestry systems in this study; (**B**) In agroforestry systems across Indonesia (primary data plus [15,49,52–54]).

We assessed whether C stocks and diversity were related across a wider range of land-use systems in Indonesia: plantations, agroforestry systems (complex and simple), and degraded and primary forest (Figure 5). The wide envelope of points appears to be bounded by an upper degradation leg and a lower restoration leg. We found a positive and relatively strong relationship between C stocks and diversity in primary and degraded forest ($R^2 = 0.50$) but weak relationships for agroforestry systems ($R^2 = 0.034$) and plantations ($R^2 = 0.032$). However, most of the agroforestry systems were in the area between the degradation and restoration leg, as agroforestry systems had higher diversity than monocultures (plantations).

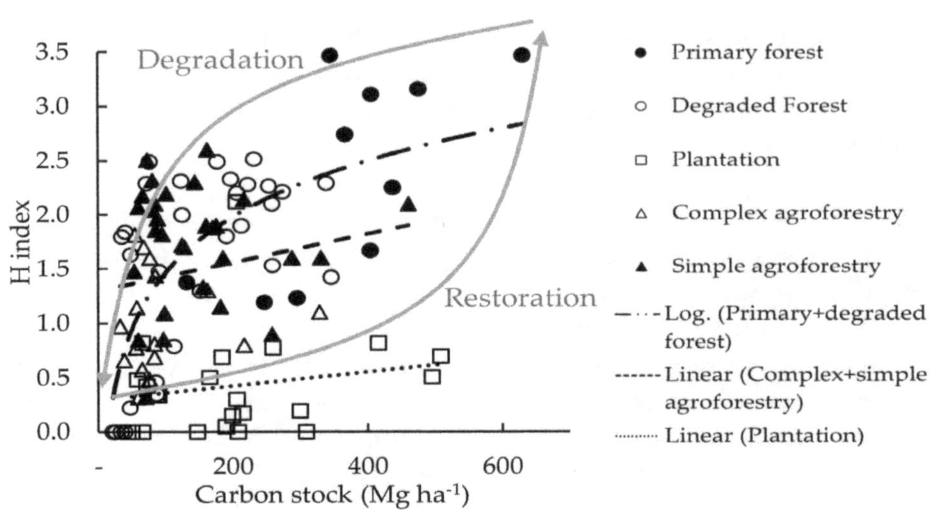


**Figure 5.** Total C stock and tree diversity (Shannon–Wiener index) in different land-use systems, n = 102 (primary data plus [15,49,52–54]).

### *3.5. Farmer Tree Preferences and Utility Value of Cacao-based Agroforestry Systems*

The major reasons for selecting tree species in the systems were economic benefits and production for subsistence, but preference for certain species differed between male and female farmers (Table 4). Female farmers preferred shade tree species that provide food, such as fruit trees and vegetables, and/or social and cultural services. Banana and coconut, for example, are not only food crops, but are also used as traditional ceremonial ornaments. Male farmers preferred species with long-term economic benefits such as timber (*Tectona grandis* and *Anthocephalus cadamba*) and fruit trees of high economic value such as *D. zibethinus*, *L. domesticum*, and *N. lappaceum*. Most of the selected tree species preferred by male and female farmers had medium to high WD (WD > 0.6 g $cm^{-3}$).

**Table 4.** Top-ranked plant species in cacao agroforestry systems that are preferred by male and female farmers in the Konawe District, Southeast Sulawesi.

| No | Species | Main Benefits | Frequency (% of Plots) | Farmer's Preference (Rank) | |
|---|---|---|---|---|---|
| | | | | Male | Female |
| 1 | *Theobroma cacao* (cacao) | Bean | 100 | 1 | 1 |
| 2 | *Pogostemon cablin* (patchouli) | Oil | 78 | 2 | 2 |
| 3 | *Gliricidia sepium* (mother of cacao) | Fodder (leave), climbed tree for pepper | 78 | 11 | 9 |
| 4 | *Musa* sp (banana) | Fruits, vegetable (flower), cultural services (leaf), toys (trunk) | 67 | - | 4 |
| 5 | *Cocos nucifera* (coconut) | Fruits, cultural services (leaf), roof (leaf), toys (trunk) | 56 | - | 5 |
| 6 | *Capsicum annum* (chili pepper) | Vegetable/spice | 56 | - | 7 |
| 7 | *Piper nigrum* (pepper) | Vegetable/spice | 44 | 10 | 6 |
| 8 | *Fagraea fragrans* (tembesu) | Timber | 44 | 3 | 10 |
| 9 | *Tectona grandis* (teak) | Timber | 44 | 4 | 11 |
| 10 | *Anthocephalus cadamba* (jabon) | Timber | 44 | 5 | - |
| 11 | *Durio zibethinus* (durian) | Fruits | 44 | 6 | 3 |
| 12 | *Lansium domesticum* (langsat) | Fruits | 22 | 7 | - |
| 13 | *Mangifera indica* (mango) | Fruits | 22 | 8 | - |
| 14 | *Nephelium lappaceum* (rambutan) | Fruits | 22 | 9 | - |

There were, in total, nine species of remnant forest trees in cacao-based agroforestry systems, 23 species of planted trees and four species of crops, according to the farmers. Almost 90% of the remnant trees that are maintained by the farmers are timber species such as *F. fragrans*, *Calophyllum* sp., *A. procera*, *Agathis* sp., and *Acacia* sp. *Arenga pinnata*, a palm species, was maintained for palm sugar production.

Farmer perspectives on the value of cacao agroforestry systems varied. The utility value of cacao-based agroforestry systems was considered to increase with time (next 30 years), especially in terms of providing food, medicine, handicraft, and recreation with an average score of 3 (Figure 6). Most farmers argued that in the future, they can no longer depend on forests because of continuous degradation. Farmers begin to recognize the importance of agroforestry systems in the provision of daily necessities. On the other hand, supply of firewood, handicraft, and decorations/ rituals were thought to become less important. Most of the respondents claimed that within the next three decades, firewood will not be required anymore because of the planned distribution of gas stoves for cooking. Likewise, some farmers predict a decrease in the production of handicrafts (woven bamboo hats and bags) due to competition from inexpensive imported products, as well as a decrease in the production of plant material for rituals and decoration because of less demand.

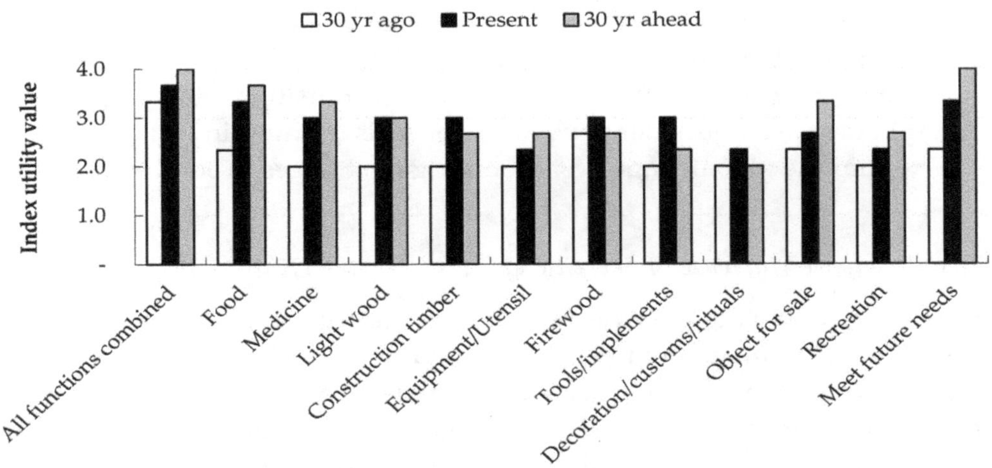


**Figure 6.** The utility value of cacao-based agroforestry systems based on its products 30 years ago, currently, and 30 years ahead based on farmer interviews.

## 4. Discussion

### *4.1. Land Use Affects Tree Diversity and C Stocks*

Forest conversion to intensive agricultural systems led to a significant decrease in biodiversity and C stocks due to the loss of trees. However, planting shade trees, thus the establishment of (simple and complex) agroforestry systems, can gradually increase tree diversity. Compared to another study of cacao-based agroforestry systems in mid-west Ghana, the diversity in Konawe was still low [55], but systems had higher tree density than cacao plantations in Talamanca, Costa Rica (149 shade trees ha$^{-1}$ and 475 cacao trees ha$^{-1}$) [35].

C stocks in annual crop systems were nearly 85% lower than in DF due to tree biomass removal (Figure 4). The higher tree density and diversity in cacao-based systems resulted in higher carbon stocks. Cacao land-use systems can increase from a total carbon stock of 56 Mg ha$^{-1}$ in a monoculture to 75–89 Mg ha$^{-1}$ if cacao is planted in a more complex agroforestry system. Shade trees played an important role in contributing to C stocks in agroforestry systems [56]. These results are similar to the C stocks of cacao-*Gliricidia* plots (56 Mg C ha$^{-1}$) in Central Sulawesi [31]. Nevertheless, we found lower C stocks than in a cacao agroforest in Mekoe, southern Cameroon, which reached a C stock of 250 Mg ha$^{-1}$ [57]. Cacao-based agroforestry systems in Konawe had also lower carbon stocks than much older agroforestry systems in Central America (117 ± 47 Mg ha$^{-1}$) [58].

Soil C stocks in the Konawe district were relatively low (38 Mg ha$^{-1}$) compared to DF and another cacao study in southern Bahia, Brazil which had more soil carbon (57 Mg ha$^{-1}$) [36]. Land-use change from forest to cropping systems can decrease belowground C through soil surface exposure, a decrease in organic matter input and soil organic C degradation. Changes due to conversion of degraded forest into agricultural systems (CR) caused a soil C loss of approximately 8.5 Mg ha$^{-1}$ in the top 10 cm. The relatively large soil C debt in Southeast Sulawesi, as we showed here, agrees with results of a recent global study by Sanderman et al. [59].

Above- and belowground C stock are linked in a virtuous cycle where organic matter accumulation (from litter and roots) increases the buffering to climate variability through a relatively stable microclimate and additional soil water storage, further enhancing plant growth. The type of tree species, and their abundance, will influence litter quality and quantity, which in turn affects soil organic matter, soil fertility and tree growth [32]. However, the estimation of the belowground part of biomass remains uncertain, with other authors estimating lower aboveground stocks for cocoa assuming an above-:below- ground biomass ratio of 7:1 [60]. We found no significant relationship between tree diversity and soil C stocks. A recent review that included our case study data suggested that increases in soil bulk density tended to conceal changes in C concentration per unit soil dry weight, as C stocks are the product of these two parameters [61]. This result aligns with Wartenberg et al. [62] who also found that there was no relation among tree diversity, soil C-content and total soil N level.

A recent summary of soil C data for Indonesian forest transitions that included the current data points concluded that increased bulk density partly conceals changes in soil organic C concentrations (per unit dry soil) within the globally agreed accounting of changes in soil C stock in the top 30 cm of the soil [61].

### *4.2. Gendered Tree Preferences Increased Diversity, C Stocks and the Utility Value of Agroforestry Systems*

Farmer preferences on shade tree species relate to the enhancement of their livelihood and are represented in their land management [8]. Gendered species choices for shade trees resulted in species-rich agroforestry systems because male and female farmers have different knowledge on, and experience with, land management and utilization [38]. Similar results were found in another study in the southern Philippines, where male smallholder farmers favoured fruit trees, while female farmers preferred food crops [63]. Overall, tree selection by farmers was influenced by how much benefit the trees provide [64,65], ease of maintenance, and their drought tolerance to avoid income loss as a result of climatic fluctuations.

Differing WD, as represented by farmers' choices, not only enriches the diversity but also contributes to C sequestration. Compared to other agricultural systems, the diversity of shade trees will remain in agroforestry systems for a long time since farmers usually plant a new tree before the old tree is harvested, which maintains C stocks. The longer trees persist in the system, the more C they will store as they increase in size [55]. Management practices geared towards high shade tree density in cacao agroforestry systems can maintain ecosystem functions that forests provide, such as aboveground C sequestration [56], enhancing the belowground C stock through organic matter addition from litter and roots, and increase resilience against changes to climate change [66].

### *4.3. Relationships between C Stocks and Tree Diversity within and Across Land-use Systems*

The relationship between C stocks and tree diversity within agroforestry systems in our research site was positive but not as strong as that within forests. A positive relationship between C stocks and biodiversity in forests has been demonstrated at different sub-climate [9] and global scales [67]. The relation between C stocks and diversity in natural forests is driven by ecological processes: high diversity allows more niches to be occupied, supporting a higher level of plant productivity [8]. In agroforestry systems, in contrast, farmers select the trees in the system, and therefore determine the diversity of agroforestry systems.

Forest degradation through logging reduces carbon stocks before affecting diversity [20,21], while a sustained loss of tree biomass leads to species loss and gradually decreases biodiversity. In terms of restoration, increases in biodiversity will usually be achieved after increases in C stocks have been achieved [23]. People tend to choose the most productive species in plantation systems to rapidly increase C stocks over time, resulting in the restoration leg with relatively high C stocks and low diversity (Figure 5). Our results suggest an intermediate position for agroforestry systems, between forest decline and restoration responses (Figures 1 and 5), with some plots demonstrating intermediate diversity and low C stocks, and some intermediate C stocks but low tree diversity, breaking the hysteresis pattern.

## 5. Conclusions

The diverse mixture of shade trees with medium to high wood density contributes to C stock maintenance in cacao-based agroforestry systems. We found a positive relationship between C stocks and tree diversity, which suggests that management practices in cacao agroforestry systems increase both C stocks and tree diversity. Gendered species preferences for shade tree species in cacao-based agroforestry systems contributed to increasing the tree diversity of the system. Farmer preferences for tree species depended on the benefit that shade tree species provide, where male and female farmers prioritized economic benefits and subsistence production, respectively.

Agroforestry systems with a mixture of diverse tree species provide more functions, both for the environment and the local people in fulfilling daily needs, including income generation. Less forest area in the future will increase the importance of trees in agroforestry systems [68]. Looking 30 years ahead, in or beyond the lifetime of the trees currently planted, farmers saw a greater need for agroforests to provide products that so far were derived from forests, and a necessary need for an increase in shade tree functional diversity. The maintenance of C stocks, biodiversity conservation and locally relevant ecosystem services (protection of the soil surface through a litter layer) will benefit from such choices.

**Author Contributions:** Conceptualization, R.R.S., M.V.N., and K.H.; methodology, R.R.S., D.D.S., K.H., and M.V.N.; writing-original draft preparation, R.R.S.; writing-review and editing, M.V.N., D.M.A.R., J.M.R., and K.H.; data analysis, R.R.S., D.M.A.R., and D.D.S.; field supervision, K.H., J.M.R.; writing supervision, M.V.N., and D.M.A.R.; funding acquisition, J.M.R. All authors have read and agreed to the published version of the manuscript.

**Acknowledgments:** We acknowledge the World Agroforestry Centre for facilitating this research. Many thanks to WCU (World Class University) Program -SAMA (Scenarios Analysis for Management of Agroforests) year 2019, Universitas Brawijaya, for supporting the writing process and Directorate General of Resources for Science, Technology and Higher Education of the Republic of Indonesia for supporting this study. We are grateful to the local community in the Wonuahoa, Lawonua, and Asinua Jaya villages for the kind help, support and warm family welcome.

## References

1. Van Noordwijk, M. Integrated Natural Resource Management as Pathway to Poverty Reduction: Innovating Practices, Institutions and Policies. *Agric. Syst.* **2019**, *172*, 60–71. [CrossRef]
2. Barrios, E.; Valencia, V.; Jonsson, M.; Brauman, A.; Hairiah, K.; Mortimer, P.E.; Okubo, S. Contribution of Trees to the Conservation of Biodiversity and Ecosystem Services in Agricultural Landscapes. *Int. J. Biodivers. Sci. Ecosyst. Serv. Manag.* **2018**, *14*, 1–16. [CrossRef]
3. Cavanaugh, K.C.; Gosnell, J.S.; Davis, S.L.; Ahumada, J.; Boundja, P.; Clark, D.B.; Mugerwa, B.; Jansen, P.A.; O'Brien, T.G.; Rovero, F.; et al. Carbon storage in tropical forests correlates with taxonomic diversity and functional dominance on a global scale. *Glob. Ecol. Biogeogr.* **2014**, *23*, 563–573. [CrossRef]
4. Van de Perre, F.; Willig, M.R.; Presley, S.J.; Andemwana, F.B.; Beeckman, H.; Boeckx, P.; Cooleman, S.; de Haan, M.; De Kesel, A.; Dessein, S.; et al. Reconciling biodiversity and carbon stock conservation in an Afrotropical forest landscape. *Sci. Adv.* **2018**, *4*, eaar6603. [CrossRef] [PubMed]
5. Loreau, M.; Hector, A. Partitioning selection and complementarity in biodiversity experiments. *Nature* **2001**, *412*, 72–76. [CrossRef] [PubMed]
6. Hicks, C.; Woroniecki, S.; Fancourt, M.; Bieri, M.; Garcia Robles, H.; Trumper, K.; Mant, R. *The Relationship between Biodiversity, Carbon Storage and the Provision of other Ecosystem Services: Critical Review for the Forestry Component of the International Climate Fund*; Department for International Development: Cambridge, UK, 2014; p. 119.
7. Cardinale, B.J.; Matulich, K.L.; Hooper, D.U.; Byrnes, J.E.; Duffy, E.; Gamfeldt, L.; Balvanera, P.; O'Connor, M.I.; Gonzalez, A. The functional role of producer diversity in ecosystems. *Am. J. Bot.* **2011**, *98*, 572–592. [CrossRef] [PubMed]
8. Swift, M.J.; Izac, A.M.N.; van Noordwijk, M. Biodiversity and ecosystem services in agricultural landscapes—Are we asking the right questions? *Agric. Ecosyst. Environ.* **2004**, *104*, 113–134. [CrossRef]
9. Lecina-Diaz, J.; Alvarez, A.; Regos, A.; Drapeau, P.; Paquette, A.; Messier, C.; Retana, J. The positive carbon stocks-biodiversity relationship in forests: Co-occurrence and drivers across five subclimates. *Ecol. Appl.* **2018**, *28*, 1481–1493. [CrossRef]
10. Healy, C.; Gotelli, N.J.; Potvin, C. Partitioning the effects of biodiversity and environmental heterogeneity for productivity and mortality in a tropical tree plantation. *J. Ecol.* **2008**, *96*, 903–913. [CrossRef]
11. Sullivan, M.J.; Talbot, J.; Lewis, S.L.; Phillips, O.L.; Qie, L.; Begne, S.K.; Chave, J.; Cuni-Sanchez, A.; Hubau, W.; Lopez-Gonzalez, G.; et al. Diversity and carbon storage across the tropical forest biome. *Sci. Rep.* **2017**, *7*, 39102. [CrossRef]

12. Poorter, L.; van der Sande, M.T.; Thompson, J.; Arets, E.J.M.M.; Alarcón, A.; Álvarez-Sánchez, J.; Ascarrunz, N.; Balvanera, P.; Barajas-Guzmán, G.; Boit, A.; et al. Diversity enhances carbon storage in tropical forests. *Glob. Ecol. Biogeogr.* **2015**, *24*, 1314–1328. [CrossRef]
13. Conti, G.; Díaz, S. Plant functional diversity and carbon storage—An empirical test in semi-arid forest ecosystems. *J. Ecol.* **2013**, *101*, 18–28. [CrossRef]
14. Mandal, R.A.; Dutta, I.C.; Jha, P.K.; Karmacharya, S. Relationship between carbon stock and plant biodiversity in collaborative forests in Terai, Nepal. *ISRN Bot.* **2013**, *2013*, 625767. [CrossRef]
15. Natalia, D.; Arisoesilaningsih, E.; Hairiah, K. Are high carbon stocks in agroforests and forest associated with high plant species diversity? *Agrivita* **2017**, *39*, 74–82. [CrossRef]
16. Kirby, K.R.; Potvin, C. Variation in carbon storage among tree species: Implications for the management of a small-scale carbon sink project. *For. Ecol. Manag.* **2007**, *246*, 208–221. [CrossRef]
17. Murray, J.P.; Grenyer, R.; Wunder, S.; Raes, N.; Jones, J.P. Spatial patterns of carbon, biodiversity, deforestation threat, and REDD+ projects in Indonesia. *Conserv. Biol.* **2015**, *29*, 1434–1445. [CrossRef]
18. Dewi, S.; van Noordwijk, M.; Zulkarnain, M.T.; Dwiputra, A.; Hyman, G.; Prabhu, R.; Gitz, V.; Nasi, R. Tropical forest-transition landscapes: A portfolio for studying people, tree crops and agro-ecological change in context. *Int. J. Biodivers. Sci. Ecosyst. Serv. Manag.* **2017**, *13*, 312–329. [CrossRef]
19. Sanchez, J.L.M.; Cabrales, L.C.; Académica, D.; Biológicas, D.C.; Juárez, U.; Tabasco, A.D. Is there a relationship between floristic diversity and carbon stocks in tropical vegetation in Mexico? *Afr. J. Agric. Res.* **2012**, *7*, 2584–2591. [CrossRef]
20. Kalaba, F.K.; Quinn, C.H.; Dougill, A.J.; Vinya, R. Floristic composition, species diversity and carbon storage in charcoal and agriculture fallows and management implications in Miombo woodlands of Zambia. *For. Ecol. Manag.* **2013**, *304*, 99–109. [CrossRef]
21. Rutishauser, E.; Hérault, B.; Baraloto, C.; Blanc, L.; Descroix, L.; Sotta, E.D.; Ferreira, J.; Kanashiro, M.; Mazzei, L.; D'Oliveira, M.V.N.; et al. Rapid tree carbon stock recovery in managed Amazonian forests. *Curr. Biol.* **2015**, *25*, R787–R788. [CrossRef]
22. Murdiyarso, D.; van Noordwijk, M.; Wasrin, U.R.; Tomich, T.P.; Gillison, A.N. Environmental benefits and sustainable land-use options in the Jambi transect, Sumatra. *J. Veg. Sci.* **2002**, *13*, 429–438. [CrossRef]
23. Martin, P.A.; Newton, A.C.; Bullock, J.M. Carbon pools recover more quickly than plant biodiversity in tropical secondary forests. *Proc. Biol. Sci.* **2013**, *280*, 20132236. [CrossRef] [PubMed]
24. Van Noordwijk, M.; Rahayu, S.; Gebrekirstos, A.; Kindt, R.; Tata, H.L.; Muchugi, A.; Ordonnez, J.C.; Xu, J. Tree diversity as basis of agroforestry. In *Sustainable Development through Trees on Farms: Agroforestry in Its Fifth Decade*; Van Noordwijk, M., Ed.; World Agroforestry Centre (ICRAF) Southeast Asia Regional Program: Bogor, Indonesia, 2019; pp. 17–44.
25. Kehlenbeck, K.; Maass, B.L. Crop diversity and classi cation of homegardens in Central Sulawesi, Indonesia. *Differences* **2004**, *63*, 53–62.
26. Dove, M.R. *The Banana Tree at the Gate: A History of Marginal Peoples and Global Markets in Borneo*; Yale University Press: New Haven, CT, USA, 2011; Volume 3, p. 332.
27. Van Noordwijk, M.; Agus, F.; Dewi, S.; Purnomo, H. Reducing emissions from land use in Indonesia: Motivation, policy instruments and expected funding streams. *Mitig. Adapt. Strateg. Glob. Chang.* **2014**, *19*, 677–692. [CrossRef]
28. Schroth, G.; Harvey, C.A. Biodiversity conservation in cocoa production landscapes: An overview. *Biodivers. Conserv.* **2007**, *16*, 2237–2244. [CrossRef]
29. Rahmanulloh, A.; Sofiyuddin, M. *Profitability of Land-Use Systems in South Sulawesi and Southeast Sulawesi*; World Agroforestry Centre: Bogor, Indonesia, 2012; p. 16. [CrossRef]
30. Dumont, E.S.; Gnahoua, G.M.; Ohouo, L.; Sinclair, F.L.; Vaast, P. Farmers in Cote d'Ivoire value integrating tree diversity in cocoa for the provision of ecosystem services. *Agrofor. Syst.* **2014**, *88*, 1047–1066. [CrossRef]
31. Smiley, G.L.; Kroschel, J. Temporal change in carbon stocks of cocoa-gliricidia agroforests in Central Sulawesi, Indonesia. *Agrofor. Syst.* **2008**, *73*, 219–231. [CrossRef]
32. Triadiati, T.S.; Guhardja, E.; Sudarsono, Q.I.; Leuschner, C. Litterfall production and leaf-litter decomposition at natural forest and cacao agroforestry in Central Sulawesi, Indonesia. *Asian J. Biol. Sci.* **2011**, *4*, 2011.
33. Deheuvels, O.; Rousseau, G.X.; Quiroga, G.S.; Franco, M.D.; Cerda, R.; Mendoza, S.J.V.; Somarriba, E. Biodiversity is affected by changes in management intensity of cocoa-based agroforests. *Agrofor. Syst.* **2014**, *88*, 1081–1099. [CrossRef]

34. Benton, T. *Sustainable Intensification*; International Plant Nutrition Institute (IPNI): Indonesia, 2016; pp. 95–109. [CrossRef]
35. Cerda, R.; Deheuvels, O.; Calvache, D.; Niehaus, L.; Saenz, Y.; Kent, J.; Vilchez, S.; Villota, A.; Martinez, C.; Somarriba, E. Contribution of cocoa agroforestry systems to family income and domestic consumption: Looking toward intensification. *Agrofor. Syst.* **2014**, *88*, 957–981. [CrossRef]
36. Monroe, P.H.M.; Gama-Rodrigues, E.F.; Gama-Rodrigues, A.C.; Marques, J.R.B. Soil carbon stocks and origin under different cacao agroforestry systems in southern Bahia, Brazil. *Agric. Ecosyst. Environ.* **2016**, *221*, 99–108. [CrossRef]
37. Jose, S. Agroforestry for ecosystem services and environmental benefits: An overview. *Agrofor. Syst.* **2009**, *76*, 1–10. [CrossRef]
38. Mulyoutami, E.; Roshetko, J.M.; Martini, E.; Awalina, D. Gender roles and knowledge in plant species selection and domestication: A case study in South and Southeast Sulawesi. *Int. For. Rev.* **2015**, *17*, 99–111. [CrossRef]
39. BPS. *Konawe in Fugures*; Konawe, B.K., Ed.; Statistics of Konawe Regency: Regency, Indonesia, 2015; p. 286.
40. Hairiah, K.; Sulistyani, H.; Suprayogo, D.; Purnomosidhi, P.; Widodo, R.H.; van Noordwijk, M. Litter layer residence time in forest and coffee agroforestry systems in Sumberjaya, West Lampung. *For. Ecol. Manag.* **2006**, *224*, 45–57. [CrossRef]
41. Shannon, C.E. A mathematical theory of communication. *Bell Syst. Tech. J.* **1948**, *27*, 379–423. [CrossRef]
42. Hairiah, K.; Dewi, S.; Agus, F.; Velarde, S.; Andree, E.; Rahayu, S.; van Noordwijk, M. *Measuring Carbon Stocks*; World Agroforestry Centre: Bogor, Indonesia, 2011; p. 95.
43. Eggleston, S.; Buendia, L.; Miwa, K.; Ngara, T.; Tanabe, K. *2006 IPCC Guidelines for National Greenhouse Gas Inventories*; Eggleston, S., Buendia, L., Miwa, K., Ngara, T., Tanabe, K., Eds.; IPCC: Geneva, Switzerland, 2006; p. 6.
44. Chave, J.; Andalo, C.; Brown, S.; Cairns, M.A.; Chambers, J.Q.; Eamus, D.; Fölster, H.; Fromard, F.; Higuchi, N.; Kira, T.; et al. Tree allometry and improved estimation of carbon stocks and balance in tropical forests. *Oecologia* **2005**, *145*, 87–99. [CrossRef]
45. Yuliasmara, F.; Wibawa, A.; Prawoto, A.A. Carbon stock in different ages and plantation system of cocoa: Allometric approach. *Pelita Perkeb.* **2009**, *25*, 86–100. [CrossRef]
46. Arifin, J. *Estimasi Cadangan Karbon pada Berbagai Sistem Penggunaan Lahan di Kecamatan Ngantang, Malang*; Brawijaya University: Malang, Indonesia, 2001.
47. Mokany, K.; Raison, R.J.; Prokushkin, A.S. Critical analysis of root: Shoot ratios in terrestrial biomes. *Glob. Chang. Biol.* **2006**, *12*, 84–96. [CrossRef]
48. Walkley, A.; Black, I.A. An examination of the degtjareff method for determining soil organic matter, and a proposed modification of the chromic acid tritation method. *Soil Sci.* **1934**, *37*, 29–38. [CrossRef]
49. Markum, M.; Soesilaningsih, E.A.; Suprayogo, D.; Hairiah, K. Plant species diversity in relation to carbon stocks at Jangkok watershed, Lombok Island. *AGRIVITA* **2013**, *35*, 45–63. [CrossRef]
50. Faranisa, R. Keanekaragaman Flora Pada Berbagai Tingkat Kepadatan Tanah di Hutan Pegunungan Taman Nasional Bromo Tengger Semeru (TNBTS). Bachelor's Thesis, Brawijaya University, Malang, Indonesia, 2017.
51. Kendom, M. Estimasi Emisi Karbon di DAS Casteel Timur Berdasarkan Perubahan Tutupan Hutan Alami di Wilayah Kabupaten Asmat, Papua. Master's Thesis, Brawijaya University, Malang, Indonesia, 2013.
52. Hairiah, K.; Suprayogo, D.; Saputra, D.D. *Reklamasi Lahan Pertanian Pascaerupsi Gunung Kelud: Efisiensi Serapan N Dalam System Agroforestry Kopi-Kakao*; Brawijaya University: Malang, Indonesia, 2017; p. 53.
53. Kurniawan, S. Bio-Physical Characteristics, Vegetation, Biomass, Carbon Stock, and Carbon Sequestration in the Bangsri Micro Watershed-East Java. In *Final Report CCCD Project*; Brawijaya University and Research group of Tropical Agroforestry: Malang, Indonesia, 2018; p. 53.
54. Prayogo, C. *Biodiversity and Ecosystem Management in the Bangsri Micro Watershed-East Java. Final Report CCCD Project*; Brawijaya University and Research Group of Tropical Agroforestry: Malang, Indonesia, 2018; p. 65.
55. Dawoe, E.; Asante, W.; Acheampong, E.; Bosu, P. Shade tree diversity and aboveground carbon stocks in Theobroma cacao agroforestry systems: Implications for REDD + implementation in a West African cacao landscape. *Carbon Balance Manag.* **2016**, *11*, 17. [CrossRef] [PubMed]
56. Middendorp, R.S.; Vanacker, V.; Lambin, E.F. Impacts of shaded agroforestry management on carbon sequestration, biodiversity and farmers income in cocoa production landscapes. *Landsc. Ecol.* **2018**, *33*, 1953–1974. [CrossRef]

57. Duguma, B.; Gockowski, J.; Bakala, J. Smallholder cacao (*Theobroma cacao* Linn.) cultivation in agroforestry systems of West and Central Africa: Challenges and opportunities. *Agrofor. Syst.* **2001**, *51*, 177–188. [CrossRef]
58. Somarriba, E.; Cerda, R.; Orozco, L.; Cifuentes, M.; Dávila, H.; Espin, T.; Mavisoy, H.; Ávila, G.; Alvarado, E.; Poveda, V.; et al. Carbon stocks and cocoa yields in agroforestry systems of Central America. *Agric. Ecosyst. Environ.* **2013**, *173*, 46–57. [CrossRef]
59. Sanderman, J.; Hengl, T.; Fiske, G.J. Soil carbon debt of 12,000 years of human land use. *Proc. Natl. Acad. Sci. USA* **2017**, *114*, 9575–9580. [CrossRef]
60. Norgrove, L.; Hauser, S. Carbon stocks in shaded Theobroma cacao farms and adjacent secondary forests of similar age in Cameroon. *Trop. Ecol.* **2013**, *54*, 15–22.
61. Hairiah, K.; van Noordwijk, M.; Sari, R.R.; Saputra, D.D.; Suprayogo, D.; Kurniawan, S.; Prayogo, C.; Gusli, S. Soil carbon stocks in Indonesian (agro) forest transitions: Compaction conceals lower carbon concentrations in standard accounting. *Agric. Ecosyst. Environ.* **2020**, *294*, 106879. [CrossRef]
62. Wartenberg, A.C.; Blaser, W.J.; Gattinger, A.; Roshetko, J.M.; Van Noordwijk, M.; Six, J. Does shade tree diversity increase soil fertility in cocoa plantations? *Agric. Ecosyst. Environ.* **2017**, *248*, 190–199. [CrossRef]
63. Ureta, J.U.; Evangelista, K.P.A.; Habito, C.M.D.; Lasco, R.D. Exploring gender preferences in farming system and tree species selection: Perspectives of JESAM smallholder farmers in southern Philippines. *J. Environ. Sci. Manag.* **2016**, *I*, 56–73.
64. Lasco, R.D.; Espaldon, M.L.O.; Habito, C.M.D. Smallholder farmers' perceptions of climate change and the roles of trees and agroforestry in climate risk adaptation: Evidence from Bohol, Philippines. *Agrofor. Syst.* **2015**, *90*, 521–540. [CrossRef]
65. Mustapha, R.I.; Jimosh, S.O. Farmer's preferences for tree species on agroforestry system in Ijebu North Local Government Area, Ogun State, Nigeria. *J. Agric. For. Soc. Sci.* **2012**, *10*, 176–187.
66. Nair, P.K.R.; Garrity, D. *Agroforestry—The Future of Global Land Use*; Springer International Publishing: New York, NY, USA, 2012.
67. Liang, J.; Crowther, T.W.; Picard, N.; Wiser, S.; Zhou, M.; Alberti, G.; Schulze, E.D.; McGuire, A.D.; Bozzato, F.; Pretzsch, H.; et al. Positive biodiversity-productivity relationship predominant in global forests. *Science* **2016**, *354*, aaf8957. [CrossRef] [PubMed]
68. Roshetko, J.M.; Snelder, D.J.; Lasco, R.D.; Van Noordwijk, M. Future challenge: A Paradigm shift in the forestry sector. In *Smallholder Tree Growing for Rural Development and Environmental Services*; Snelder, D.J., Lasco, R.D., Eds.; Springer: Berlin/Heidelberg, Germany, 2008; pp. 453–485.

# 6

# Tree Roots Anchoring and Binding Soil: Reducing Landslide Risk in Indonesian Agroforestry

**Kurniatun Hairiah [1,*], Widianto Widianto [1], Didik Suprayogo [1] and Meine van Noordwijk [1,2,3]**

1 Faculty of Agriculture, Brawijaya University, Jl Veteran, Malang 65145, Indonesia; widianto@gmail.com (W.W.); suprayogo@ub.ac.id (D.S.); m.vannoordwijk@cgiar.org (M.V.N.)
2 World Agroforestry (ICRAF), Bogor 16155, Indonesia
3 Plant Production Systems, Wageningen University & Research, 6700 AK Wageningen, The Netherlands
* Correspondence: kurniatun_h@ub.ac.id or k.hairiah@cgiar.org

**Abstract:** Tree root systems stabilize hillslopes and riverbanks, reducing landslide risk, but related data for the humid tropics are scarce. We tested fractal allometry hypotheses on differences in the vertical and horizontal distribution of roots of trees commonly found in agroforestry systems and on shear strength of soil in relation to root length density in the topsoil. Proximal roots of 685 trees (55 species; 4–20 cm stem diameter at breast height, dbh) were observed across six landscapes in Indonesia. The Index of Root Anchoring (IRA) and the Index of Root Binding (IRB) were calculated as $\Sigma D_v^2/dbh^2$ and as $\Sigma D_h^2/dbh^2$, respectively, where $D_v$ and $D_h$ are the diameters of vertical (angle > 45°) and horizontal (angle < 45°) proximal roots. High IRA values (>1.0) were observed in coffee and several common shade trees. Common fruit trees in coffee agroforestry had low medium values, indicating modest 'soil anchoring'. Where root length density ($L_{rv}$) in the topsoil is less than 10 km $m^{-3}$ shear strength largely depends on texture; for $L_{rv}$ > 10 shear strength was > 1.5 kg $m^{-2}$ at the texture tested. In conclusion, a mix of tree species with deep roots and grasses with intense fine roots provides the highest hillslope and riverbank stability.

**Keywords:** coffee; fruit trees; index of root anchoring; slope stability; soil shear strength; root length density; root tensile strength

---

## 1. Introduction

Watersheds usually provide water but occasionally they generate mudflows. Intense seasonal precipitation during monsoons in mountainous watersheds can trigger landslides, debris flows, and flash floods. On steep slopes, landslides can destroy houses, villages, or any vegetation encountered in the downhill path. Soil and debris flow can also be a major contributor of sediment load in the river systems. Extreme rainfall events generate shallow slope failures by elevating pore pressures and decreasing effective stress, but numerous site-specific factors, such as preferential hydrologic flow paths, slope steepness, soil thickness, and existing plants root systems influence the potential for slope instability [1,2]. Agroforestry can reduce risks by having appropriate trees (species, age, diversity, management) at strategic locations at hillslope and landscape scale [3]. Higher plant diversity was found [4] to enhance soil stability in disturbed alpine ecosystems.

Shear strength is a term used in soil mechanics to describe the resistance against structural failure, or the magnitude of the shear stress that a soil can sustain before submitting to a sliding failure along a plane that is parallel to the direction of the force. The shear resistance of soil is a result of friction and interlocking of particles, and possibly cementation or bonding at particle contacts, strengthened by roots depending on their tensile strength. Tensile strength is a measure of the force required to pull something such as rope, wire, or a structural beam to the point where it breaks. Soil shear strength of rooted soil $s_r$ exceeds that of unrooted soil $s$ according to $s_r = s+1.2 \sum T_a$, where $T_a$ it the tensile strength per unit cross-sectional area of all roots at the plane sheared. The factor 1.2 is an approximation derived for common slope angles [5]. A review of experimental methods and procedures to test root-strength and rooted-soil shear-strength behavior, focused on obtaining data to parameterize root reinforcement models. [6]. Common methods either determine the tensile strength of individual roots by loading the root in a pulling device until it breaks or determine the shear-strength of rooted soil in comparison to non-rooted soil in a Coulomb-type shear-box test. Laboratory shear-box tests encounter a difficulty in that the roots are not generally fixed or constrained at the base of the shear-box and are pulled rather than break.

The mechanics of 'failure' are broadly understood in terms of weight of the (wet) soil plus vegetation, slope, and critical soil shear strength. A higher soil clay content and lower water content lead to higher soil shear strength [7–9]. Intensifying soil management after forest conversion can lead to soil compaction meaning lower pore space and soil infiltration, enhancing shear strength and overland flow; along with a reduction of the weight of the vegetation such changes can reduce the chances of slope failure with time after forest conversion [10]. However, loss of 'root anchoring' associated with gradual decay of tree roots in the first few years after conversion will enhance the risk of landslides [11,12]. Conversion affecting only part of a hill slope, e.g., by road construction, can modify subsurface flows, leading to concentrated flows and locally enhanced landslide risk. The deepest and most destructive landslides, however, may be hardly influenced by vegetation. Still, a combination of deep-rooted trees for anchoring and shallow rooted grass (for stabilizing topsoil) is hypothesized to stabilize slopes prone to mass movement. At least four system levels are involved: the vegetation (species, stem diameters, spatial distribution of trees), the distribution of roots of an individual tree over the soil profile (shoot/root allometry, vertical root distribution), the effects of root length density on soil shear strength throughout the profile (it is the weakest plane on which the soil will shear), and the tensile strength of roots determining when individual roots will break. Most studies focus on only part of this range of scales.

Several studies on how woody species, mostly planted forests, influence slope stability have been reported in temperate areas such as Alaska [13,14], Canada [15], China [16–18], New Zealand [19], Australia [20], and the Mediterranean region [21,22]. A study in Australia [23] quantified increases in soil strength due to root reinforcement, on the basis of root strength, interface friction between the roots and the soil, and the distribution of roots within the soil for two common riparian species: river red gum (*Eucalyptus camaldulensis*) and swamp paperbark (*Melaleuca ericifolia*). They found that interspecies differences in the strength of living roots had less significance for bank reinforcement than interspecies differences in root distribution. A study in New Zealand found that shear strength contributions per tree did not differ between the native kanuka tree (*Kunzea ericoides*) and the species replacing it (*Pinus radiata*) on hillslopes prone to storm-generated land sliding, but higher stand density explained a slope protection advantage of the native vegetation [24]. A comparison between 9, 20, and 30-year-old stands of *Cryptomeria japonica* stands in China found that root density was highest in the 9-year-old stand, but that root tensile strength was lowest at this stage. As older stands had been thinned, however, the 9-year old stands contributed most to slope stability [25]. A study in Italy with three local shrub species found that these complement each other in different parts of the landscape in stabilizing steep hillslopes which are seasonally affected by storm-induced shallow landslides, by anchoring into the soil mantle. Root tensile force, at which a root breaks, increased with increasing root diameter, but tensile strength (per unit cross-sectional area) decreased with increasing root diameter following a power law curve [26]. Other studies found that when root distribution has a wide range of diameters, the root reinforcement results are controlled by large roots, which hold much more force than small roots [27].

Measurements from tropical vegetation, with higher tree diversity, on shear strength in relation to tree properties are scarce, however [28]. Interspecific variation in root strength in the tropics remains largely undescribed, with the exception of a study in southern Thailand [29] that compared roots of seven tree species in their effects on slope stabilization in the context of biotechnical slope protection.

The coarse root system of trees may differ in mixtures from that in monocultures [30], with shorter, more dense roots in focal trees growing with conspecific neighbors. A three-dimensional view on tree coarse roots is needed to verify this [31].

The mechanical effect of root systems in enhancing soil stability is based on at least three mechanisms [32]: (a) fine root systems in the surface layers bind soil particle strongly increasing cohesion [33]; through the more stable soil structure they reduce the entrainment of soil particles in overland flow of water; (b) the tensile strength of roots in the surface layers enhance shear strength and the risks that small blocks of soil break away on river banks, roadsides, gullies, or natural channels; (c) deep tree root systems support tree trunks and act as an anchor to the soil [34] resulting in a high root resistance to storm force and reducing the chance that larger soil masses are flushed away once channels are formed. Depending on the relative importance of those mechanisms, the choice of vegetation and its associated rooting pattern can influence slope stability. The susceptibility of cut slopes to land sliding can under certain circumstances be increased rather than reduced by the establishment of a vegetation cover, if the increase in infiltration rate offsets the mechanical benefits of soil reinforcement by roots.

Not all tree roots are the same. A study of 10 European perennials [35] found that root tensile strength–diameter relationships depend strongly on taxa. Root characteristics including root length density, diameter, lignin, and polyphenolic content are known to affect the soil cohesion [36]. Lignin content of wood is the primary determinant of strength, but also affects the rate of decomposition [18]. The role of polyphenolics and protein (linked to N content) is less clear in root strength, but these factors certainly influence the rate of decomposition and hence decay rate of strength of dead roots [37–39]. Root tensile strength, which is supposedly an intrinsic property that does not depend on the dimensions of the test specimen, can increase with decreasing root diameter, associated with a higher percentage of cellulose [40]. While trees differ in their spatial distribution of woody roots over the soil profile, their branching pattern generally adheres to fractal branching rules and allometric relations on the basis of proximal root diameters at the stem base can be used to predict total size of a root system, similar to the allometric equations that relate aboveground biomass to stem diameter [41,42].

Scaling up from a bundle of roots to tree stands [43] needs to consider counteracting processes. Tree roots not only influence tensile strength that reduces, but also infiltration and subsurface flows that contribute to landslide risk [44]. In the context of a broad evaluation of the role of agroforestry in maintaining or restoring watershed functions in the humid tropics [45], we tested hypotheses at four scales (individual) woody roots, rooted volumes of soil, tree root systems, and riparian vegetation:

1. The critical load where individual roots break is related to the root's lignin content,
2. The shear strength of a volume of soil increases proportionally to the number and strength of individual roots,
3. With a similar overall ratio of woody root to stem cross sectional area, there are consistent differences between tree species in root development in the topsoil and at depth, that contribute to differences in soil binding and soil anchoring, respectively,
4. Differences in the distribution of tree roots between species can be used to reduce landslide risks in the context of productive coffee agroforestry systems.

## 2. Materials and Methods

### 2.1. Location

The research was carried out in six sites: (a) Waybesai watershed, Sumberjaya (W. Lampung), (b) Ciherang and Cibadak sub-watershed, Sentul (West Java), (c) Kalikonto sub-watershed, Pujon and Ngantang-Malang (East java), (d) the UB-forest on the slopes of the Arjuna mountain (Malang), (e) Bangsri watershed Wajak-Malang (East java), (f) Upper Bedadung watershed, Jember, (East Java) (Figure 1 and Table 1). The measurements were performed in three steps: (a) individual root strength in relation to root properties; (b) root length density in relation to soil shear strength in the surface layer in Sumberjaya, West Lampung; (c) inventory of the potential of tree root systems as an anchor to maintain hillslope and river bank stability in all six sites.

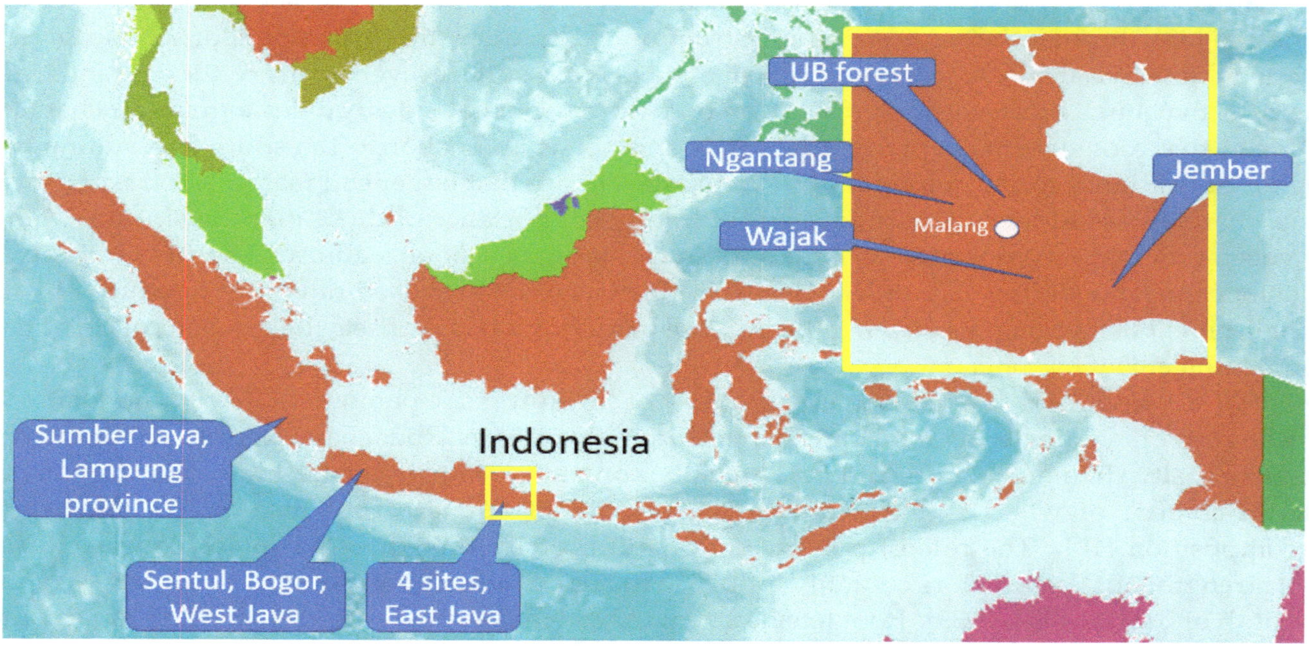


**Figure 1.** Location of six study sites in East Java, West Java, and Sumatra.

**Table 1.** Geoposition and characteristics of five study sites.

| Watershed and Time of Study | Geoposition | Altitude, m above Sea Level | Annual Rainfall, mm | Soil Type | Main Tree Crops |
|---|---|---|---|---|---|
| Way Besai, Sumber Jaya, West Lampung, (2005–2006) | 5°01′29.9″–5°02′34.2″ S, 104°25′46.5″–104°26′51.4″ E | 700–1700 | 2500–2614 | Inceptisol and Andisols | *Coffea, Gliricidia, Persea, Durio, Artocarpus, Aleurites* |
| Ciherang and Cibadak sub-watershed, Sentul Bogor, West Java (2007) | 6°35′37″–6°35′29″ S, 106°54′58″–106°55′47″ E | 1529 | 4000 | Inceptisol | *Maesopsis, Pangium, Eugenia, Sandoricum* |
| Kali Konto sub-watershed (Pujon and Ngantang village), Malang, East Java (2006–2007) | 7°46′03″–7°56′54″ S, 112°19′20″–112°29′55″ E. | 1150 | 1797–3151 | Inceptisol | *Coffea, Durio, Persea, Gliricidia, Pinus, Maesopsis, Toona, Agathis* |
| Kalisari sub-watershed (UB Forest), Malang, East Java (2019) | 7°49′30″–7°51′36″ S, 112°34′38″–112°36′53″ E | 900–1300 | 2005 | Inceptisol, Andisol | *Pinus, Swietenia, Coffea, Gliricidia, Calliandra, Hibiscus, Parkia, Toona, Erythrina* |
| Bangsri sub-watershed, Wajak, Malang, East Java (2018) | 8°7′29″–8°7′32″ S, 112°47′30″–112°48′30″ E | 660–720 | 2220–2314 | Inceptisol, Entisol | *Persea, Durio, Paraserianthes, Swietenia, Pinus, Michelia* |
| Upper Bedadung watershed, Jember, East Java (2006) | 7°59′66″–8°33′56″ S, 113°19′–114°02′30″ E, | 300–3000 | 2699 | Inceptisol and Andisols | *Coffea, Leucaena, Swietenia, Hevea, Ochroma,* |

### 2.2. Individual Root Strength in Relation to Root Properties

Root strength, the tensile force per unit area of root needed to break the root, was measured by clamping a weight onto a vertically hanging root and increasing the downward force until rupture. A tensile strength apparatus [46,47] was developed locally. Five tree species (all 5 years old) were selected for this study: mahogany (*Swietenia mahogani*), gmelina (*Gmelina arborea*), suren (*Toona sureni*), coffee (*Coffea canephora*), and bamboo (*Bambusa arundinacea*). Root samples were collected from three trees of each species from farmers' plots in Singosari village (Malang). Fresh roots with a length of about 25 cm and a diameter of close to 2 mm were placed in the tensile strength apparatus. The mean of six replicate root measurements of the force required to shear the root was used as indicator of root strength. A composite dry root sample of about 10 g from each species was used for analysis of chemical root characteristics: total carbon content C measured with the wet oxidation method of Walkley and Black [48], total nitrogen content (N) measured using the Kjeldahl distillation method, concentration of lignin by boiling plant sample with a sulfuric acid solution of Cetrimethyl Ammonium Bromide (CTAB) under controlled temperature [49], and polyphenolics by extracting plant tissues with methanol and subsequently colorimetrically measuring absorbance at a wavelength of 760 nm by the Folin–Denis method [50].

### 2.3. Role of Root Length Density in Soil Shear Strength

Roots add shear strength to soil when the root network penetrates a potential failure surface. The amount of tensile root force contributed to a potential slide mass should increase with increasing area of root intersection. Hence, we explored how soil shear strength was related to the root length density ($L_{rv}$) at different distances to tree trunks. Tree roots can increase the shear strength of shallow soils through mechanical reinforcement [9].

Measurement on tree root length density was done in two study sites, focused on coffee (*Coffea canephora*) and shade tree species commonly grown in coffee agroforestry system of each site. In Sumberjaya, root measurements thus included coffee (*Coffea canephora*), *Gliricidia sepium* ('kayu hujan'), *Maesopsis eminii* ('pohon afrika'), *Artocarpus heterophyllus* (jackfruit), and *Bambusa arundinacea* (bamboo), and mixed 'fallow vegetation' (dominated by *Trema orientalis* and *Melastoma*; local names are, respectively, 'anggrung' and 'harendong') and the edges of rice fields. In Malang, the measurement was performed on five tree species i.e., *Coffea canephora* (coffee) and *Bambusa arundinacea* (bamboo), *Swietenia mahogany* (mahogany), *Gmelina arborea* (gmelina), *Toona sureni* (suren). The trees selected for measurements were 5 years old, with three replications.

Soil and root samples were taken using two PVC rings of 5 cm height and 10 cm in diameter, temporarily connected, for measuring soil shear strength (Figure 2; [51]), and subsequently root length density. Soil and root samples were taken from the topsoil layer of 0–10 cm, at different distances to trees i.e., 50, 100, 150, 200 cm to obtain variation in root length density. For rice, samples were taken along a bund at 0–5 cm soil depth. Soil shear strength was measured by holding the lower and pulling the upper ring, recording the weight that had to be added to cause a break between the two rings. Soil texture and soil bulk density were measured in these same samples using the pipette and gravimetric technique. Subsequently, roots were separated from the soil by wet sieving.

Roots of selected trees were separated from other roots of different crops. Root length density ($L_{rv}$) was measured using the line intersect method [52] on washed soil samples. Root samples were dried in the oven at 80 °C for 48 h to estimate their dry weight ($D_{rv}$). The GENSTAT 8.0 software [53] was used to carry out ANOVA (analysis of variance) including a *t*-test to separate means.

**Figure 2.** Taking soil samples in the study site in a double ring and measuring soil shear resistance by increasing the pull by adding water to a bucket until the upper ring moves.

## *2.4. Inventory of Potential Tree Root System as an Anchor to Maintain Slope and Riverbank Stability*

At each of the six sites saplings (<5 cm dbh) and trees (>5 cm dbh) of common trees species were selected in multistrata coffee and other agroforestry systems, with a minimum age of 5 years (according to local farmer assessment). High economic value trees were prioritized for the observations and compared to bush fallow vegetation. Proximal roots (close to the tree stem) were exposed [41], with their diameter measured at 20–30 from the trunk, classified as either 'horizontal' or 'vertical' (using an angle of 45° as threshold; Figure 3). Two root indices were calculated i.e., Index of Root Anchoring (IRA) and Index of Root Binding of Soil (IRB). IRA was calculated as $\Sigma D_v^2/dbh^2$ where dbh is tree diameter at breast height (1.3 m height) and $D_v$ is the diameter of vertical roots [54,55]; IRB was calculated as $\Sigma D_h^2/dbh^2$, where $D_h$ is diameter of horizontal roots. Some vertical roots were sampled for measurement of wood density, lignin content, and polyphenolic content using the same method as described in Section 2.2.

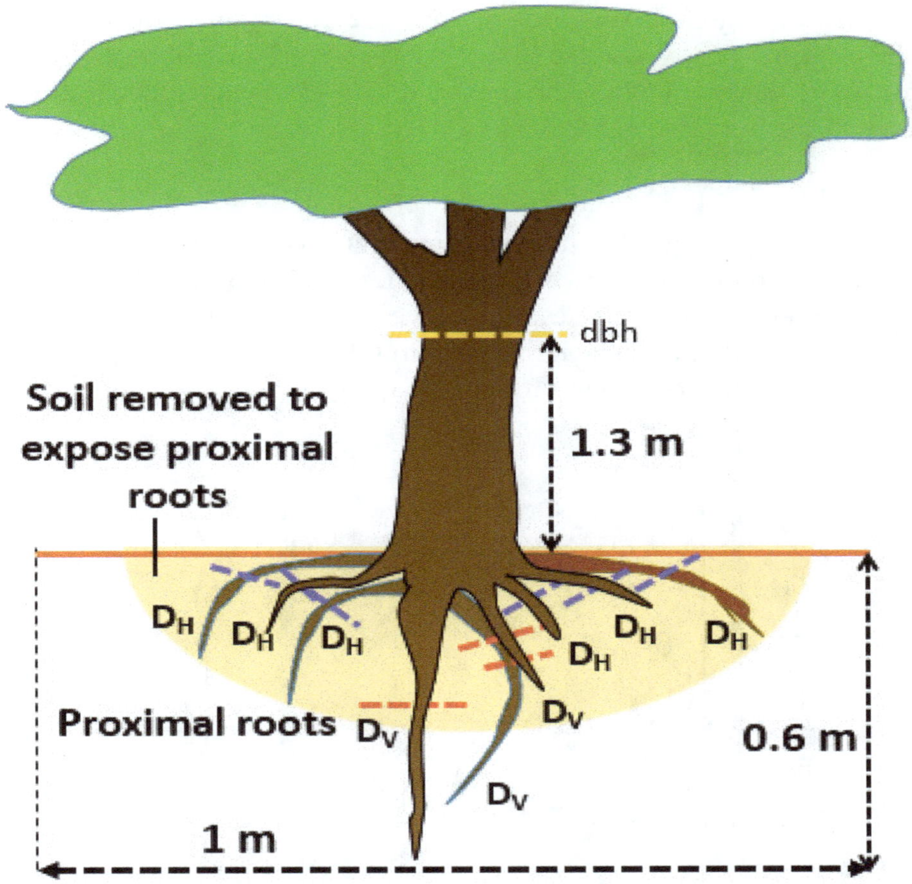


**Figure 3.** Schematic diagram of the distribution of proximal roots [41]; horizontal (H) roots descend at an angle of <45°, vertical (V) roots descend at an angle of >45°; D = root diameter; dbh = diameter at breast height.

## 3. Results

### *3.1. Individual Root Strength in Relation to Root Properties*

Tree roots of approximately 2 mm in diameter (fine roots) were found to break when a weight of 2–15 kg was applied, which corresponds with 64–478 kPa (kg $cm^{-2}$) or 0.06–0.48 MPa of root tensile strength. There was considerable variation in lignin content of the roots (13–30%, with high contents in mahogany and coffee; Table 2) and this was associated with root tensile strength (Figure 4A). About 70% of the variation in root strength was associated with variation in lignin content. The ratio of (Lignin + Polyphenol)/N that generally associate with decomposition rate [38] had less predictive power on root strength than lignin content alone. A multiple regression shows that root strength was positively related with the N content, but negatively with the polyphenol content, with about 80% of variation accounted for (Figure 4B).

**Table 2.** Chemical characteristics of tree root system.

| Litter Type | N (%) | L (%) | P (%) | L/N | P/N | (L + P)/N |
|---|---|---|---|---|---|---|
| Mahogany | 0.6 | 29.2 | 26.4 | 45.0 | 40.7 | 85.7 |
| Toona | 1.2 | 18.9 | 7.7 | 16.4 | 6.7 | 23.1 |
| Gmelina | 1.3 | 13.5 | 8.1 | 10.1 | 6.0 | 16.1 |
| Coffea | 5.4 | 20.1 | 5.4 | 3.7 | 1.0 | 4.7 |
| Bamboo | 1.0 | 16.0 | 1.6 | 16.5 | 1.6 | 18.2 |

N: total nitrogen content, L: lignin content, P: polyphenol content.

The specific root length (specrol = $L_{rv}/D_{rv}$, m $g^{-1}$ of roots) differed between tree species; coffee had finer roots as shown by a high specrol value of about 22 m $g^{-1}$, but it declined rapidly with time to 9 m $g^{-1}$ (Figure 5C); a similar case was also found in *Maesopsis*. Jackfruit roots, by contrast, increased in specrol with time up to 19 m $g^{-1}$, while the specrol of *Gliricidia* was not changing with time at about 8 m $g^{-1}$. Bamboo had the lowest specrol (5.5 m $g^{-1}$).

### 3.3. Soil Shear Strength

The average soil shear strength in our measurements was 3.37 MPa (Figure 6). There appeared to be ($p = 0.055$) a significant interaction between tree species and age on soil shear strength, accompanying a significant difference between tree species. The highest soil shear strength was found in the plot with bamboo (average of 4.82 MPa). Soil shear strength in the bund of a paddy rice field (2.38 MPa) was higher than that for all trees tested, except for coffee.

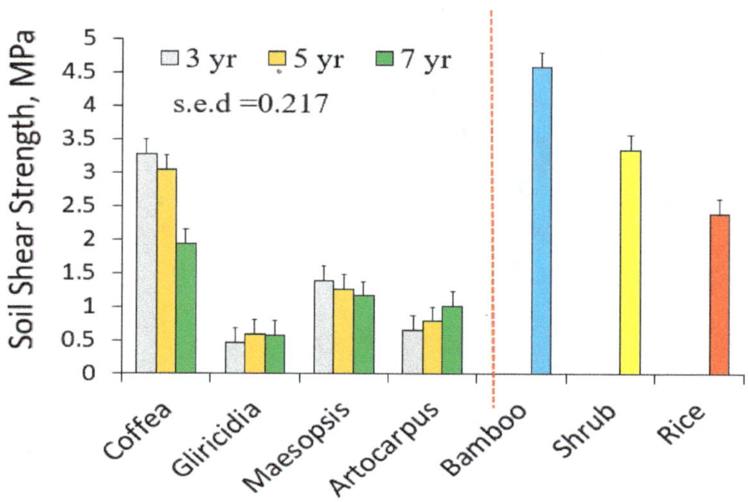


**Figure 6.** Soil shear strength (SS) at 5 cm depth for samples taken close to four species of trees and three types of non-trees.

### 3.4. Soil Shear Strength in Relation to Root Length Density in the Topsoil

The high shear strength close to bamboo stands was clearly correlated to the high root length density (Figure 7). In other trees and non-trees such relation was not clear, possibly due to the scarcity of soil samples with $L_{rv}$ values above 10 km $m^{-3}$.

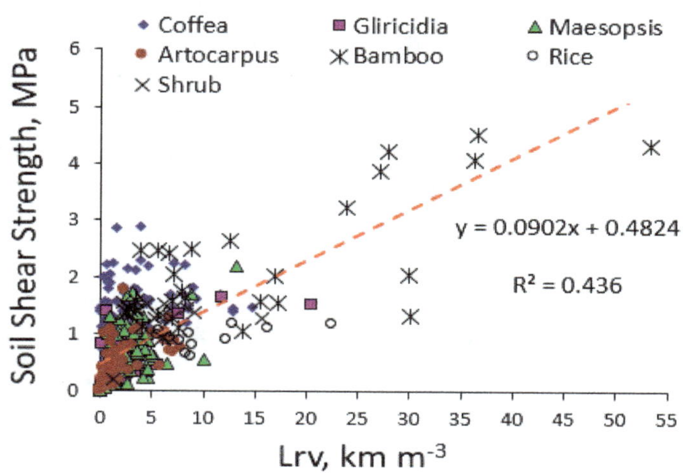


**Figure 7.** Regression of soil shear strength at 5 cm depth on root length density ($L_{rv}$) of various plant species in the layer 0–10 cm.

*3.5. Allometry of Proximal Tree Root Systems*

Results of the root survey of various tree species in six locations suggested difference among locations, with proximal root diameters relative to the stem diameters high in Bangsri and low in Ngantang (Figure 8). Our data allow a partial disentangling of site (e.g., soil or climate related), species, and age effects.

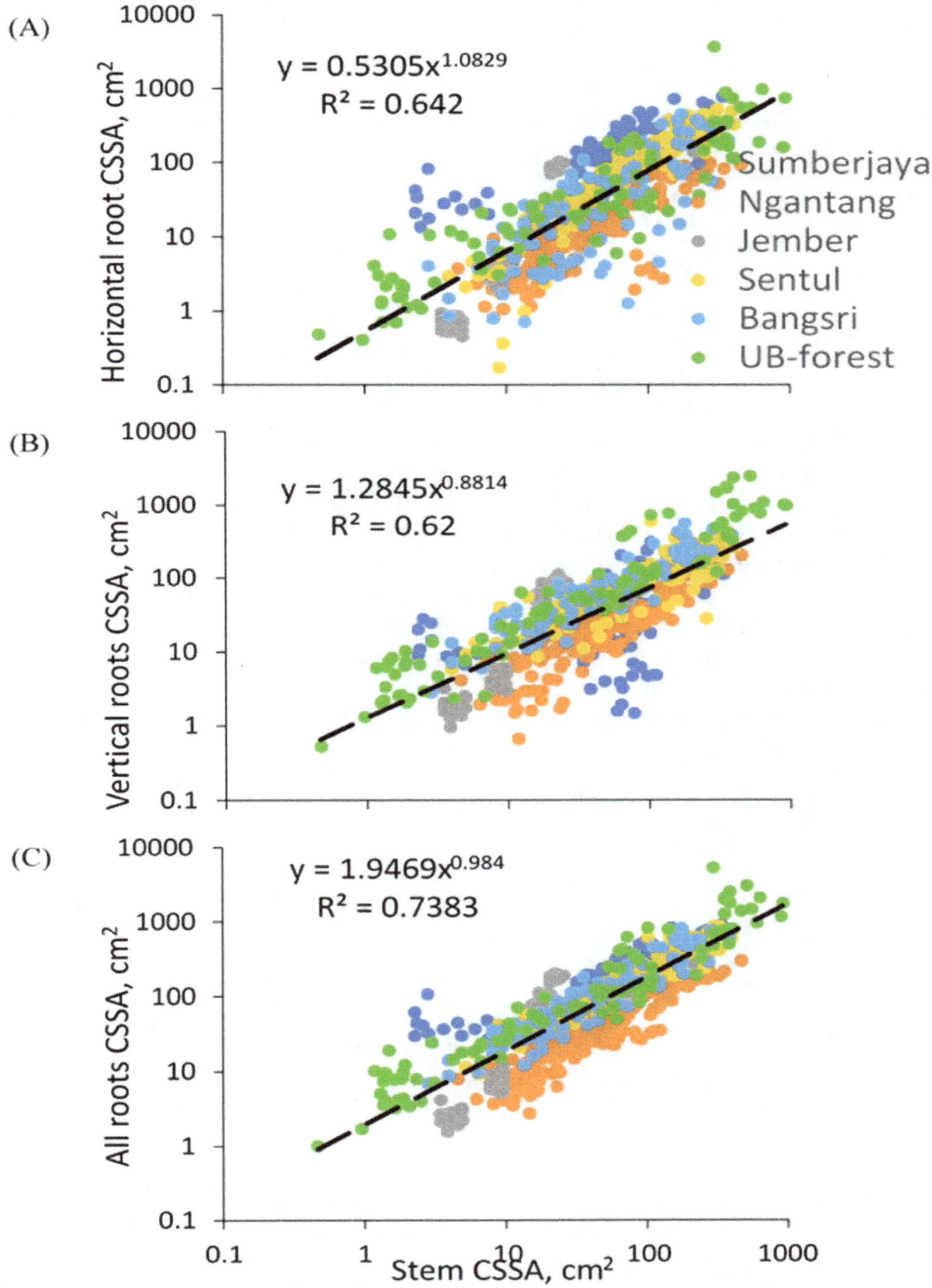


**Figure 8.** Relationship between cross-sectional area (CSSA) of tree stems and (**A**) horizontal roots ($D_h^2$), (**B**) vertical roots ($D_v^2$), and (**C**) total proximal roots for 685 tree samples of 60 tree species across five study sites.

Across all tree observations, variation in tree diameter accounted for 74% of variation in the sum of proximal root CSSA (Figure 8C), 62% of variation in vertical root CSSA (Figure 8B), and 64% of variation in horizontal root CSSA (Figure 8A). Average root-to-shoot ratio in terms of CSSA was around 2.39 (mean 2.21), average IRA 1.30 (median 1.11), and average IRB 1.09 (median 0.816).

Across all species the relative importance of vertical roots decreased with increasing stem diameter, and that of horizontal roots increased, with only a slight decline in root/shoot ratio (in terms of cross-sectional area) from 1.95 to 1.74 in the observed range of tree sizes (Table 3). The ratio of vertical to horizontal roots declined from 2.42 to 0.60, with lateral roots taking over from the tap root in most species.

**Table 3.** Average impact of increasing stem diameter (and cross-sectional area, CSSA) on vertical and horizontal proximal root CSSA, based on regression equations in Figure 8.

| Stem CSSA, $cm^2$ | Vertical (IRA) | Horizontal (IRB) | Root/Shoot CSSA | Vertical/Horizontal Ratio |
|---|---|---|---|---|
| 1 | 1.28 | 0.53 | 1.95 | 2.42 |
| 10 | 0.98 | 0.64 | 1.88 | 1.52 |
| 100 | 0.74 | 0.78 | 1.81 | 0.96 |
| 1000 | 0.57 | 0.94 | 1.74 | 0.60 |

The ANOVA showed that tree species differed significantly ($p < 0.05$) in IRA (Index of Root Anchoring) and IRB (Index of Root Binding) values but given the overall relationship between stem diameter and both IRA and IRB values (Table 3), differences in the diameters at which tree species were observed could be (partially) responsible for this finding.

Appendix A lists the 50+ species to which the 685 measured trees belong with, a minimum of four, a median of 9.5, an average of 12.7, and a maximum of 58 observations per tree species (with robusta coffee as most studied tree). The mean diameters per species of the observed trees had a minimum of 1.5 cm, a median of 8.5 cm, an average of 8.0 cm, and a maximum of 19.9 cm. Across all tree species, there was no significant relationship (as evident from the low fraction of variance accounted for or $R^2$ value) between the average stem diameter at which a species was observed, and the average IRA (or IRB) value recorded (Figure 9). We can thus assume that differences in observed IRA and IRB values between species cannot be simply attributed to the differences in stem diameters for the trees observed.

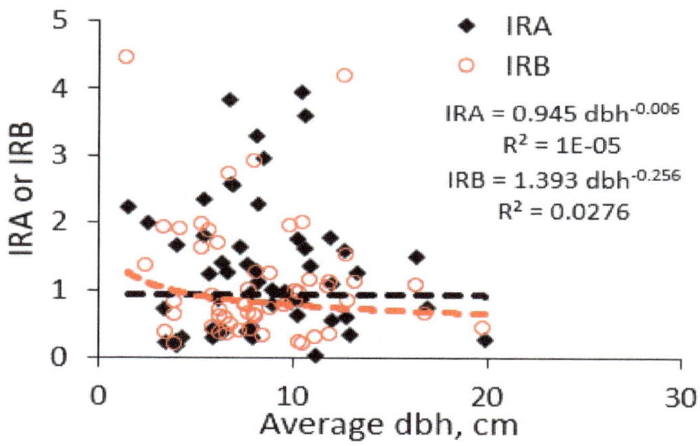


**Figure 9.** Relationship between average value of the IRA and IRB indices and average stem diameter at which a tree species was observed for the 50+ tree species in the survey.

Based on the observed IRA and IRB distributions at species level (rather than at the observed tree level, as the number of observations per species was uneven), the lower and upper quartile were 0.5 and 1.8 for IRA, and 0.5 and 1.5 for IRB, respectively (Figure 10). Using these thresholds, Table 4 shows a two-way classification of the tree species observed in the survey.

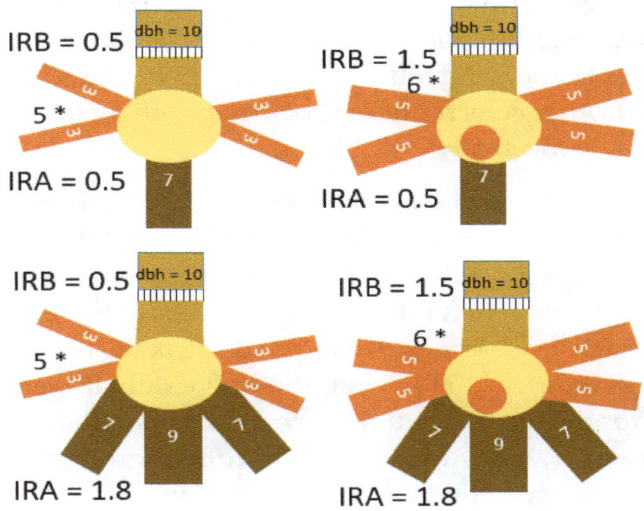


**Figure 10.** Examples of root-to-shoot relations at the four classification thresholds in Table 4. The IRB for a tree with stem diameter of 10 and 6 horizontal roots of 5 cm diameter is $6 * 5^2/10^2 = 1.5$, while that for a tree with 5 roots of 3.16 cm is $5 * 3.16^2/10^2 = 0.5$.

**Table 4.** Two-way classification of trees based on rooting indices IRA (Index of Root Anchoring) and IRB (Index of Root Binding soil particles) classified by lower and upper quartiles of their distributions

| INDEX | IRB_Low (<0.5) | IRB_ Medium (0.5–1.5) | IRB_High (>1.5) |
|---|---|---|---|
| IRA_Low <0.5 | *Mangifera indica, Citrus × sinensis, Malus domestica, Cassia fistula, Agathis dammara, Ficus benjamina, Annona muricata* | *Spathodea campanulata, Pinus merkusii, Melia azedarach, Jatropha curcas, Ceiba pentandra, Trema orientalis* | *Litsea garcia* |
| IRA_Medium 0.5–1.8 | *Eucalyptus deglupta, Cinnamommum burmannii* | *Parkia speciosa, Sandoricum koetjape, Hevea brasiliensis, Aleurites moluccana, Durio zibethinus, Erythrina subumbrans, Ochroma pyramidale, Syzygium polyanthum, Swietenia mahagoni, Persea americana, Calliandra calothyrsus, Pangium edule, Maesopsis eminii, Nephelium lappaceum, Psidium guajava, Toona sureni, Coffea canephora var. robinson, Artocarpus communis, Tectona grandis* | *Homalantus populneus, Artocarpus heterophyllus, Hibiscus tiliaceus, Pterocarpus indicus, Paraserianthes falcataria, Leucaena leucocephala, Artocarpus elasticus* |
| IRA_High >1.8 | *Quercus lineata, Macarangga triloba, Syzygium aqueum, Croton argyratus, Ficus padana* | *Coffea canephora var. robusta, Piper aduncum, Gmelina arborea* | *Michellia alba, Coffea arabica, Gliricidia sepium, Senna spectabilis* |

All nine cells of Table 4 had at least one tree species, with the largest number (obviously) in the mid-range values for both IRA and IRB. The seven species in the upper left cell (low IRA and low IRB) had a low overall root-to-shoot ratio, while the four species in the lower right cell (high IRA and high IRB) a high overall one. The other cells represent a partial tradeoff between vertical and horizontal roots (or IRA and IRB values, respectively).

The two main coffee species (arabica and robusta) had a relatively high IRA value (strong taproot system), with average or high lateral root development as well (relative to stem diameters). Most of the fruit trees commonly found in coffee agroforestry (such as *Artocarpus elasticus* (jackfruit), *Parkia speciosa* (petai), and *Durio zibethinus* (durian)) had intermediate values for both IRA and IRB. The same holds for common timber species such as *Maesopsis eminii* (pohon afrika), *Toona sureni* (suren), *Tectona grandis* (teak), *Swietenia mahagoni* (mahogany). However, their generally larger stem diameters, relative to coffee, still imply a larger absolute contribution to vertical root development in the plot as a whole. Legume trees commonly used as shade trees in current coffee agroforestry systems, *Gliricidia sepium* and *Senna spectabilis* (ramayana), provided a higher anchoring per unit stem diameter than the more traditional shade tree, *Erythrina subumbrams* (dadap).

A further comparison between pruned and unpruned robusta coffee in Sumberjaya (data not shown) found the highest IRA (7.7) for unpruned coffee, suggesting that unpruned coffee is a good species for anchoring riverbanks. This characteristic disappears with pruning, confirming earlier research where frequent pruning of other tree species also led to the formation of more roots in the surface layer [56].

## 4. Discussion

All four hypotheses were confirmed in this study. The break strength of woody roots across five species was positively related to lignin and nitrogen content and negatively to polyphenol content. A linear model explained 81% of the variation (Figure 4B). Lignin content alone accounted for 70% of the variation. Mahogany (*Swietenia mahogani*) and coffee (*Coffea canephora*) had the strongest roots, gmelina (*Gmelina arborea*) and suren (*Toona sureni*) the weakest, and giant bamboo (*Bambusa arundinacea*) had an intermediate root strength.

Soil shear strength depends on soil texture and soil water content, and we found a considerable spread in shear strength with low root length densities (Figure 7). The only species with sufficiently high root length densities to increase shear strength was bamboo. However, our test was based on the top 10 cm of soil, testing a possible break at 5 cm depth, which is unlikely to happen. The shallow landslides of most interest would slide at depths of 30 cm to 1 m. From our data it is unlikely that root length densities at such depth would be high enough to have a substantial effect on shear strength. However, as landslides are binary (yes or no) events, a small change in the threshold value may still be relevant. In terms of land use change, it is the change in root presence that matters, especially a decrease, as a preceding (forest) vegetation may have prevented landslides to occur that were otherwise 'due' (as the thickness and weight of soil layers exceeded what can be prevented from sliding by the soil properties as such). To judge absolute contributions to anchoring of the rooted soil, absolute tree size may be at least as important as the relative allocation over vertical and horizontal roots indicated by the IRA and IRB values.

The combination of root quality characteristics determining tensile strength according to our results in Figure 4 is different from that determining decomposition rate [39], although high lignin content is associated with strong and slowly decomposing roots. Tree species having specific rooting habits (high root strength and root diameter) can be used to control erosion and when linked with extreme flood probability can be used to indicate the risk of a storm likely to cause slope instability in the period between clear-felling and tree re-growth [19]. Our root tensile strength measurements of

64–478 kPa were lower than values reported in the literature, such as the 17.6 MPa for woody roots of 5.5 mm diameter measured for *Pinus radiata* [24]. More measurements on larger-diameter woody roots are still needed to clarify the difference between these results.

Trees with high IRA (Index of Root Anchoring) do not always have a low IRB (Index of Root Binding of soil particles) value or vice versa. Results in Table 4 can be interpreted as guide for suitability of trees as riverbank stabilizer. Coffee (especially without pruning) is potentially suitable for anchoring and soil surface holding at the riverbank, but it has a low root length density (Figure 5). Combination with other trees such as Gliricidia (high IRB as well as IRA) and other grasses is probably better to increase slope stability. High economic value trees (fruit and timber trees) in Sumberjaya (West Lampung) [57] are mostly grouped in the medium class (Table 4). Indeed, tree roots contribute to soil shear strength, but only in the upper 0.4 m of the soil. Most failures, however, occur at greater depths as anchorage by deeper roots was not effective or absent [58]. Evaluation of suitable trees for stabilizing riverbank should also consider the root strength differences [59].

Compared to other methods of observing tree root characteristics the proximal root method, based on fractal branching theory [42], allowed a much wider part of the field-level variation among sites, tree species, tree management, and tree size to be sampled, with an investment of 1–2 person-hours per tree for preparation and observation. As no roots are cut and soil can be returned to the original position the method is essentially 'non-destructive'. Among the limitations of the classification of proximal roots is that tree species with large horizontal roots may have 'sinker' roots further away from the stem that can have a major effect on strengthening the rooted soil as a whole. Root studies on Indonesian trees in the 1930s showed a wide range of rooting patterns along vertical and horizontal spread, including sinker roots [60].

A further caveat is that measurements in the dry season might be different; as all our measurements were during the rainy season. Landslide risk can be higher at the start of the rainy season (when roots are not yet fully active), as well as towards the end (when the whole soil system tends to be water-saturated).

The interest of farmers or watershed managers in deep roots for possible landslide risk reduction, may need to be balanced by considerations from the tree's perspective [61]. In terms of plant strategies, deep roots help first of all in water acquisition during dry (and sunny etc.) periods [62]. Horizontal roots provide nutrients and water during rainy seasons. It is tempting to interpret the root-to-sheet ratio in terms of cross-sectional area (and likely xylem vessel transport capacity) as a fail-safe strategy to be able to support the leaf canopy on either surface or deeper soil water, depending on conditions. Any impact on slope stabilization is at best a 'co-benefit' for the tree, not a major selection force.

Conceptually the different scales (root, tree, and vegetation) can be combined to understand landslide risk at landscape level and to target interventions to reduce risk, but a fully quantitative model that includes all relations of dynamic tree root system growth and decay along with the probability of rainfall intensities and slope geo-hydrology is not feasible at this stage. While most practices that try to reduce erosion target an increase of water infiltration [45], such increase can effectively increase landslide risk unless it is accompanied by deep and strong tree root systems. The collapse of terraces on steep slopes can be interpreted this way. Increased risk of landslides after reduction in tree cover is likely to peak once tree root system decay has reduced soil shear strength and before water infiltration into the soil profile has been reduced. Gradual replacement of trees on hill slopes, as practiced in the 'sisipan' management style of agroforests [63], will have lower risk than a clear felling-replanting. Maintaining and promoting mixed tree vegetation will reduce (but not necessarily minimize) risk, while providing other benefits to local livelihoods. Based on this study, we suggest a mix of tree species with deep roots and grasses with intense and strong fine roots will provide the highest riverbank and hillslope stability, especially if drastic tree pruning or clearing-replanting events can be avoided.

## 5. Conclusions

Our investigations confirmed that woody roots play a role in holding soil together and resisting mass movement in landslides. The critical load where individual roots break is related to the root's lignin content (its woodiness). The shear strength of a volume of soil over any plane increases proportionally with the number and strength of individual roots. With a similar overall ratio of woody root to stem cross sectional area, there are consistent differences between tree species in root development in the topsoil and at depth, that contribute to differences in soil binding and soil anchoring, respectively. Differences in the distribution of tree roots between species can be used to reduce landslide risks in the context of productive coffee agroforestry systems. Tree root architecture varies with species, location, and age. On average the cross-sectional area of proximal roots is twice that of the stem, with only a small reduction in the ratio with increasing stem diameter. The fraction of proximal roots in the vertical and horizontal categories varies more between species than the root-to-shoot ratio of cross-sectional areas. Using the Index of Root Anchoring (based on vertical roots) and the Index of Soil Binding (based on horizontal roots), tree species in the upper and lower quartile of each indicator were identified. However, the strong location effect that was observed implies that site-specific observations will be needed for selecting trees that are locally deep-rooted, taking current observations only as first indication. A mixed vegetation is the simplest way of ensuring that both soil binding and soil anchoring functions are provided.

**Author Contributions:** Conceptualization, K.H. and M.V.N; Investigation, K.H., W.W. and D.S.; Methodology, K.H., W.W., D.S. and M.V.N.; Validation, D.S.; Writing—original draft, K.H. and M.V.N.; Writing—review and editing, K.H. and M.V.N. All authors have read and agreed to the published version of the manuscript.

**Acknowledgments:** Ari Santosa, Mohadi Nurhada, Veronika Kurniasari, Emanuel Ardian Kristanto, Astrid Ferera, Arya Putra Widyatmadya, and Tsarwah Al Sausan offered valuable contributions during their field work as undergraduate students of University of Brawijaya, Malang. We appreciated the valuable inputs of Rachmat Mulia, Fahmuddin Agus, and Jianchu Xu to earlier drafts.

## Appendix A

**Table A1.** Root observations across six sites, with number of trees observed (N), average dbh, IRA, and IRB, and quartiles in IRA and IRB distribution.

| Species | N | Dbh Avg | IRA Avg | IRB Avg | Quartile | Quartile |
|---|---|---|---|---|---|---|
| *Mangifera indica* | 6 | 11.2 | 0.04 | 0.29 | IRA Q1 | IRB Q1 |
| *Citrus × sinensis* | 6 | 4.0 | 0.18 | 0.19 | IRA Q1 | IRB Q1 |
| *Malus domestica* | 6 | 3.5 | 0.23 | 0.36 | IRA Q1 | IRB Q1 |
| *Cassia fistula* | 6 | 19.9 | 0.26 | 0.42 | IRA Q1 | IRB Q1 |
| *Agathis dammara* | 6 | 5.9 | 0.29 | 0.43 | IRA Q1 | IRB Q1 |
| *Ficus benjamina* | 6 | 7.9 | 0.29 | 0.41 | IRA Q1 | IRB Q1 |
| *Annona muricata* | 6 | 6.3 | 0.37 | 0.35 | IRA Q1 | IRB Q1 |
| *Spathodea campanulata* | 6 | 4.0 | 0.25 | 0.61 | IRA Q1 | IRB Q23 |
| *Pinus merkusii* | 11 | 12.9 | 0.34 | 0.83 | IRA Q1 | IRB Q23 |
| *Melia azedarach* | 11 | 7.6 | 0.40 | 0.75 | IRA Q1 | IRB Q23 |
| *Jatropha curcas* | 6 | 6.3 | 0.41 | 0.68 | IRA Q1 | IRB Q23 |
| *Ceiba pentandra* | 6 | 5.9 | 0.41 | 0.89 | IRA Q1 | IRB Q23 |
| *Trema orientalis* | 10 | 7.8 | 0.47 | 0.64 | IRA Q1 | IRB Q23 |
| *Litsea garcia* | 5 | 4.3 | 0.30 | 1.88 | IRA Q1 | IRB Q4 |
| *Eucalyptus deglupta* | 6 | 12.0 | 0.56 | 0.33 | IRA Q23 | IRB Q1 |
| *Cinnamommum burmannii* | 10 | 7.3 | 1.65 | 0.38 | IRA Q23 | IRB Q1 |
| *Parkia speciosa* | 19 | 12.8 | 0.59 | 1.50 | IRA Q23 | IRB Q23 |
| *Sandoricum koetjape* | 20 | 10.2 | 0.62 | 0.96 | IRA Q23 | IRB Q23 |

**Table A1.** *Cont.*

| Species | N | Dbh Avg | IRA Avg | IRB Avg | Quartile | Quartile |
|---|---|---|---|---|---|---|
| *Hevea brasiliensis* | 10 | 16.9 | 0.75 | 0.64 | IRA Q23 | IRB Q23 |
| *Aleurites moluccana* | 15 | 8.9 | 0.80 | 0.71 | IRA Q23 | IRB Q23 |
| *Durio zibethinus* | 15 | 10.4 | 0.87 | 0.95 | IRA Q23 | IRB Q23 |
| *Erythrina subumbrans* | 20 | 7.8 | 0.95 | 0.98 | IRA Q23 | IRB Q23 |
| *Ochroma pyramidale* | 7 | 9.6 | 0.99 | 0.76 | IRA Q23 | IRB Q23 |
| *Syzygium polyanthum* | 5 | 8.9 | 1.00 | 1.218 | IRA Q23 | IRB Q23 |
| *Swietenia mahagoni* | 45 | 12.0 | 1.09 | 1.11 | IRA Q23 | IRB Q23 |
| *Persea americana* | 21 | 8.2 | 1.13 | 1.26 | IRA Q23 | IRB Q23 |
| *Calliandra calothyrsus* | 15 | 6.6 | 1.26 | 0.57 | IRA Q23 | IRB Q23 |
| *Pangium edule* | 23 | 13.3 | 1.27 | 1.11 | IRA Q23 | IRB Q23 |
| *Maesopsis eminii* | 31 | 10.9 | 1.36 | 1.14 | IRA Q23 | IRB Q23 |
| *Nephelium lappaceum* | 32 | 7.6 | 1.37 | 0.77 | IRA Q23 | IRB Q23 |
| *Psidium guajava* | 10 | 6.3 | 1.40 | 0.58 | IRA Q23 | IRB Q23 |
| *Toona sureni* | 25 | 16.4 | 1.50 | 1.07 | IRA Q23 | IRB Q23 |
| *Coffea canephora* var. *robinson* | 16 | 4.0 | 1.66 | 0.80 | IRA Q23 | IRB Q23 |
| *Artocarpus communis* | 4 | 10.2 | 1.76 | 0.83 | IRA Q23 | IRB Q23 |
| *Tectona grandis* | 9 | 11.9 | 1.79 | 1.03 | IRA Q23 | IRB Q23 |
| *Homalantus populneus* | 5 | 3.4 | 0.73 | 1.90 | IRA Q23 | IRB Q4 |
| *Artocarpus heterophyllus* | 35 | 9.9 | 0.82 | 1.93 | IRA Q23 | IRB Q4 |
| *Hibiscus tiliaceus* | 20 | 6.2 | 0.86 | 1.68 | IRA Q23 | IRB Q4 |
| *Pterocarpus indicus* | 5 | 5.7 | 1.25 | 1.86 | IRA Q23 | IRB Q4 |
| *Paraserianthes falcataria* | 15 | 8.1 | 1.28 | 2.91 | IRA Q23 | IRB Q4 |
| *Leucaena leucocephala* | 15 | 12.7 | 1.59 | 4.17 | IRA Q23 | IRB Q4 |
| *Artocarpus elasticus* | 4 | 10.6 | 1.62 | 1.99 | IRA Q23 | IRB Q4 |
| *Quercus lineata* | 4 | 6.9 | 2.57 | 0.48 | IRA Q4 | IRB Q1 |
| *Macarangga triloba* | 4 | 8.5 | 2.96 | 0.30 | IRA Q4 | IRB Q1 |
| *Syzygium aqueum* | 4 | 10.6 | 3.60 | 0.19 | IRA Q4 | IRB Q1 |
| *Croton argyratus* | 4 | 6.7 | 3.84 | 0.34 | IRA Q4 | IRB Q1 |
| *Ficus padana* | 4 | 10.4 | 3.96 | 0.22 | IRA Q4 | IRB Q1 |
| *Coffea canephora* var. *robusta* | 58 | 2.5 | 1.99 | 1.35 | IRA Q4 | IRB Q23 |
| *Piper aduncum* | 4 | 8.2 | 2.28 | 0.58 | IRA Q4 | IRB Q23 |
| *Gmelina arborea* | 4 | 8.1 | 3.28 | 0.62 | IRA Q4 | IRB Q23 |
| *Michellia alba* | 5 | 5.4 | 1.81 | 1.96 | IRA Q4 | IRB Q4 |
| *Coffea arabica* | 10 | 1.5 | 2.23 | 4.43 | IRA Q4 | IRB Q4 |
| *Gliricidia sepium* | 14 | 5.4 | 2.35 | 1.60 | IRA Q4 | IRB Q4 |
| *Acacia auriculiformis* | 16 | 6.8 | 2.58 | 2.72 | IRA Q4 | IRB Q4 |

## References


1. Sidle, R.C.; Dhakal, A.S. Recent advances in the spatial and temporal modeling of shallow landslides. In Proceedings of the 2003 MODSIM Conference, Townsville, Australia, 14–17 July 2003; pp. 602–607.
2. Stokes, A.; Norris, J.E.; Van Beek, L.P.H.; Bogaard, T.; Cammeraat, E.; Mickovski, S.B.; Jenner, A.; Di Iorio, A.; Fourcaud, T. *How vegetation reinforces soil on slopes. In Slope Stability and Erosion Control: Ecotechnological Solutions*; Springer: Dordrecht, The Netherlands, 2008; pp. 65–118.
3. Van Noordwijk, M.; Hairiah, K.; Lasco, R.; Tata, H.L. How can agroforestry be part of disaster risk management. In *Sustainable Development through Trees on Farms: Agroforestry in Its Fifth Decade*; Van Noordwijk, M., Ed.; World Agroforestry (ICRAF): Bogor, Indonesia, 2019; pp. 215–229.
4. Pohl, M.; Alig, D.; Körner, C.; Rixen, C. Higher plant diversity enhances soil stability in disturbed alpine ecosystems. *Plant Soil* **2009**, *324*, 91–102. [CrossRef]
5. Alam, S.; Banjara, A.; Wang, J.; Patterson, W.B.; Baral, S. Novel Approach in Sampling and Tensile Strength Evaluation of Roots to Enhance Soil for Preventing Erosion. *Open J. Soil Sci.* **2018**, *8*, 330. [CrossRef]
6. Giadrossich, F.; Schwarz, M.; Cohen, D.; Cislaghi, A.; Vergani, C.; Hubble, T.; Phillips, C.; Stokes, A. Methods to measure the mechanical behaviour of tree roots: A review. *Ecol. Eng.* **2017**, *109*, 256–271. [CrossRef]
7. Iverson, R.M. Landslide triggering by rain infiltration. *Water Resour. Res.* **2000**, *36*, 1897–1910. [CrossRef]

8. Micheli, E.R.; Kirchner, J.W. Effects of wet meadow riparian vegetation on streambank erosion. 2. Measurements of vegetated bank strength and consequences for failure mechanics. *Earth Surf. Process Landf.* **2002**, *27*, 687–697. [CrossRef]
9. Roering, J.J.; Schmidt, K.M.; Stock, J.D.; Dietrich, W.E.; Montgomery, D.R. Shallow landsliding, root reinforcement, and the spatial distribution of trees in the Oregon Coast range. *Can. Geotech.* **2003**, *40*, 237–253. [CrossRef]
10. Wu, T.H.; Swanston, D.N. Risks of landslides in shallow soils and its relation to clearcutting in Southeastern Alaska. *For. Sci.* **1980**, *26*, 495–510. [CrossRef]
11. O'Loughlin, C.L.; Watson, A.J. Root wood strength deterioration in radiate pine after clearfelling. *N. Z. J. For. Sci.* **1979**, *9*, 284–293.
12. Wilkinson, P.L.; Anderson, M.G.; Lloyd, D.M. An integrated hydrological model for rain-induced landslide prediction. *Earth Surf. Processes Landf.* **2002**, *27*, 1285–1297. [CrossRef]
13. Ziemer, R.R.; Swantson, D.N. *Root Strength Changes after Logging in Southeast Alaska*; USDA Forest Service Research Note PNW; Department of Agriculture, Forest Service, Pacific Northwest Forest and Range Experiment Station: Portland, OR, USA, 1997; Volume 306.
14. Abe, K.; Ziemer, R.R. *Effect of Tree Roots on Shallow-Seated Land Slides*; USDA Forest Service Gen. Tech. Rep. PSW-GT130; International Union of Forestry Research: Vienna, Austria, 1991; pp. 11–20.
15. Schiechtl, H.M. *Bioengineering for Land Reclamation and Conservation*; University of Alberta Press: Edmonton, AL, Canada, 1980.
16. Shen, P.; Zhang, L.M.; Chen, H.X.; Gao, L. Role of vegetation restoration in mitigating hillslope erosion and debris flows. *Eng. Geol.* **2017**, *216*, 122–133. [CrossRef]
17. Wang, B.; Zhang, G.H. Quantifying the binding and bonding effects of plant roots on soil detachment by overland flow in 10 typical grasslands on the Loess Plateau. *Soil Sci. Soc. Am. J.* **2017**, *81*, 1567–1576. [CrossRef]
18. Zhu, J.; Wang, Y.; Wang, Y.; Mao, Z.; Langendoen, E.J. How does root biodegradation after plant felling change root reinforcement to soil? *Plant Soil* **2020**, *446*, 211–227. [CrossRef]
19. Watson, A.; Phillips, C.; Marden, M. Root strength, growth, and rates of decay: Root reinforcement changes of two tree species and their contribution to slope stability. *Plant Soil* **1999**, *217*, 39–47. [CrossRef]
20. Hubble, T.C.T.; Docker, B.B.; Rutherfurd, I.D. The role of riparian trees in maintaining riverbank stability: A review of Australian experience and practice. *Ecol. Eng.* **2010**, *36*, 292–304. [CrossRef]
21. Bischetti, G.B.; Chiaradia, E.A.; Simonato, T.; Speziali, B.; Vitali, B.; Vullo, P.; Zocco, A. Root Strength and Root Area Ratio of Forest Species in Lombardy (Northern Italy). *Plant Soil* **2005**, *278*, 11–22. [CrossRef]
22. De Baets, S.; Poesen, J.; Reubens, B.; Wemans, K.; De Baerdemaeker, J.; Muys, B. Root tensile strength and root distribution of typical Mediterranean plant species and their contribution to soil shear strength. *Plant Soil* **2008**, *305*, 207–226. [CrossRef]
23. Abernethy, B.; Rutherfurd, I.D. The distribution and strength of riparian tree roots in relation to riverbank reinforcement. *Hydrol. Processes* **2001**, *15*, 63–79. [CrossRef]
24. Ekanayake, J.C.; Marden, M.; Watson, A.J.; Rowan, D. Tree roots and slope stability: A comparison between Pinus radiata and kanuka. *N. Z. J. For. Sci.* **1997**, *27*, 216–233.
25. Genet, M.; Kokutse, N.; Stokes, A.; Fourcaud, T.; Cai, X.; Ji, J.; Mickovski, S. Root reinforcement in plantations of *Cryptomeria japonica* D. Don: Effect of tree age and stand structure on slope stability. *For. Ecol. Manag.* **2008**, *256*, 1517–1526. [CrossRef]
26. Tosi, M. Root tensile strength relationships and their slope stability implications of three shrub species in the Northern Apennines (Italy). *Geomorphology* **2007**, *87*, 268–283. [CrossRef]
27. Giadrossich, F.; Cohen, D.; Schwarz, M.; Ganga, A.; Marrosu, R.; Pirastru, M.; Capra, G.F. Large roots dominate the contribution of trees to slope stability. *Earth Surf. Processes Landf.* **2019**, *44*, 1602–1609. [CrossRef]
28. Collison, A.J.C.; Anderson, M.G. Using a combined slope hydrology/stability model to identify suitable conditions for landslide prevention by vegetation in the humid tropics. *Earth Surf. Processes Landf.* **1996**, *21*, 737–747. [CrossRef]
29. Nilaweera, N.S.; Nutalaya, P. Role of tree roots in slope stabilisation. *Bull. Eng. Geol. Environ.* **1999**, *57*, 337–342. [CrossRef]

30. Madsen, C.; Potvin, C.; Hall, J.; Sinacore, K.; Turner, B.L.; Schnabel, F. Coarse root architecture: Neighbourhood and abiotic environmental effects on five tropical tree species growing in mixtures and monocultures. *For. Ecol. Manag.* **2020**, *460*, 117851. [CrossRef]
31. Danjon, F.; Reubens, B. Assessing and analyzing 3D architecture of woody root systems, a review of methods and applications in tree and soil stability, resource acquisition and allocation. *Plant Soil* **2008**, *303*, 1–34. [CrossRef]
32. Collison, A.; Pollen, N. The effects of riparian buffer strips on streambank stability: Root reinforcement, soil strength and growth rates. In *Roots and Soil Management: Interaction between Roots and the Soil*; Zobel, R.W., Wright, S.F., Eds.; American Society of Agronomy: Madison, WI, USA, 2005; Volume 48, pp. 15–56.
33. Reubens, B.; Poesen, J.; Danjon, F.; Geudens, G.; Muys, B. The role of fine and coarse roots in shallow slope stability and soil erosion control with a focus on root system architecture: A review. *Trees* **2007**, *21*, 385–402. [CrossRef]
34. Stokes, A.; Atger, C.; Bengough, A.G.; Fourcaud, T.; Sidle, R.C. Desirable plant root traits for protecting natural and engineered slopes against landslides. *Plant Soil* **2009**, *324*, 1–30. [CrossRef]
35. Boldrin, D.; Leung, A.K.; Bengough, A.G. Root biomechanical properties during establishment of woody perennials. *Ecol. Eng.* **2017**, *109*, 196–206. [CrossRef]
36. Chen, H.; Harmo, M.E.; Griffiths, R.P. Decomposition and nitrogen release from decomposing woody roots in coniferous forests of the Pacific Northwest: A chronosequence approach. *Can. J. For. Res.* **1999**, *31*, 246–260. [CrossRef]
37. Schroth, G. Decomposition and nutrient supply from biomass. In *Trees, Crops and Soil Fertility: Concepts and Research Methods*; Schroth, G., Sinclair, F.L., Eds.; CABI Publishing: Wallingford, UK, 2003; pp. 131–150.
38. Hairiah, K.; Sulistyani, H.; Suprayogo, D.; Widianto Purnomosidhi, P.; Widodo, R.H.; Van Noordwijk, M. Litter layer residence time in forest and coffee agroforestry systems in Sumberjaya, West Lampung. *For. Ecol. Manag.* **2006**, *224*, 45–57. [CrossRef]
39. Palm, C.A.; Sanchez, P.A. Nitrogen release from some tropical legumes as affected by lignin and polyphenol contents. *Soil Biol. Biochem.* **1991**, *23*, 83–88. [CrossRef]
40. Genet, M.; Stokes, A.; Salin, F.; Mickovski, S.B.; Fourcaud, T.; Dumail, F.; van Beek, R. The Influence of Cellulose Content on Tensile Strength in Tree Roots. *Plant Soil* **2005**, *278*, 1–9. [CrossRef]
41. Van Noordwijk, M.; Purnomosidhi, P. Root architecture in relation to tree-soil- crop interactions and shoot pruning in agroforestry. *Agrofor. Syst.* **1996**, *30*, 161–173. [CrossRef]
42. Van Noordwijk, M.; Mulia, R. Functional branch analysis as tool for scaling above- and belowground trees for their additive and non-additive properties. *Ecol. Model.* **2002**, *149*, 41–51. [CrossRef]
43. Schwarz, M.; Lehmann, P.; Or, D. Quantifying lateral root reinforcement in steep slopes—From a bundle of roots to tree stands. *Earth Surf. Processes Landf. J. Br. Geomorphol. Res. Group* **2010**, *35*, 354–367. [CrossRef]
44. Nespoulous, J.; Merino-Martin, L.; Monnier, Y.; Bouchet, D.C.; Ramel, M.; Dombey, R.; Viennois, G.; Mao, Z.; Zhang, J.L.; Cao, K.F.; et al. Tropical forest structure and understorey determine subsurface flow through biopores formed by plant roots. *Catena* **2019**, *181*, 104061. [CrossRef]
45. Van Noordwijk, M.; Agus, F.; Verbist, B.; Hairiah, K.; Tomich, T.P. Managing Watershed Services in Ecoagriculture Landscapes. In *Farming with Nature: The Science and Practice of Ecoagriculture*; Scherr, S.J., McNeely, J.A., Eds.; Island Press: Washington, DC, USA, 200; pp. 191–212.
46. Burroughs, E.R.; Thomas, B.R. *Declining Root Strength in Douglas-Fir after Felling as a Factor in Slope Stability*; US Department Agriculture Forest Services Research Paper INT-190; Department of Agriculture, Forest Service, Intermountain Forest and Range Experiment Station: Ogden, UT, USA, 1977.
47. Ziemer, R.R. An apparatus to measure the crosscut shearing strength of roots. *Can. J. For. Res.* **1978**, *8*, 142–144. [CrossRef]
48. Walkley, A.; Black, I.A. An examination of the Degtjareff method for determining soil organic matter, and a proposed modification of the chromic acid titration method. *Soil Sci.* **1934**, *37*, 29–38. [CrossRef]
49. Goering, H.K.; Van Soest, P.J. *Forage Fiber Analysis*; Agricultural Handbook no 379; USDA: Washington, DC, USA, 1970; p. 20.
50. Anderson, J.M.; Ingram, J.S.I. *Tropical Soil Biology and Fertility: A Handbook of Methods*; CAB. International: Wallingford, UK, 1989; p. 171.
51. Rose, C. *An Introduction to the Environmental Physics of Soils, Water and Watershed*; Cambridge University Press: Cambridge, UK, 2004; p. 65.

52. Tennant, D. A test of a modified line intersect method of estimating root length. *J. Ecol.* **1975**, *63*, 995–1001. [CrossRef]
53. Payne, R.W.; Lane, P.W.; Ainsley, A.E.; Bicknell, K.E.; Digby, P.G.N.; Harding, S.A.; Leech, P.K.; Simpson, H.R.; Todd, A.D.; Verrier, P.J.; et al. *Genstat 5 Reference Manual*; Clarendon Press: Oxford, UK, 1987.
54. Van Noordwijk, M. Functional branch analysis to derive allometric equations. In *Modelling Global Change Impacts on the Soil Environment: Report of Training Workshop 5–13 May 1998*; IC-SEA Report No., 6; Murdiyarso, D., van Noordwijk, M., Suyamto, D.A., Eds.; BIOTROP-GCTE: Bogor, Indonesia; ICSEA: Lisbon, Portugal, 1999; pp. 77–80.
55. Akinnifesi, F.K.; Rowe, E.C.; Livesley, S.J.; Kwesiga, F.R.; Van Lauwe, B.; Alegre, J.C. Tree root architecture. In *Below-Ground Interactions in Tropical Agroecosystems. Concepts and Models with Multiple Plant Components*; Van Noordwijk, M., Cadisch, G., Ong, C.K., Eds.; CABI Publishing: Wallingford, UK, 2004; pp. 61–82.
56. Hairiah, K.; Widianto, W.; Utami, S.R.; Sunaryo, D.S.; Sitompul, S.M.; Lusiana, B.; van Noordwijk, M.; Cadish, G. *Pengelolaan Tanah Masam Secara Biologi: Refleksi Pengelaman dari Lampung Utara*; World Agroforestry Centre (ICRAF): Bogor, Indonesia, 2000; p. 184.
57. Budidarsono, S.; Wijaya, K. Praktek Konservasi dalam Budidaya Kopi Robusta dan Keuntungan Petani. *Agrivita* **2004**, *26*, 107–117.
58. Cammeraat, E.; van Beek, R.; Kooijman, A. Vegetation succession and its consequences for slope stability in SE Spain. *Plant Soil* **2005**, *278*, 135–147. [CrossRef]
59. Stokes, A.; Mattheck, C. Variation of wood strength in tree roots. *J. Exp. Bot.* **1996**, *47*, 693–699. [CrossRef]
60. Coster, C. Wortelstudien in de tropen. V. Gebergtehoutsoorten [Root studies in the Tropics. V. Tree species of the mountain region]. *Tectona* **1935**, *28*, 861–878.
61. Van Noordwijk, M.; Lawson, G.; Hairiah, K.; Wilson, J. *Root Distribution of Trees and Crops: Competition and/or Complementarity. Tree-Crop Interactions: Agroforestry in a Changing Climate*; CABI: Wallingford, UK, 2015; pp. 221–257.
62. Van Noordwijk, M.; Martikainen, P.; Bottner, P.; Cuevas, E.; Rouland, C.; Dhillion, S.S. Global change and root function. *Glob. Change Biol.* **1998**, *4*, 759–772. [CrossRef]
63. Wibawa, G.; Hendratno, S.; van Noordwijk, M. Permanent smallholder rubber agroforestry systems in Sumatra, Indonesia. In *Slash and Burn: The Search for Alternatives*; Palm, C.A., Vosti, S.A., Sanchez, P.A., Ericksen, P.J., Juo, A.S.R., Eds.; Columbia University Press: New York, NY, USA, 2005; pp. 222–232.

# 7

# Sustainable Agroforestry Landscape Management: Changing the Game

**Meine van Noordwijk [1,2,3,*], Erika Speelman [4], Gert Jan Hofstede [5,6], Ai Farida [7], Ali Yansyah Abdurrahim [8,9], Andrew Miccolis [10,11], Arief Lukman Hakim [3,11,12], Charles Nduhiu Wamucii [13], Elisabeth Lagneaux [1,14], Federico Andreotti [4,15], George Kimbowa [16,17], Gildas Geraud Comlan Assogba [1], Lisa Best [4,18], Lisa Tanika [11], Margaret Githinji [5], Paulina Rosero [5], Rika Ratna Sari [1,3], Usha Satnarain [19], Soeryo Adiwibowo [9], Arend Ligtenberg [4], Catherine Muthuri [20], Marielos Peña-Claros [11], Edi Purwanto [21], Pieter van Oel [16], Danaë Rozendaal [1,22], Didik Suprayogo [3] and Adriaan J. Teuling [13]**

1 Plant Production Systems, Wageningen University and Research, 6700 AK Wageningen, The Netherlands; elisabeth.lagneaux@wur.nl (E.L.); gildas.assogba@wur.nl (G.G.C.A.); rika.sari@wur.nl (R.R.S.); danae.rozendaal@wur.nl (D.R.)
2 World Agroforestry (ICRAF), Bogor 16155, Indonesia
3 Agroforestry Research Group, Faculty of Agriculture, Brawijaya University, Jl. Veteran no 1, Malang 65145, Indonesia; arief.hakim@wur.nl (A.L.H.); suprayogo@ub.ac.id (D.S.)
4 Laboratory of Geo-Information Science and Remote Sensing, Environmental Sciences, Wageningen University & Research, 6708 PB Wageningen, The Netherlands; Erika.speelman@wur.nl (E.S.); federico.andreotti@wur.nl (F.A.); lisa.best@wur.nl (L.B.); arend.ligtenberg@wur.nl (A.L.)
5 Information Technology Group, Social Sciences, Wageningen University & Research, 6706 KN Wageningen, The Netherlands; Gertjan.hofstede@wur.nl (G.J.H.); Margaret.githinji@wur.nl (M.G.); paulina.roseroanazco@wur.nl (P.R.)
6 Centre for Applied Risk Management (UARM), North-West University, Potchefstroom 2520, South Africa
7 Applied Climatology, IPB University, Bogor 16680, Indonesia; farida_ai@apps.ipb.ac.id
8 Research Centre for Population, Indonesian Institute of Sciences (LIPI), Jakarta 12190, Indonesia; aliy002@lipi.go.id
9 Rural Sociology, IPB University, Bogor 16680, Indonesia; adibowo3006@gmail.com
10 World Agroforestry (ICRAF), 66095-903 Belém/PA, Brazil; a.miccolis@cgiar.org
11 Forest Ecology and Forest Management Group, Wageningen University & Research, 6708 PB Wageningen, The Netherlands; lisa.tanika@wur.nl (L.T.); marielos.penaclaros@wur.nl (M.P.-C.)
12 Faculty of Science and Technology, Universitas Islam Raden Rahmat, Malang 65163, Indonesia
13 Hydrology and Quantitative Water Management Group, Wageningen University & Research, 6700 AA Wageningen, The Netherlands; charles.wamucii@wur.nl (C.N.W.); ryan.teuling@wur.nl (A.J.T.)
14 iES Landau, Institute for Environmental Sciences, University of Koblenz-Landau, 76829 Landau, Germany
15 CIRAD, UPR GREEN, F-34398 Montpellier, France
16 Water Resources Management, Wageningen University & Research, 6700AA Wageningen, The Netherlands; George.kimbowa@wur.nl (G.K.); Pieter.vanoel@wur.nl (P.v.O.)
17 Faculty of Engineering, Busitema University, 236 Tororo, Uganda
18 Tropenbos Suriname, University Campus (CELOS Building), Paramaribo, Suriname
19 Environmental Policy, Department of Social Sciences, Wageningen University & Research, 6706 KN Wageningen, The Netherlands; usha.satnarain@wur.nl
20 World Agroforestry (ICRAF), 00100 Nairobi, Kenya; C.Muthuri@cgiar.org
21 Tropenbos Indonesia, Bogor 16163, Indonesia; edipurwanto@tropenbos-indonesia.org
22 Centre for Crop Systems Analysis, Wageningen University & Research, 6700 AK Wageningen, The Netherlands
* Correspondence: m.vanNoordwijk@cgiar.org

**Abstract:** Location-specific forms of agroforestry management can reduce problems in the forest–water–people nexus, by balancing upstream and downstream interests, but social and ecological finetuning is needed. New ways of achieving shared understanding of the underlying ecological and social-ecological relations is needed to adapt and contextualize generic solutions. Addressing these challenges between thirteen cases of tropical agroforestry scenario development across three continents requires exploration of generic aspects of issues, knowledge and participative approaches. Participative projects with local stakeholders increasingly use 'serious gaming'. Although helpful, serious games so far (1) appear to be ad hoc, case dependent, with poorly defined extrapolation domains, (2) require heavy research investment, (3) have untested cultural limitations and (4) lack clarity on where and how they can be used in policy making. We classify the main forest–water–people nexus issues and the types of land-use solutions that shape local discourses and that are to be brought to life in the games. Four 'prototype' games will be further used to test hypotheses about the four problems identified constraining game use. The resulting generic forest–water–people games will be the outcome of the project "Scenario evaluation for sustainable agroforestry management through forest-water-people games" (SESAM), for which this article provides a preview.

**Keywords:** boundary work; ecohydrology; forest–water–people nexus; landscape approach; participatory methods; scenario evaluation; social-ecological systems; tropical forests

---

## 1. Introduction

### *1.1. Agroforestry and the Forest–Water–People Nexus*

Current understanding of the term agroforestry links plot, landscape and policy aspects of the ways farmers interact with trees [1]. We present early results and planned next steps of a new interdisciplinary research program on scenario evaluation for sustainable agroforestry management, focusses on the (agro) forest–water–people nexus and exploring how serious games can help bridge the science–policy divide, as adequate supplies of clean water are a key development challenge while the roles of trees and forest in modifying ecohydrology are context specific and often contested. A top-down policy perspective needs to be reconciled with a bottom-up farmer understanding, while social, cultural aspects may be as important as ecological, technical ones. To unpack the context dependence of past research results, a comprehensive diagnosis has to be the basis for any comparative study. We restrict ourselves here to a pan-tropical perspective.

The 2030 agenda for Sustainable Development, adopted by all United Nations member states in 2015, provides a shared blueprint for peace and prosperity for people and the planet, now and into the future [2]. As desirable as this sounds, there was a long way to go to achieve these goals even before the emergence of the SARS-CoV-2 virus showed the fragility [3] of the increasingly globalized and interdependent world. As part of the Sustainable Development agenda, land use, often a mix of forests, agroforestry, agriculture and built-up areas, has to meet many and partly conflicting needs, including the provision of food, energy, water as well as environmental protection from floods, droughts and biological extinctions. The interconnected global social-ecological system needs to be understood as a dynamic feedback system before 'leverage points' [4] can be used wisely to shift development to more sustainable trajectories understanding the forces working in opposite directions. By understanding the landscape as a social-ecological system [5], the ecological pattern of 'land cover', its hydrological consequences and the social overlay of 'land use' and 'water use' can be analyzed as part of multiple feedback loops [6].

The nexus of forest–water–people interactions is central to the landscape as social-ecological system; it is not restricted to institutional forest definitions, but relates to actual tree cover of different quantity, quality, age and spatial pattern [7]. Many authors have noted that current drivers and pressures lead to partial-interest decisions, suboptimal and often contested consequences, including disturbances of hydrological cycle and flow regimes [8,9]. Diagnosing the issue, from plot-level soil effects to continental hydro-climates [10], is a relevant first step, but unless it contributes to decisions and actions that correct the system, it remains an academic exercise. Socially, the ecological impacts can increase conflict, aggravating land mismanagement (Figure 1). Rather than trying to redress the symptoms one by one, a more incisive analysis is needed of the social-ecological system as a whole. Change will require a concurrence of local, national and international decisions translated into actions, and as such will be helped by a shared understanding of issues, commitments to achievable goals and action at appropriate temporal (immediate improvement plus long-term results) and spatial scales.

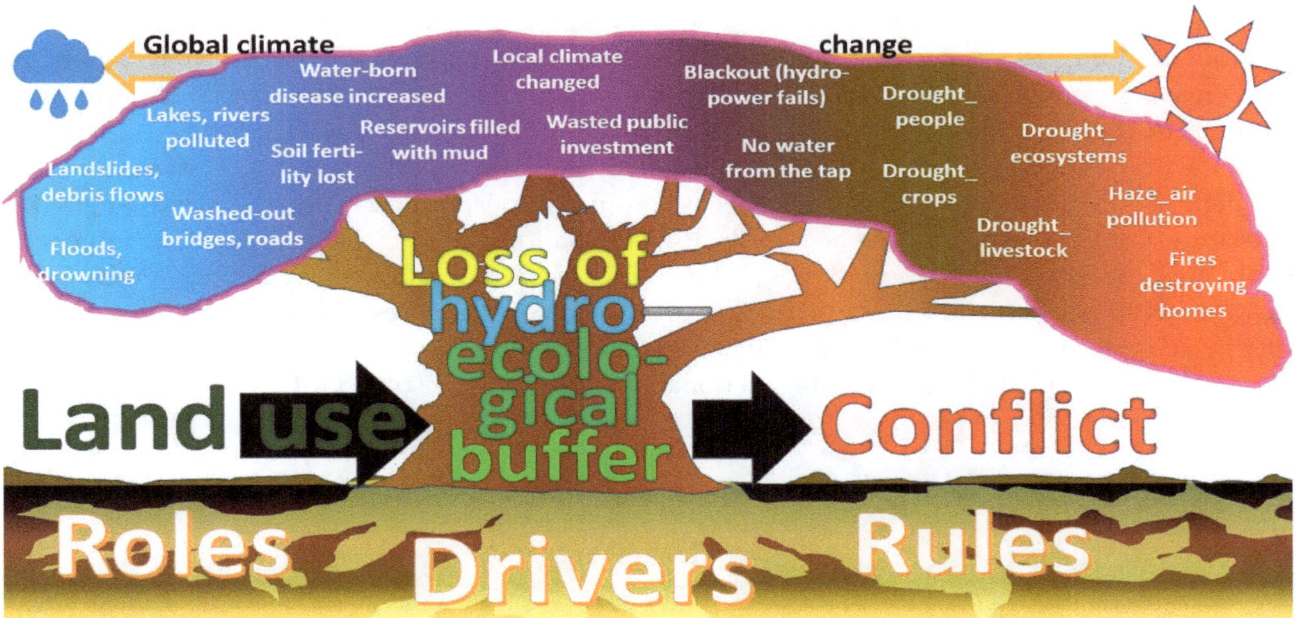


**Figure 1.** Problem tree of symptoms of disturbed hydrology (ranging from 'too wet' to 'too dry') and its underlying causes internal to the landscape, and external (including climate change).

Regional planning for resilient landscape forest-water management that can adapt to change and that can incorporate a diversity of knowledge has been shown to be more effective when collective approaches for problem-solving are employed [11]. However, while active stakeholder participation has become a main approach in problem solving and solution exploration [12–15], methods are less commonly used for supporting stakeholders to explore and evaluate alternative scenarios and facilitate (social) learning [16].

### *1.2. Use of Serious Games: Issues Arising*

An innovative participatory approach to learn about, discuss and explore the complexity of the various dimensions of complex contested landscapes to facilitate social learning among stakeholders can be best described by the keyword 'serious gaming' [17]. Serious games are emerging as a valid possible intervention tool [18–20]. In multi-stakeholder settings, games function as (social) learning tools and boundary objects to discuss local voices and concerns. A game-based approach can stimulate participants, in a safe space, to explore system behavior through scenario evaluation and to support negotiations in local contexts. Shared experience and jointly acquired knowledge in scenario evaluation games may help in the emergence of coalitions for change in the real world. In this context, playing games has shown to provide information that can support better-informed decision making [21,22].

While the use of serious games has become increasingly popular in research and development projects, there still are relevant methodological questions that remain unanswered [23].

The development and use of these games appear to suffer from a number of drawbacks that will be addressed in this paper:

(1). Games are commonly ad hoc, case dependent with poorly defined extrapolation domains for responsible use, and therefore less relevant of applicable in other contexts;
(2). Games often require heavy research investment from intervention experts to be constructed in ways that are relevant for important local discussions;
(3). Games have untested cultural limitations in where and how they can be used [24];
(4). Game users lack clarity on where and how games relate to policy making in local and/or global issue cycles, negotiations and reforms of governance instruments.

Here, we review these four identified drawbacks of serious gaming as background to the Scenario Evaluation for Sustainable Agro-forestry Management (SESAM) program (2019–2025), which includes 15 PhD research projects, geared at credible, salient and legitimate action research (Figure 2). The overall SESAM program and the individual PhD projects within it are designed to (1) be systematic in their coverage of the pantropical forest–water–people nexus in its main manifestations and issues, using generic forest and tree cover transitions as continuum description rather than forest–agriculture dichotomies; (2) use well-established hydrological, ecological, social and economic concepts to complement local empirical knowledge; (3) be cognizant of the main dimensions in which cultural contexts of inter-human and human–nature relations vary to guide responsible game use; and (4) be explicitly adapted (or adaptable) to different stages of local and global issue (policy) cycles [25], where issues become part of a political agenda, get debated and (partly) resolved (often sowing seeds for the next issue to emerge).

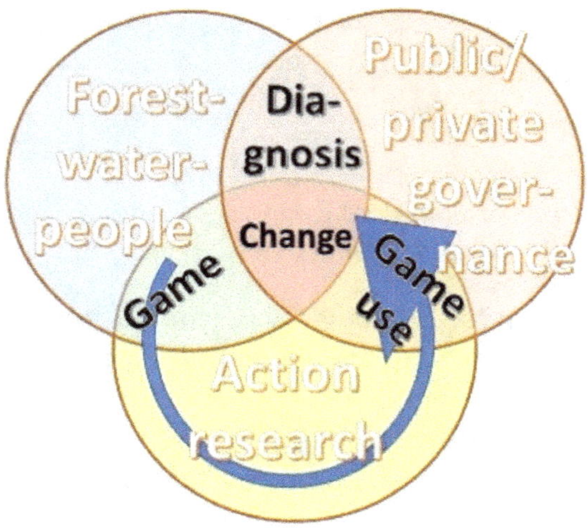


**Figure 2.** The three spheres in which SESAM will operate to ensure credible (based on current understanding), salient (actionable in terms of public and/or private governance) and legitimate (aligned with stakeholder interests) games, used appropriately.

These four targets are highly ambitious when taken one by one but tackling them jointly may prove to tap into synergies. In this early stage of confronting reality and our ambitions, we will describe (A) Steps to make games less 'ad hoc' by a selection of existing frameworks that can be used in a systemic understanding of forest–water–people nexus issues, human decision making in cultural contexts, and constructing games, (B) Link generic understanding of pantropical variation in forest–water–people configurations to the current set of landscapes/sub-watersheds that form the primary contexts and a list of emerging issues for more efficient linking of scientific, local and policy-oriented knowledge, (C) Place the set of landscapes in the known geography of cultural variation as background for exploring cultural limitations in game re-use, (D) Present a set of game prototypes that can match stages of issue cycles in policy making in local to global issue cycles, negotiation support and reforms of governance instruments.

## 2. Frameworks for Understanding Social-Ecological System Change

A central issue in todays' Sustainability Science is how knowledge, power and values interact in decision making and how long-term impacts and planetary responsibility can be woven into existing institutions. While there are hard trade-offs that lead to difficult choices in prioritizing one goal over another, existing space for multifunctionality is missed due to incomplete understanding of positive and negative consequences of land and water management decisions, oversimplified ideas of a forest vs. agriculture dichotomy (missing out on intermediate land uses with trees), 'free and prior informed consent' and inadequate stakeholder involvement. Therefore, 'game-changing' ways to involving diverse types of knowledge, plural values, multiple voices and heterogeneous stakes are needed. Science has always contributed to 'environmental issue cycles', but current concepts of "Boundary work' across the science–policy interface include roles in agenda setting, understanding patterns and processes, exploring options and scenarios, commitment to principles and implementation decisions in a complex reality of issues and concerns. Boundary work aims for salient, credible and legitimate information and understanding [26], avoiding being 'normative' in presenting insights and results. Three research traditions and their tools may need to be combined for effective boundary work (Figure 3):

1. The environmental science tradition of analyzing Drivers, Pressures, System states, Impacts and Responses (DPSIR) as multi-scale phenomena that can be studied one by one, but need to be understood jointly [27–29].
2. The natural resource governance 'issue cycle' concept [30,31] in five boundary work steps that clarify the R of DPSIR: (a) Agenda setting, (b) Better and widely shared understanding of what is at stake, (c) Commitment to principles, (d) Details of operation, devolved to (newly created or existing) formal institutions that handle implementation and associated budgets, and (e) Efforts to monitor and evaluate effects ('outcomes'); it thus relates to the 'Responses' part of DPSIR, and
3. The social side of human decision making based on Groups, Rituals, Affiliation, Status, Power (GRASP) [32], as will be further discussed in Section 5.

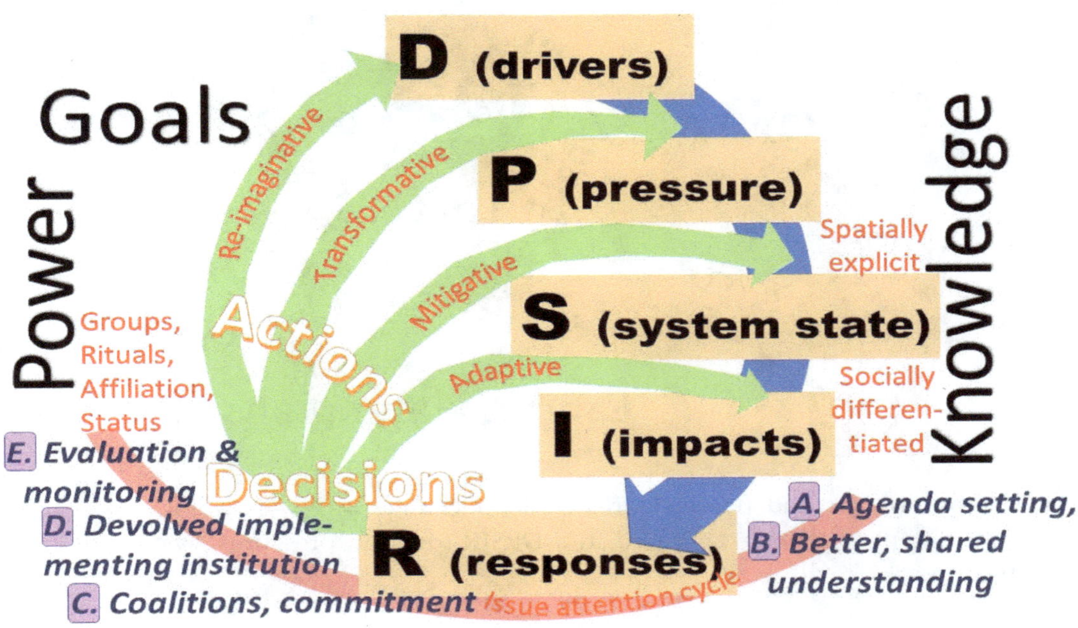


**Figure 3.** Drivers, pressures, system state, impacts and responses (also known as DPSIR) of landscape level change, with a decision cycle (**A**–**E**) closing various feedback loops interacting with knowledge, power and human sociality and its GRASP determinants.

With the 'system state' (S) of land cover/use as starting point, one can analyze the 'pressures' (P) that the landscape actors respond to (e.g., food production, income generation, health, water and energy requirements, protection from disasters), or look at underlying drivers (D) that lead to the pressures (e.g., demographic trends, international markets, national policies). One can also focus on the impacts (I) that the system state has on social and ecological well-being, or on the responses (R) that are triggered. These responses can lead to a change in the S→I relationship, the often spatially explicit P→S relationship, the often-generic D→P relationships, or challenge the drivers themselves. Such decisions can be labelled as adaptive, mitigative, transformative and re-imaginative, requiring increasingly drastic change in the existing social-ecological system (deeper roots of the problem tree in Figure 1). While the 'adaptive' decisions can be taken individually or in small groups, the others depend on collective action (especially the mitigative ones) and policy-level institutional change.

Basic questions in any landscape approach [33] are the 'Who?', 'What?' and 'Where?' of current land use, in a temporal perspective ('When?'), as a start for exploring 'Why?' questions of drivers and pressures. Impacts ('So what?') are the entry point to 'Who cares?' stakeholder analysis, and opportunities to close the loop if those who care can directly change the who/what/where of land use, or at more fundamental level the why of drivers and pressures. These questions lead to a further elaboration of the DPSIR framework defining the breadth of the type of drivers and pressures that we may need to understand on the human side of the eco-hydrological-social system that deals with the forest–water–people nexus issues. (Figure 4).

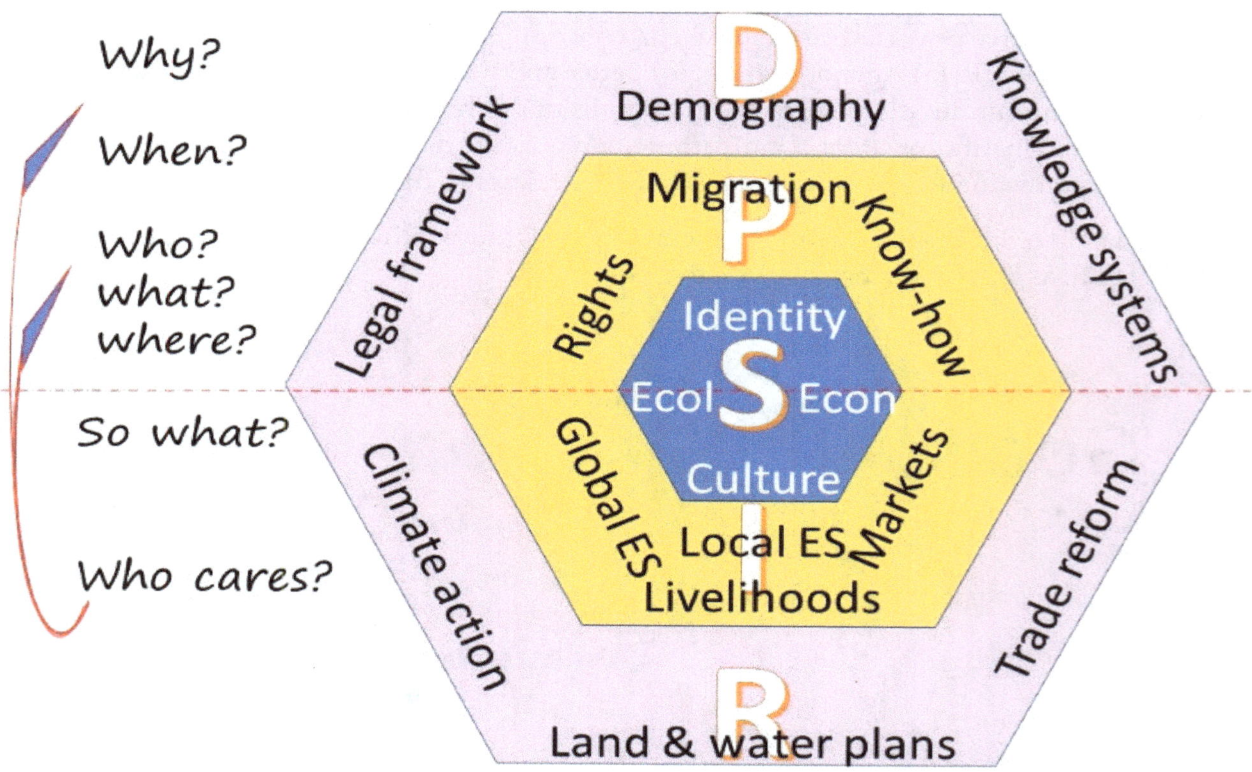


**Figure 4.** State variables and processes relevant for change in eco-hydrological-social systems in the forest–water–people nexus grouped in a DPSIR (drivers, pressures, system state, impacts, responses) framework.

The System-state variables (at the heart of DPSIR) deal with ecosystem structure and function on the (hydro-) ecological side, and with identity, (collective) culture, and economic indicators on the social side (Figure 4, center of the diagram). On the Impacts side, the most relevant indicators can be grouped as livelihoods and local ES, Markets and global ES. Among the pressures, migration (rather than birth rates), rights and know-how, relate to underlying drivers that include legal frameworks, demography and knowledge systems. Responses can include land and water use plans (linked to rights within a legal framework), trade reform and climate action, emphasizing relations between the local system and global change. Part of the latter are attempts to make incentive structures at the land-user level more aligned with 'downstream' impacts, via Payments for Ecosystem Services (PES) [34]. The recent PRIME (Productivity, Rights, complementary Investments, Market access and Ecosystem services) framework of the World Bank that has helped conceptualize forests' contribution to poverty reduction and guide intervention design [35] is compatible with the middle hexagon in Figure 4. Further conceptual framing will be presented after the portfolio of landscapes are briefly introduced.

## 3. Representativeness and Diagnostic DPSIR Analysis of the SESAM Landscape Portfolio

Brief descriptions of the 13 landscapes in which SESAM will diagnose issues, develop games and test their relevance for local decision making are provided in Appendix A. In this section, we will consider the degree to which the portfolio can be expected to include major variation in pantropical manifestation of forest–water–people nexus issues and apply a DPSIR analysis to identify the major issues at stake.

A pantropical typology of forest–people interactions [36] has proposed six recognizable stages of a 'forest transition', starting with landscapes with >80% of core forest and human population densities below 1 to over 1000 $km^{-2}$. All forest transition stages can be found in any of five ecoclimatic zones, that link to a forest water analysis, and are based on the ratio of rainfall (precipitation, P) and potential evapotranspiration ($E_{pot}$). This typology separates drylands and semi-arid zones ($P/E_{pot} < 0.5$; 35% of tropical area, 20% of people) from a dry–sub-humid zone ($0.5 < P/E_{pot} < 0.65$; 10% of area, 11% of people). On the wetter side, we can distinguish (relatively) high 'water towers' with $P/E_{pot} > 0.65$ and generating streamflow to lower, drier parts of the same watershed (11% of area; 15%of people), a lowland (non-water tower) humid forest zone ($0.65 < P/E_{pot} < 0.9$; 19% of area, 22% of people) and a per-humid lowland forest zone ($P/E_{pot} > 0.9$) with 25% of pantropical land area and 32% of its human population.

Human life in the drier zones depends on rivers (or groundwater flows) that originate in wetter areas, especially in the 'water tower' configuration where the wetter part is higher in the landscape. All eco-climatic zones can contain all stages of the forest transition, although in the drier parts of the tropics climate and human agency are not easily disentangled as causes of low tree cover vegetation. Forest transitions are defined by a phase of declining cover of (natural) forest, an inflection point and a phase of increasing forest cover (secondary forest in landscapes with rural land abandonment and urbanization, and, more commonly, planted forests).

The pantropical classification at sub-watershed level [35] forms a basis for judging pantropical representativeness of our SESAM landscapes. Summary statistics of the various SESAM landscapes (Table 1) show that the landscapes include all five ecoclimatic zones (dryland/semiarid; dry-sub-humid; water tower; humid per-humid lowland) and all six forest transition stages. On further analysis, the forest–water–people nexus issues in each of the landscapes involve plot-level aspects that relate to lack or presence of trees, hillslope and watershed hydrological relations (and associated water rights), and policies for agriculture, forestry, land-use rights (Table 2). These three issues match the three current agroforestry paradigms at plot/farm, landscape and policy scales, within a common definition of agroforestry as the interaction of agriculture and trees, including the agricultural use of trees [37].

**Table 1.** Basic properties of SESAM landscapes; W = water-shed, R = Region, C = coastal zone.

| Context | Location | Coordinates | Hydro-Climate | Mean Annual Rainfall, mm | Human Population Density, km$^{-2}$ | Forest Cover, % | Forest Transi-tion Stage | Scale: | Area of Focus, km$^2$ |
|---|---|---|---|---|---|---|---|---|---|
| Core forests, upriver | 1. Suriname upriver | 3–4°N, 54–56° W | Per-humid | 2700 | <1 | >70 | 1 | W, R | 7860 |
| | 2. Madre de Dios, Peru | 12°36′ S, 69°11′ W | Per-humid | 2221 | 1.3 | 95 | 1 | W | 85300 |
| Mangrove coast | 3A. Nickerie, Suriname | 5°51′N, 55°12′W | Per-humid | 1800 | 6.4 | 90 | 1 | C, District | 5353 |
| | 3B. Paramaribo, Suriname | 5°56′N, 57°01′W | Per-humid | 2210 | 1297 | 19 | 3 | C, District | 182 |
| Agroforestry mosaic | 4. Tomé Açu, Pará, Brazil | 2°25′S, 48°09′W | Per-humid | 2371 | 10 | 37 | 2 | Municipality | 5145 |
| Coastal peatland | 5. Ketapang, Indonesia | 1°27′–2°0′S, 110°4′–110°8 E | Per-humid | 3169 | 20 | 34 | 2 | Peatland Hydrolo-gical Units | 948 and 1048 |
| Mountain lakes | 6. Singkarak, Indonesia | 0°30′–°45′N 100°20′–100°43′E; | Water-tower | 1700–3200 | 338 | 16 | 3 | W | 1135 |
| Watertower/Semi-arid gradient | 7. Mount Elgon, Uganda | 01°07′ 06″ N, 34°31′ 30″ E | Water-tower | 1600 | 355 (up),606 (md),870 (lw) | 25 (up), 63(md) 36(lw) | 4 | W | 4200 |
| | 8. Ewaso Ng'iro, Mt Kenya | 0°15′S–1°00′N, 36° 30′– 37°45′ E | Water-tower | 600 (lw), 1600 (up) | 150 (up), 12 (lw) | 18 | 4 or 6 | W | 15,200 |
| Mountain farms | 9. Andes, Peru | 15°50′S, 70°01′W | Dry-sub humid | 700 | 5.1 | 0.1 | 5 | W | 8490 |
| Dryland farming | 10. Mossi plateau, Burkina Faso | 12°45′– 13° 06′N, 0°.99–1°33′W | Semi-arid | 400–700 | 148 | 10 | 5 or 6 | Village territory | |
| Agroforestry mosaic | 11. Kali Konto, Indonesia | 7°45′–7°57′S, 112°19′–112°29′E | Water tower | 2995–4422 | 453 | 20 | 6 | Sub-watershed | 240 |
| Water towers under pressure | 12. Rejoso, Indonesia | 7°32′–7°57′S, 112°34′–113°06′E | Water tower | 2776 | 414 (up), 693 (md), 1925 (lw) | 11 | 6 | W | 628 |
| | 13. Upper Brantas, Indonesia | 7°44′–8° 26′S, 112°17′–112°57′E | Water tower | 875–3000 | 1042 | 24 | 6 | Sub-watershed | 180 |

**Table 2.** Drivers, pressures/system state and impacts across the SESAM landscapes, based on descriptions in Appendix A.

| Location | Drivers of Land-use Change | Pressure/System State: Plot-Level Land Use | Pressure/System State: Landscapes, Watersheds | Pressure/System State: Policy Interactions | Impacts on Forest–Water–People Nexus |
|---|---|---|---|---|---|
| 1. Suriname upriver | Pressures: Shifting cultivation, logging, road infrastructure, encroaching gold mining. Drivers: income generation, weak law enforcement. | Agriculture in the landscape is basically all shifting cultivation. Traditionally within the plot there are more crops and only a few trees (usually palm-fruit trees). | A multifunctional landscape with agriculture, forestry, nature-based tourism Good water quality, availability of drinking water and maintaining water levels in the streams and rivers for agriculture. | Forestry is the better regulated sector than agriculture sector No sector-coordinated policy. Low capacity in district government institutions. | High vulnerability of rain-fed agriculture and water security, forest degradation due to logging and shifting cultivation, increasing deforestation in the Guiana Shield. |
| 2. Madre de Dios, Peru | Illegal gold mining and informal agricultural expansion along the recently paved Inter-Oceanic Highway. | Diverse agroforestry systems are expected to increase hydraulic redistribution, soil macroporosity and in general to represent a sustainable alternative to gold mining. | Degraded or deforested areas, when dedicated to farming, can be agro(re-)forested to increase the tree cover at landscape level while improving socioecological resilience. | Regional governments are supporting market-oriented low-diversity agroforestry systems as an alternative to the illegal gold mining and slash-and-burn farming. Land-use planning does not take the impact on watershed functions into account, and vulnerability to droughts (fires). | Increase in drought and flooding episodes, drought-related fires, mercury contamination of rivers resulting from gold-mining activities. |
| 3. Mangrove coast, Suriname | Drivers: Population growth (urbanization), income generation (for different actors such as fishermen, tour operators), poor law enforcement. | The mangrove forest along the coast prevent erosion and its ecosystems also serves as a potential water source for the nearby agriculture land mainly used for rice cultivation. | Unsustainable use of the mangrove ecosystem services and the removal of mangrove trees for various purposes (building infra-structure, housing, etc.) resulting in saltwater intrusion, reducing water quality land inwards. | There is poorly integrated coordination among the different stakeholders (direct and indirect users) to foster mangrove conservation. | Mangrove forest degradation affecting coastal resilience. Excessive and unsustainable use of mangrove ecosystem service |
| 4. Tomé Açu, Pará, Brazil | Poor environmental governance and enforcement, migration, rural poverty. | Logging, extensive grazing, slash and burn farming (cassava and annual crops), monocrop oil palm. | Oil palm drives contamination of waterways through chemical fertilizer/pesticide use and mill effluents. Fire, low-yielding extensive grazing and land degradation influence water flow regimes. | Lack of coordination between various levels of government (municipal, state and federal) and lack of landscape-level land-use planning. | Oil palm drives contamination of waterways through chemical fertilizer/pesticide use and mill effluents. Different land-use types have different implications on water flows at the plot and landscape scale. |

**Table 2.** *Cont.*

| Location | Drivers of Land-use Change | Pressure/System State: Plot-Level Land Use | Pressure/System State: Landscapes, Watersheds | Pressure/System State: Policy Interactions | Impacts on Forest–Water–People Nexus |
|---|---|---|---|---|---|
| 5. Ketapang, Indonesia | Increase in number of migrant communities and expansion of plantations in peat-swamp forest that led to massive canal construction. | Maintain soil infiltration for water level in the agricultural area in the peatland. Control drainage (smart 'canal blocking') for more constant water levels. | Converting the burnt areas surrounding protected forests into agro-forestry system will restore the functions as well as improving local peoples' livelihood. Saltwater intrusion in coastal zone when peat domes are drained. | Managing the land-use planning in peatland area based on peat depth and characteristics: Areas with peat depth > 3 m have protected forest status; with peat depth < 3 m plantation or agroforestry systems are allowed; with peat depth < 0.5 m and sapric peat (open-field) agriculture is allowed. | Haze episodes from the forest fire of the degraded peatland ecosystem, as an impact of decreasing groundwater level during dry season. |
| 6. Singkarak, Indonesia | Highly intensified agriculture, population increment, urbanization. | Soil erosion occurs in the agricultural land (highly intensive horticulture) which is located in the upstream and hillslope area. | Land-use change into highly intensive agri-culture and residential in upper basin impact on water quantity and quality of the lake. | There is no single integrated authority for watershed and lake management. | Water resource, forest and land-use change impact on the lake (water quantity, quality and biodiversity), impact of climate variability (dry years) on river basin. |
| 7. Mount Elgon, Uganda | High population, High poverty levels (low income generating activities), Land fragmentation (land tenure system), Favorable Climate, Urban extension, food gap, Conflicting policies. | Due to the topography of the region, soil erosion is common; Planting shade trees in existing coffee fields/systems controls soil erosion and boosts coffee production hence enhancing and sustaining crop yield and food security. | Fragmented forests due to population growth and increased agricultural activities; Subsistence farmers cultivating wooded areas and practicing agroforestry (with other crops and coffee); The degraded soil/land needs to be rehabilitated in order to promote ecosystem services of the mountainous forests (East Africa's water tower). | Empowering the local agroforestry communities and cooperatives to plant more trees; Supporting payments to local communities to avoid deforestation and restore forest inside the park; Joint environment policy Implementation—as community motivation to encroach into park forest is dependent on policy, commodity prices, law enforcement and political interests; Enforcement of forest/environment bylaws, and resource use agreements. | LULC changes—High deforestation levels (urban area extension and agricultural land expansion); Upstream–downstream conflicts (decreasing rivers base flow); Human encroachment (national park, riparian zones of riverbanks and swamps); Riverbank degradation and Land degradation (soil erosion and declining soil fertility); Seasonal downstream floods and landslides |

**Table 2.** *Cont.*

| Location | Drivers of Land-use Change | Pressure/System State: Plot-Level Land Use | Pressure/System State: Landscapes, Watersheds | Pressure/System State: Policy Interactions | Impacts on Forest–Water–People Nexus |
|---|---|---|---|---|---|
| 8. Ewaso Ng'iro, Mt Kenya | Increasing irrigation water demands, increasing human population, changing weather patterns (erratic rainfall, prolonged droughts), poor governance, political interference. | Sustainable Land Management practices that reduces soil erosion and increase water infiltration. The area under water-demanding crops affects irrigation de-mands. Increasing rainwater harvesting and storage at plot level would reduce water demands. | Land cover/use changes at watershed level affects water retention/water yields Climate change and variability (P, PET) affects distribution of water resources Excessive abstraction of river water upstream affects river flows downstream. | Watershed planning and governance needs to be improved by capacity building community level structures such as Water Resources Users Association (WRUA) in Kenya. Irrigation water management efficiency is constrained by lack of knowledge, understanding and technology access for farmers. | Uncontrolled abstraction of water, human encroachment (e.g., farming in riverbank protected/riparian areas, limited downstream flows (dry riverbeds); Poor enforcement of policies (metering, riparian corridors), local politics versus national government interests, climate change/variability |
| 9. Andes, Peru | Agricultural boom: quinoa production for export supported by cooperatives and NGOs. | Highland agroecological zone with good soil organic matter. Depending on the on exposure, soils and drainage, agricultural activity remains in the high altitude. | A multifunctional landscape with agroecological practices quinoa with their crop wild relatives and diverse activities as silvo-pastoral systems. | This landscape is recognized from the United Nations Organization as Globally Important Agricultural Heritage Systems (GIAHS) valorizing the ancestral systems of cultivation of quinoa. | Increased drought, soil fertility loss, loss of cultivated diversity—the export market only includes few quinoa varieties. |
| 10. Mossi plateau, Burkina Faso | Demography, climate variability and land degradation | Onset of rainy season, soil quality, infiltration, crop water use, yield formation. | Overland flow capture, Village level transfers of biomass, (fodder), farmer groups. | Land and water use rights, local institutions for collective action, grazing management. | Low and unstable biomass production. |
| 11. Kali Konto, Indonesia | Agricultural and dairy products demand (market), poor land management, and population increment. | Intensive agricultural farming with minimum tree cover in hillslope increase soil surface exposure, increase on livestock. | Land-use change from forested area to open field (including grassland for fodder), reservoir siltation due to high sedimentation which affect water quality and quantity. | There is lack of integrated coordination among stake holder (government, farmers, and water beneficiary) for achieving sustainable watershed conservation effort. | Increase on horticultural area, soil fertility lost, landslide (debris flows), reservoir filled with muds. |

**Table 2.** *Cont.*

| Location | Drivers of Land-use Change | Pressure/System State: Plot-Level Land Use | Pressure/System State: Landscapes, Watersheds | Pressure/System State: Policy Interactions | Impacts on Forest–Water–People Nexus |
|---|---|---|---|---|---|
| 12. Rejoso, Indonesia | Population growth (all), changes in forest and agricultural crop commodity prices (all), deforestation for horticulture (upland), rock and sand mining (midland), groundwater exploitation (lowland) | Upper zone erosive Andisols used for intensive vegetable production with low tree cover. Local communities are not allowed to use forest resources or use forest areas for farming and raising livestock; Community only carries out agricultural activities outside the state forest area. | Local communities are involved in planting trees allowed to farm among tree stands (intercropping) in production forests with the rules of cooperation for the results. Negative impacts of forest conversion to agri-culture; increasing tourism, stone mining, uncontrolled water use for paddy rice | The implementation of recent Community Based Forest Management Programs Collaboration between stakeholders A PES (Payment for Ecosystem Services) scheme began to be implemented | Land conversion and commodity change without planning, reduced flowing springs, drought in the dry season, flooded and landslides in the rainy season, an explosion of agricultural pests almost every year, overexploiting downstream use of bore wells and potential conflicts, environmental damage due to central rock mining and potential conflicts. |
| 13. Upper Brantas, Indonesia | Population growth (0.95%/year), urbanization; upland vegetable markets, tourism industry. | Upland vegetable area, degraded state-owned forest soil erosion and sedimentation, intensive use of fertilizer and chemical pesticides. | Gradient land-use change from grand forest protected area, production forest, upland vegetable, settlement and tourism are. Deficit water balance in dry season triggering water conflict. | There is no coordination and comprehend Policy of Land and (Ground) Water Resource. | Land-use change, ground water and deep well extraction, water user conflict, flood, water supply and demand. |

## 4. Unpacking the Forest–Water–People Nexus

A large number of 'issues' in the case studies relate to water and need to be understood in their landscape context (Figure 5). They all relate to the way water flows from mountains ultimately back to oceans, unless it follows the atmospheric 'short cycle' downwind route back to rainfall over land. Along the streamflow pathway, water changes in quality (sediment load, pollution), quantity (annual water yield to reservoirs or lakes), and flow regime (regularity of flow, flood risks), affecting settlements, agriculture, fisheries and human health. A major theme is the upstream versus downstream (with the 'water tower' configuration of freshwater sources for lowlands specifically challenging), complemented by upwind versus downwind for rainfall recycling, and specific features such as riparian wetlands and forests, peat swamp forests and mangroves protecting the coastal zone, while supporting marine fisheries. The scientific debate on the mechanisms and patterns of 'biological rainfall infrastructure' is undecided [6,38]. While the 'biotic pump' theory suggests forests cause the flow of moist air towards them, observations of windspeeds show that rainforests are associated with relatively low wind speeds that imply the 'short cycle' can remain relatively local, with transport distances during the mean atmospheric residence time of 8 days of hundreds rather than thousands of kilometers [39].

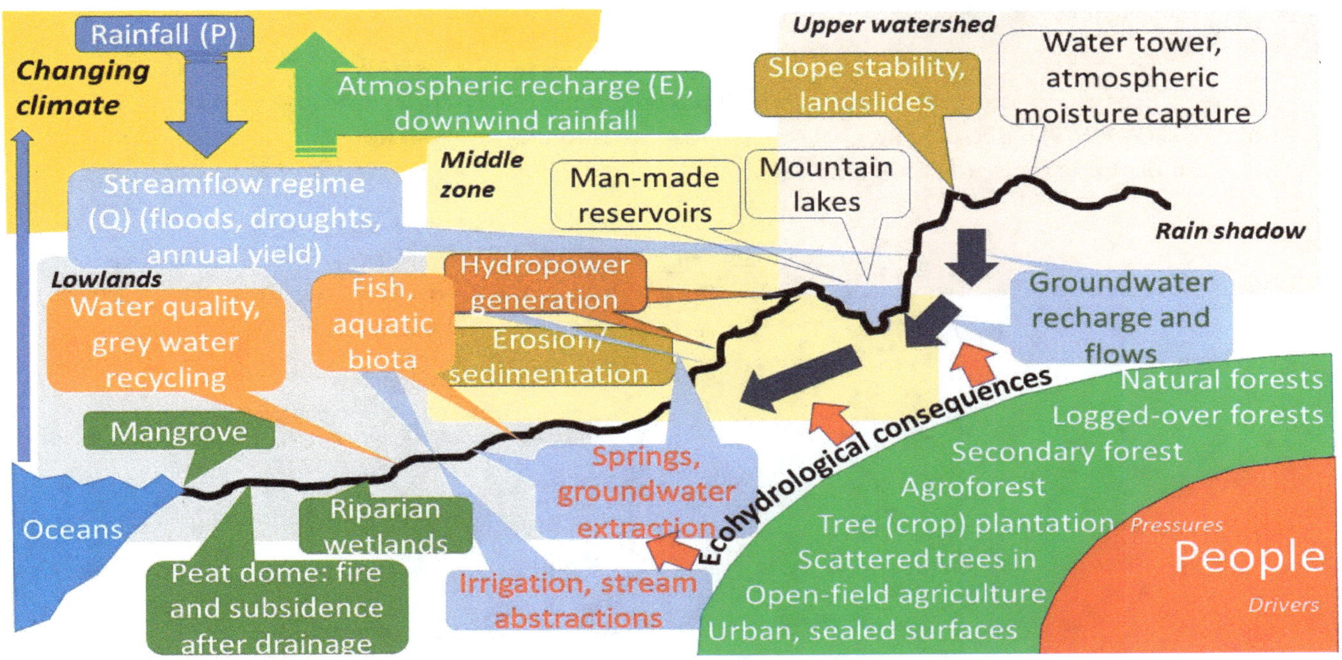


**Figure 5.** A first synthesis of eco-hydrological relations across the various landscapes in Table 2.

Where issues relate to forests and/or trees outside the forest, the cascade of processes that start with rainfall, are followed by canopy interception, infiltration into the soil or overland flow and contribution to local water storage and/or direct response of streams and rivers (Figure 6). Ecosystem structure interacts with hydro-ecological processes, jointly shaping a set of 'ecosystem services' that relate to benefits humans derive from well-functioning systems. A typology of nine such 'watershed functions' are indicated in Figure 6 and described in Table 3. These functions (or 'services') can be used to analyze site-specific differences between land cover types, relating the actions of roots in soil to landscape-level streamflow [40].

The principal concept in all of ecohydrology is that of the water balance, where input (precipitation P or snowmelt in cooler parts of the world) is related to two main pathways out of the system: back to the atmosphere via evapotranspiration (E) and streamflow Q (surface or groundwater flows) and to changes in the 'buffer' of water held inside the system (Figure 7). At timescales of a year or more this buffering can be ignored (except for interannual climate variability), but at the timescale of a rainstorm it is key to reduce downstream flooding risk [41,42].

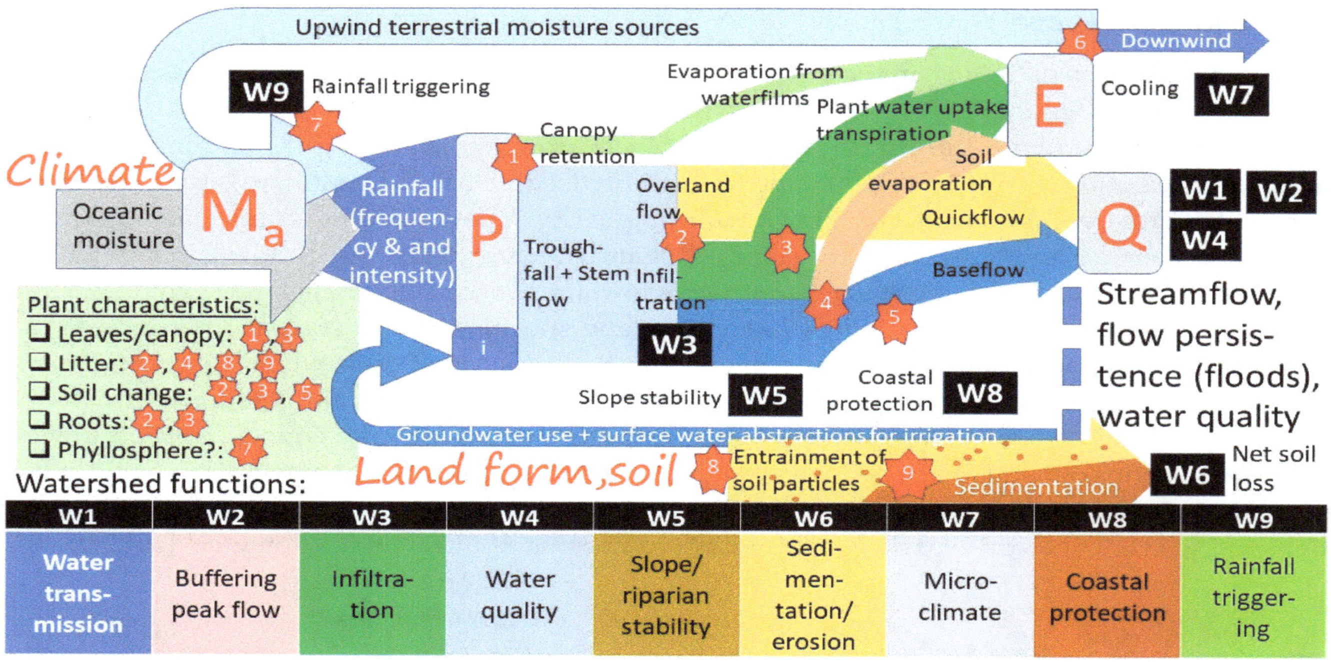


**Figure 6.** Ecosystem structure (landform, soil, plant characteristics), processes (**1–9**), and 'watershed functions' (**W1–W9**) based on water balance: P = precipitation, Q = stream flow, E = evapotranspiration, $M_a$ = atmospheric moisture, i = irrigation.

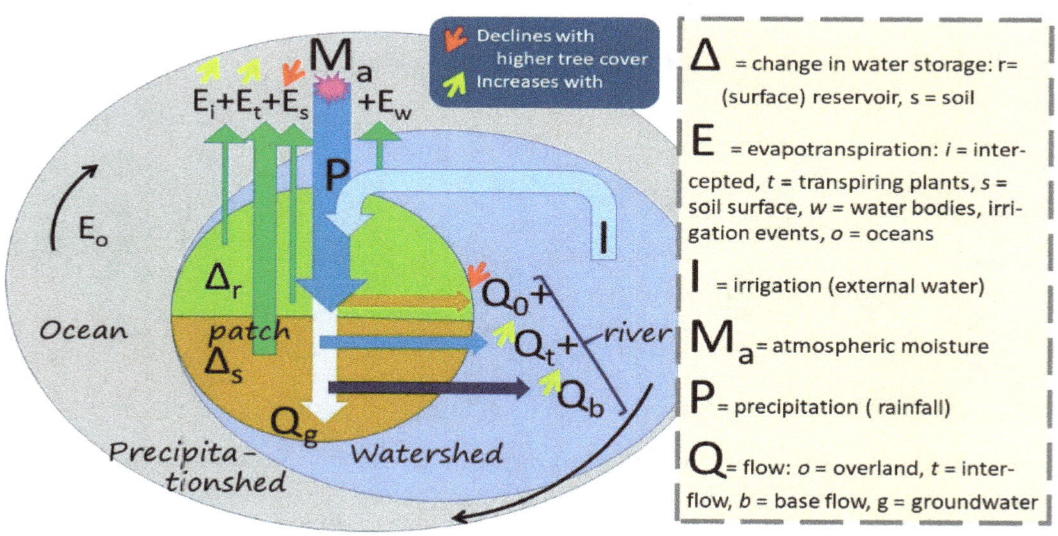


**Figure 7.** Basic elements of a water balance at patch, watershed and precipitationshed scale levels, that can form a biophysical basis for forest–water–people nexus games, with expected changes in terms with increasing/decreasing forest cover.

We can now relate a typology of 'watershed functions' (**W1–W9**) to an understanding of ecosystem structure and function, across the various landscapes (Table 3). Decisions on adding landscapes to the SESAM portfolio were made with this representation in mind. Probably the most contested in this list is W9, the influence of vegetation on rainfall regimes at landscape and continental scales [38,43–46]. Three of the Latin America landscapes are at the start of the Amazon rainfall recycling system, one is at the receiving end, and one in the much drier Andes range that depends on left-over atmospheric moisture. The two East African water towers are part of the complex hydro-climatic system, influencing rainfall further North (including to some extent that in the Burkina Faso landscape, along with water coming in from the Atlantic Ocean).

**Table 3.** Reference (V, major, or v, relevant) to nine 'watershed issues' across the various landscapes [47,48].

| Locations | Watershed Functions | | | | | | | | |
|---|---|---|---|---|---|---|---|---|---|
| | W1 | W2 | W3 | W4 | W5 | W6 | W7 | W8 | W9 |
| | Water Trans-mission | Buffering Peak Flow | Infiltra-tion | Water Quality | Slope/ Riparian Stability | Sedi-men-tation/ Erosion | Micro-Climate | Coastal Protec-tion | Rainfall Trigger-ing |
| Upper Suri-name River | v | | V | | | v | V | | v |
| Madre de Dios, Peru | v | | V | V | | | V | | V |
| Mangrove coast, Suriname | | | | V | | | V | V | |
| Para, Brazil | | v | | V | | v | V | | v |
| Ketapang, peatland | v | V | V | V | | | V | v | |
| Singkarak Lake | v | V | | V | V | v | | | |
| Mount Elgon | | V | | V | V | V | | | |
| Mount Kenya | v | V | V | | | v | V | | |
| Peruvian Andes | v | | V | | | v | | | |
| Mossi plateau | | | V | V | | | V | | |
| Kali Konto | | v | V | V | | v | V | | |
| Rejoso Watershed | v | V | V | V | V | v | | | |
| Upper Brantas | | V | V | V | V | V | | | |

Beyond the 'provisioning' and 'regulating' services captured in W1 ... W9, there also are 'cultural' services based on 'relational values' between Humans and Nature. The two interact when visits by domestic (or international) tourists to the mangroves of Nickerie district in Suriname or the mountain resorts in East Java increase pressure on freshwater resources. At the global scale the 'Ecosystem Service' (ES) language is currently embedded as 'instrumental values' in a wider 'relational values' paradigm on 'Nature's Contributions to People' (NCP) [49]. The way cultural relations with a landscape and its forest and water aspects is expressed varies with history and religion. The springs in the upper Brantas still are sacred places, where mountain spirits are brought offerings to secure the gift of fresh water continues. On densely populated Java, the *wayang* tradition re-tells stories of the past. All stories start with the *gunungan* or mountain symbol (Figure 8), a tree of life connecting forest animals and creatures to people's homes and lives. The flip-side, shown occasionally for dramatic effect, shows demons and fire: social conflict and mismanaging human–nature relations destroys human livelihoods. Throughout human history, perspectives on spirits, deities, personified nature or a single Almighty have been described in metaphors of words that also describe human–human relations in terms of (A) family, such as ancestors, (grand-) parents (Mother Nature), siblings, offspring, partners, in-laws), (B) neighbors, friends, business partners, (C) adversaries, competitors, armed attackers and defenders, (D) servants or (E) educators. A subset of these relations (the providing mother of NCPs and ES servants) can be interpreted as 'instrumental', directly supporting human goals and objectives, but even those imply that there will have to be a two-way (rather than unidirectional) relationship to maintain or support what is relevant to people because it cannot be taken for granted [50].

Across the various landscapes we can now understand that debates about increasing/decreasing forest cover and/or tree cover in agroforestry or other land uses, can have both instrumental and wider relational aspects. Technical solutions to local issues may increase problems if they do not lead to shared understanding and negotiated trade-offs. Such effects will have to be represented in applicable games at a reasonable level of detail and accuracy, and in their cultural context.

**Figure 8.** Instrumental as part of relational values of nature from a human perspective as exemplified in the Javanese *gunungan* representation with its idyllic harmony and fearful disaster side, and the UN Sustainable Development Goals; NCP = Nature's contributions to people, PCN = People's contributions to nature.

The immediate issues of public concern to which landscape management contributes (or can help solve) are very diverse among the SESAM portfolio, even though they all relate to the same water balance. Upstream on the Suriname river, increasing deforestation (gold mining) and forest degradation (logging) in the Guiana Shield will need to be managed in a better way. Increasing food security, and water security during the long dry season, depends on maintaining and restoring the flow buffering aspect of the river, along with sustained biodiversity as a basis for other forest ecosystem services beneficiaries depend on for subsistence and income. In Madre de Dios (Peru) focus is on reduced water shortages for agriculture and household use (including fruit production) in the dry season, but also on fire prevention during the dry season. In Pará State, water recycling is reduced by deforestation and land degradation and water quality undermined by chemical inputs used on large-scale monocrop plantations, pointing to the need for restorative and climate-smart agroecological practices including agroforestry. In the Ketapang peatlands (Indonesia), reducing public health impacts of haze episodes due to peat-swamp forest fire may align with an interest in reducing agricultural damage due to flooding during the rainy season by restoring the water storage capacity of peat domes and reducing the speed of drainage. Restoring the mangrove forest in Suriname will protect the coastal zone from saltwater intrusion and floods, while the mangrove ecosystem provides services to coastal fisheries, beyond its direct products.

Water resources management in water towers of Kenya and Uganda involves better crop selection, caution in increasing middle-zone tree densities and attention to water distribution for downstream users. It also involves reducing landslides, minimizing population displacement and deaths, loss of fertile land and famine. Reducing flooding in the downstream area during the rainy season can go hand in hand with reducing drought (severe dry Spells).

In the densely populated Brantas river basin in Indonesia, the targets are water security and water quality for domestic use and tourism, and irrigation water for dry-season paddy rice production (three or even four crops per year). Similarly, around Lake Singkarak there's a need for increasing water availability for irrigation and hydropower during the dry season, but also a need for maintaining the local endemic fish habitat due to maintained water quality. In the Rejoso Watershed, beyond these points, reducing water use in lowland rice paddies may be needed to secure groundwater for a nearby metropole. Before using these insights, however, the cultural context needs to be considered as it may call for different ways of constructing games, despite hydro-ecological similarity.

## 5. Cultural Diversity in Response to Forest–Water–People Nexus Issues

The response options chosen in any society and any point in time are under the influence of culture, as a layer between generic human nature and individual personality. Research on geographic variation in culture at any point in time (Figure 9A) and cultural change as part of economic development can be reconciled in identifying at least two (but up to six) main axes of variation [51]. In the simplest portrayal (Figure 9B) a collectivism–individualism axis aligns with distance to the equator, and a monumentalism-flexibility axis on which Latin America and East Asia are the bookends.

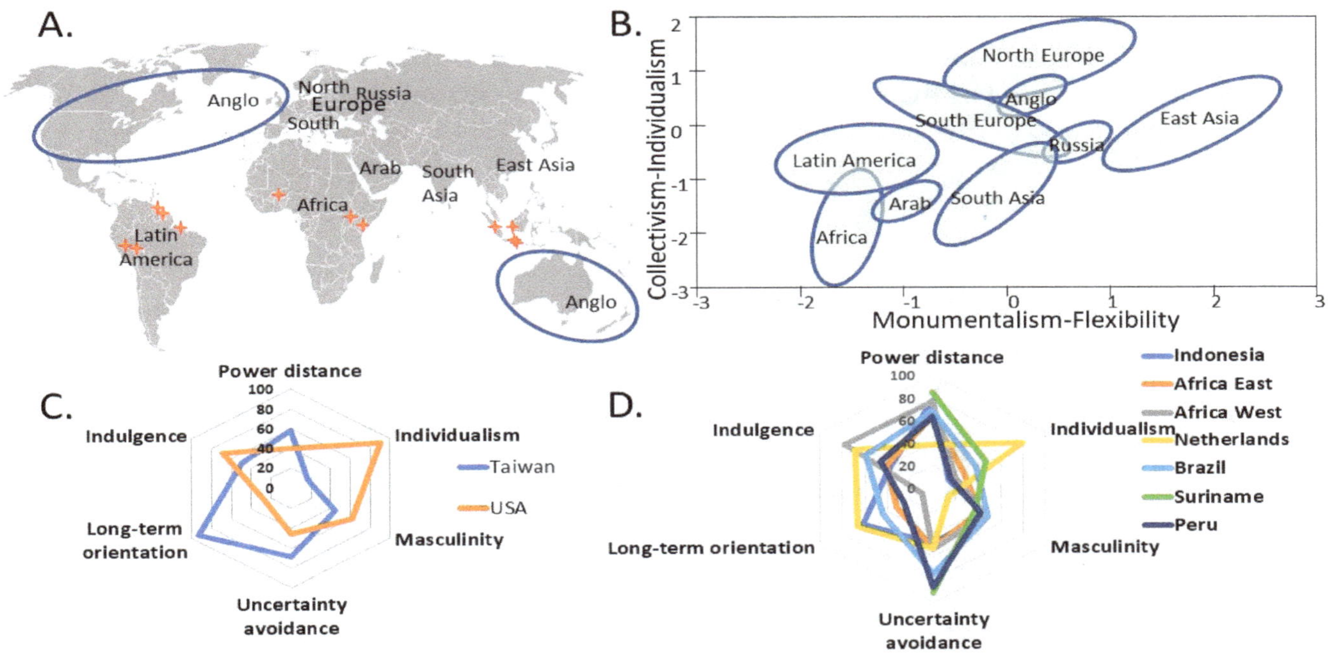


**Figure 9.** Geography of cultural dimensions: (**A**) Global map with red signs indicating landscapes where SESAM is active; (**B**) Two main axes of cultural variation and the relative position of countries (based on national surveys) and country groupings [52]; (**C**) Comparison of six Hofstede dimensions for two contexts in which the 'so long sucker' game was played; (**D**) Hofstede dimensions for the countries (or where such data are lacking regions) involved in SESAM [53].

People follow any of four directions in addressing any (new) issue, with preferences depending on culture [54]: hierarchy (clarity on power in decision making), unleashing private initiative (aiming for efficiency and public benefits via an 'invisible hand'), (perceived) fairness of social outcomes and (public) transparency (accountability, anticorruption). Forestry issues follow this general pattern around the globe. The high degrees of private (agro)forest ownership in Scandinavian countries reflect a low sense of hierarchy and belief in private initiative as opposed to a history of centralized control in much of the (formerly colonized) tropics. Recent responses to the COVID-19 pandemic may confirm basic cultural patterns of relying on central authority versus citizen responsibility, and in orientation on long-term goals versus immediate gratification.

A simulation game of an agroforestry landscape is a model of a socio-ecological system, clarifying components, actors, interactions, roles and rules. A run of such a game puts that system into action. Some of the components will have been designed, while others are implicitly embodied in the participants, making the game dependent on the players. This mixed composition is crucial for how the game unfolds. In doing so, we may need to refer to the six culture dimensions identified in [52,55], although the axes are only partially independent of each other in a statistical sense.

As an illustration of the way this matters in how a single game can be interpreted differently, consider some intercultural experiences with the game 'So Long Sucker' [56]. This board game allowed four players to form coalitions, rapidly eliminating one another until one remained. It had been designed in the USA to show how an incentive structure can lead to selfish behavior. When a class of Taiwanese came for a visit and groups from this class played the game, they managed to turn the dynamics around, using exactly the same rules to create a sustained group of participants that played for hours until the facilitator intervened. The reason for this is that the Taiwanese used different unwritten rules, implicit in their culture, than the US students were using, or the US designers of the game had anticipated. This story reminds us that one game may not fit all cultures [57]. Let us look at these cultural differences, since they may hold a message for SESAM. The USA and Taiwan have cultures that are widely apart (Figure 9C) on the six dimensions of culture [52].

For the US players, of individualistic and short-term-oriented mindset, the spirit of the game was "each one for himself and the devil take the hindmost". There was no moral penalty on kicking out your fellow players. For the Taiwanese partners, excluding someone from the game would be morally wrong and also imply a destroyed future of the game. This made the difference in behaviors of both types of players rational from their perspective. For SESAM we need to distinguish the inherent properties of a game from the multiple ways it can be used (re-interpreted) in various contexts.

A further example of the way games can be understood in different ways depending on cultural context emerged when researchers adopted a game originally developed for university students in Sudan to understand land degradation due to interaction of grazing pressure and erratic rainfall to farmers in N. Ghana [58]. The game centers on cattle, grass and watering points, but missed out on an important consideration of managing grazing pressure in Ghana: Guinea fowl like short grass, and especially female farmers found that the lack of these animals in the game missed opportunities for their land management ideas to emerge. The game became a 'boundary object' for debating gender balance in the local context, as much as discussing climate change adaptation.

## 6. Action Orientation: Game Typology and Prototypes

The game development process is conventionally divided in four major steps: Step 1—Baseline study of the local landscape and context, diagnosing key issues, Step 2—Game development, Step 3—Game implementation, Step 4—Game impact assessment. In each of the SESAM landscapes, research will be strongly embedded in participatory action research and executed in close collaboration with (key)stakeholders. SESAM will facilitate and guide research in the four steps by providing a toolbox of relevant methods for each of the research steps. The toolbox will be filled with a collection of existing tools and methods as well as potentially newly developed methods. Inspiration will be taken from relevant well-established gaming communities, namely companion modelling network [59] and the International Simulation and Gaming Association [60].

In step 1, a thorough understanding of the complexity and context of the studied landscape will be developed by identifying and exploring the current social-ecological system. Proposed methods for this include stakeholder assessment [61], (P)ARDI [62], and fuzzy cognitive mapping [63]. Through these methods a conceptual understanding of the system will be developed. In addition, the Q methodology is recommended to identify the current perceptions of various stakeholder groups. In Step 2, the game will be developed based on the conceptual understanding of the systems developed in Step 1. Key actors, elements, of the system and their interactions will be represented in the game to allow for relevant dynamics and patterns to be reproduced. From the initial stakeholder assessment, stakeholders will be selected to be actively involved in the development and implementation of the game. Step 3 will be part of an initial learning loop with Step 2 in the target area, although it is common to pre-test games on other target audiences (e.g., students...). If games are re-used beyond their initial place of origin, Step 3 can follow directly from Step 1. Step 4, the impact of using serious games on (social) learning and stakeholder opinions will be assessed through qualitative, and semi-quantitative analysis. Some of the existing innovative methods, e.g., Q-method, to distinguish between concurrent discourses and ways of explaining observed phenomena [64,65] are currently being explored in the context of game impact (before vs. after) assessment. Existing natural resource management frameworks [29] are also relevant in this context as a way to assess stakeholder understanding, perception, and willingness to act and adopt improved management options.

Reflecting on the types of games that we may need to develop and test, we may need at least four types of games (Figure 10): A. games that share the discovery process of a diagnostic stage (Table 4), B. games that focus on land-use decisions (Table 5), C. games that add hydrological consequences (with their human impacts) to land-use decisions (Table 6), and D. games that also include responses, where stakeholders outside the landscape try to influence land-use decisions (Table 7).

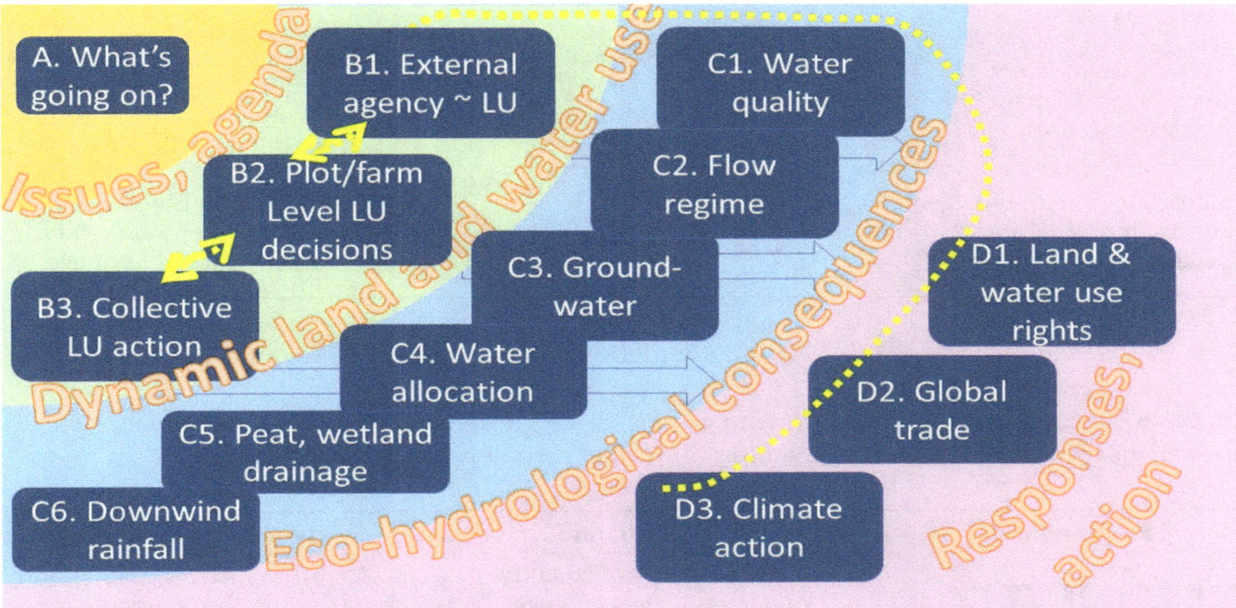


**Figure 10.** Initial game typology based on scope (A for agenda setting, B for better understanding land-use decisions, C for consequences of land-use decisions for water flows, D. for driver level responses by stakeholders external to the landscape considered in B).

**Table 4.** Prototype of 'What's going on' agenda-setting game (journalist/detective quest), with targeted/expected endpoints as hypothesis for further corroboration; type A in Figure 10.

| Starting Point | Dynamic | Targeted/Expected End Point |
|---|---|---|
| An issue of public concern that involves water and people, and in which forests and trees may play a role. | One or more journalist/detective teams are formed and have opportunity to interact with stakeholders. | The various pieces of the puzzle come together and start to give an 'emergent' perspective on what's going on. |
| Multiple stakeholders of the issue have diverse interpretations of what is at stake, how it works, what are (alternative) facts. | Stakeholder groups have their own interpretation, e.g., deforestation, climate change, technical failures, water grabs, of underlying causes of the issue and possible solutions. | A first 'agenda setting' conclusion may well be that the issue is indeed an important one, that it is 'wicked' (no easy way out), requiring deeper analysis. |
| There is no consensus on 'what's going on', tensions may be rising, conflicts emerging. | If the journalists/detectives interact appropriately with a stakeholder group, they may get 'a piece of the puzzle'. | Depending on how the process is managed, an overarching 'framing' of the issue may emerge that is shared by all. |

**Table 5.** Prototype of a who? what? where? land and water use game, with targeted/expected endpoints as hypothesis for further corroboration; type B in Figure 10

| Starting Point | Dynamic | Targeted/Expected End Point |
|---|---|---|
| A locally recognizable functional terminology of land uses along the forest transition curve. | Land users (farmers, communities) make choices with direct consequences for their livelihoods, leading to an emerging 'land-use mosaic'. | Patterns of change in land (and water) use mosaic that are made visible along with the multiple 'causes' that were at play. |
| A spatial representation of topography, soils and water flows as interaction 'arena'. | B1: Focus on plot- and farm-level decisions, including trees and tree diversity. | Reported experience of players in various roles, (partially) achieving their goals. |
| Characterization of local livelihoods, on-farm and off-farm, leading to 'land use'. | B2: Focus on external agency, pulling and pushing local land-use decisions according to various agendas. | Clarity, within the game, on what are 'externalities' for the various actors and how this contributes to an overall result. |
| Identification of external agents, influences and pressures that shape land-use decisions. | B3: Focus on collective action in land and/or water use and the decision making that can enhance synergy. | Depending on physical landscape context, a better understanding of its role in shaping land use. |

**Table 6.** Prototype of a so what? who cares? game with eco-hydrological consequences, with targeted/expected endpoints as hypothesis for further corroboration; type C in Figure 10.

| Starting Point | Dynamic | Targeted/Expected End Point |
|---|---|---|
| A climate plus topography description of abiotic context. | **C1**: Focus on water quality ('pollution'), consequences for health, sedimentation. | As follow-up to games A and B, clarify the consequences for a wider range of 'stakeholders. |
| Pre-human vegetation interacting with abiotic context. | **C2**: Focus on water quantity and flow regime: water yield, floods, droughts. | C1–C5: Identify downstream people influenced by decisions made upstream: 'who cares?' |
| Human land use modifying vegetation, soils, drainage pat-terns (as shaped in games B). | **C3**: Focus on (blue) water availability and its allocation to (appropriation by) competing users. | C6: Identify downwind people influenced by decisions made upwind: 'who cares?' |
| Awareness (based on game A) of the 'down-stream' issues land-use change can influence. | **C4**: Focus on groundwater recharge and availability through springs and wells. | Identify vulnerability to climate change (trend, increased variability). |
| A technical water balance model that may stay in the background, but provides 'ballpark' rules for the game. | **C5**: Focus on wetland and peat drainage and its consequences for subsidence and/or fire risk. | Identify the contributors to 'buffering' that reduce impacts of external variability. |
| Climate variability and climate change scenarios that provide challenges to existing land use. | **C6**: Focus on atmospheric moisture recycling, downwind effects on rainfall. | Reflect on the 'wicked' nature of the underlying issue (game A). |

**Table 7.** Prototype of a 'closing-the-loop' game, with targeted/expected endpoints as hypothesis for further corroboration; type D in Figure 10.

| Starting Point | Dynamic | Targeted/Expected End Point |
|---|---|---|
| Current land (and water) use is a direct cause of problems downstream/downwind. | D1: Land-use planning and water use rights negotiations modify future land-use change and incentives (incl. PES?). | Unexpected winners and losers of various 'feedback loops', deepening the sense of 'wicked' problems. |
| Current land use is a resultant of local + external forces that expect to benefit from their choices, but don't take 'externalities' into account. | D2: Global trade as driver of land-use change becomes aware of its social and ecological 'footprint' and starts to take responsibility, e.g., by standards and 'certification'. | Deeper understanding of 'common but differentiated responsibility' in resolving issues at landscape, national and/or global scale. |
| Those affected by 'externalities can take action, depending on power relations, political and cultural context. | D3: Global climate action expands from its current carbon focus to concerns over water cycles and downwind effects of tree cover change. | Need to balance *'efficiency'* and *'fairness'* in interacting with social-ecological systems in a given cultural context. |

## 7. Discussion: the Four Challenges to Use of Serious Games

Returning to the four issues restricting use of 'serious games' in the context of forest–water–people nexus issues, we need to take stock of how the SESAM program, in our current preview, will address the critique that games (1) are ad hoc with poorly defined extrapolation domains; (2) require heavy research investment from intervention experts; (3) have untested cultural limitations in where and how they can be used; and (4) lack clarity on where and how they can be used in policy making in local or global issue cycles. In doing so, we will need to return to the perspective of Figure 2, articulating it further in Figure 11. The figure shows (A) real-world social-ecological systems and their forest–water–people issues, (B) games as simplified representations of such systems, focused on specific aspects, and (C) use of these games in real-world contexts (potentially beyond where they were initially constructed). The match can be viewed from an ecological, a social or policy-oriented perspective.

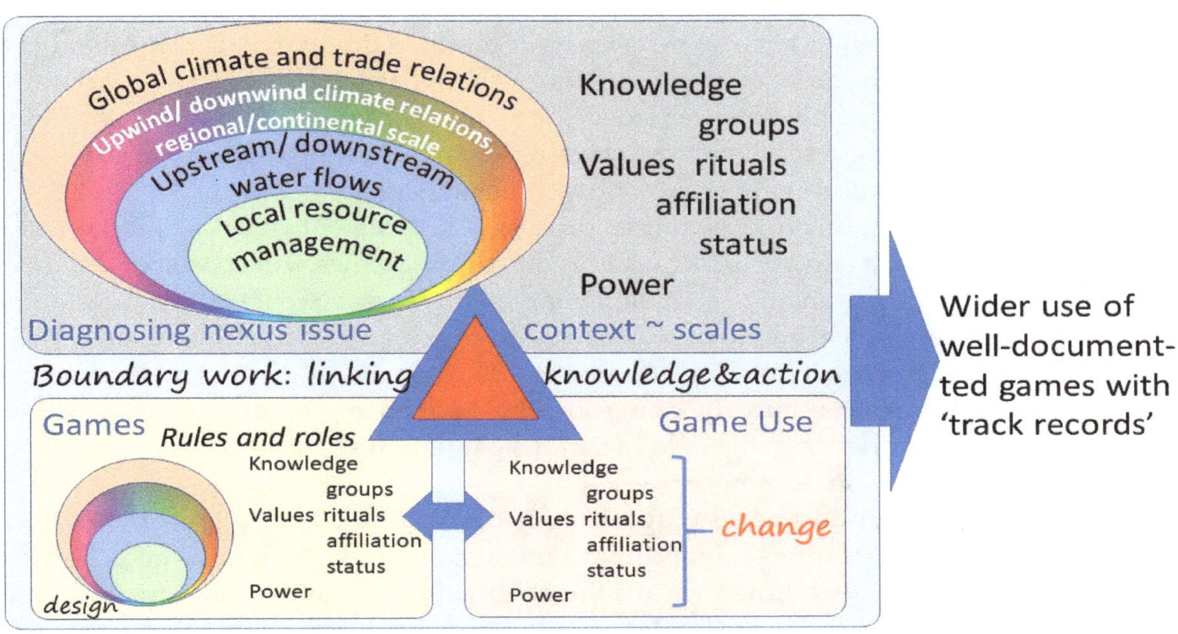


**Figure 11.** Real-world forest–water–people nexus issues across scale levels, games that reflect these, and game use that can modify the trajectory of real-world issues, where the same game can be used in different settings, in different ways, yielding a track record that can be documented.

### *7.1. Providing a Scaffold for Scenario Evaluation Games in the Forest–Water–People Nexus*

In order to address the current rather ad hoc nature of serious gaming studies in agroforestry, SESAM aims to develop a scaffold for scenario evaluation games in the forest–water–people nexus that could form an example in other fields in which games have been increasingly used. We will build up a library of games, describing games based on scope, content, format, track record (experiences) and identify games (e.g., as subtitle) by generic issue-in-context (e.g., tropical mountain lake). By creating this overview of games, starting from the games developed within SESAM, but not limited to the set of SESAM games, the program aims to be able to draw grounded conclusions on the interplay at the core of the forest–water–people nexus and the relevance and contribution of a serious gaming approach in this context. Through a cross-case study comparative analysis, we aim to identify how different interpretations of forest–water interaction influences the understanding, awareness and action related to landscape change and management at different levels of governance in different locations on a pantropical forest transition curve.

SESAM will perform a comparative analysis of all games developed and their impact on actor learning in different contexts. This will provide insights into larger methodological gaming-related questions contributing to the wider application of serious games as a generic approach for decision making under high levels of complexity and uncertainty. SESAM will match case-by-case games with other landscapes with (a) similar issue, (b) similar ecosystem and social characteristics, (c) same country or region, while starting from a 'User demand' point of view ('what is the question?' rather than 'supply' 'we want to disseminate our research results'). The portfolio of games developed within the SESAM program should enable a more in depth understanding of (1) The relative value of different type of games, e.g., simple vs. complex, or fully defined vs. open, (2) game comfort zones linked to cultural context, age, gender and other social stratifiers, and, (3) learning effects (through before–after comparison) of game session participants who enter with different levels of knowledge. Classifying games and assessing the relative learning impact of gaming participants experienced will allow comparing learning from different game sessions with different games in distinct systems and contexts around the world. Based on this, recommendations will be developed to facilitate a wider use of well-documented games with track records.

### *7.2. Optimizing Research Investment in Game Development*

Games currently rely on long-term, expensive prior research involvement. Optimizing research investment is key to support the broader use of serious games in participatory social-ecological systems research on the forest–water–people nexus and development of agroforested landscapes. SESAM will contribute towards optimizing research investments by (i) streamlining the serious gaming process from start to finish, and (ii) exploring options for the re-using games.

In order to streamline the serious gaming process, SESAM will develop guidelines for the development, implementation and analysis of games. These guidelines will consist of a standardized step-wise approach based on a number of complementary methods (see Section 6). By developing these SESAM guidelines and implementing them in our SESAM case studies, we aim to offer a streamlined game development process as well as to facilitate communication about games and between research teams in the serious gaming community. By doing so, SESAM aims to provide a scaffold to build upon existing knowledge and experience in the field of serious gaming in scenario evaluation and allow for research teams to connect and share experiences.

In addition, SESAM will address the question of transposability of games by exploring the use of games in contexts and settings beyond the original case study they have been developed in and assessing the impact of co-designed vs. off-the-shelf games. Some authors have already been successfully developing and exploring the implementation of games beyond their place or origin [66,67]. The development of such games that allow for more generic applications, requires the concepts used to be 'valid' and 'robust' in a wider range of circumstances, without claiming high accuracy in any specific location. Within SESAM we will cross-test games between case study areas within the comprehensive coverage of the forest–water–people nexus among the sites. SESAM will develop means of communication to assist in finding existing games and finding out whether they could be of use in a specific (new) situation. In addition, SESAM will describe how 'long-term' adaptability goals can be reconciled with 'short-term' match with data (e.g., by shifting from data-driven heuristics to first-principles models underlying the game).

SESAM will also develop Agent-Based Models (ABM) in parallel and interacting with game development, as this provides valuable insights in the adequacy of current process-level understanding of the social-ecological systems represented in the game [68,69].

### 7.3. Culture-Sensitive Gaming

Scenario games are artefacts that embody certain perceptions and values, but also allow various usages, e.g., through selection of player groups, setting of incentives, or usage of group discussions surrounding game rounds. The larger cultural setting influences these issues [52]. In SESAM we will have to create games that involve the future of agroforestry and the sustained livelihood of people. Based on existing national-level data on culture [51], there is a considerable 'band-width' for each of the six culture dimensions across the SESAM countries (Figure 9D), especially if the Netherlands is included as well. Within countries, there will undoubtedly be further variation, for example between the mountain and lowland forest people of Peru, the upriver and coastal zone people in Surinam and the islands (Sumatra, Kalimantan, Java) of Indonesia, within the SESAM portfolio.

The knowledge that unwritten rules of culture play a role implies that SESAM participants will be playing similar games across sites and looking for potential differences in game dynamics even if the pre-designed rules and incentives are identical. At design time, cultural differences between the designers and the intended users can cause blind spots [52]. We will be careful to design games that are meaningful and acceptable to participants. The local knowledge of our PhD candidates, as well as intensive contacts with stakeholders, should guarantee this. Dimensions of culture that we expect to be important for game design and dynamics are individualism and power distance—mentioned in one breath because they tend to be strongly correlated with tropical countries typically more hierarchical and collectivist than Western ones; and long-term orientation, which varies across countries in both the Tropics and the West. In conclusion, while a game is just an artefact and has no cultural awareness, its design and its usage can be done in culture-aware ways. SESAM intends to do this and document the results.

### 7.4. Game Relevance in the Policy Domain

One of the most confounding elements of policy issues is that there is a can of worms (and other parts of belowground biodiversity) of actors and interests, and nobody has an overview. This is precisely where games are strong: they provide a shared system boundary and show how actions of stakeholders impact the overall behavior of the system. In doing so, they also allow emotional responses to events, as well as joint (collective) action. In terms of system patterns, games allow to experience patterns such as the Tragedy of the Commons, or a Fix that Fails.

A review of 43 serious games and gamified applications related to water [70], covering a diversity of serious games, noted the still unsettled terminology in the research area of gamification and serious gaming and discussed how existing games could benefit early steps of decision making by problem structuring, stakeholder analysis, defining objectives, and exploring alternatives. Behavioral games on common pool resources may be used to facilitate self-governance [71], with groundwater a particularly challenging common pool resource to govern due to its low visibility, resulting in resource depletion in many areas. Serious games can promote values that transcend self-interest (transcendental values), based on the contributions of social psychology, but to do so, their design should incorporate the many value conflicts that are faced in real life water management and promote learning by having players reflect on the reasoning behind value priorities across water management situations [72,73].

This implies that potentially, not only playing games but also designing or adapting them can be very valuable for stakeholders in a policy-setting context [74]. A scenario game of one of the types in Figure 10 can be contextualized for and by local stakeholders; this exercise will generate relevant discussion and important learning. Being inclusive in which actors to select, as well as having support from important local persons, is essential [75]. Games also fit in a movement towards more plural and participatory approaches to 'valuation' of natural capital and ecosystem services [76].

All SESAM games will build on the boundary work tradition of taking the three 'ways of knowing' (local ecological knowledge, public/policy knowledge and science-based knowledge) as potentially complementing each other [32,77], with scope for new solutions to emerge at interfaces.

How can game use trigger policy change? From the five steps mentioned in Figure 3, it is relatively easy to see how games (including those of type 1) can be used for 'Agenda setting' and raising awareness of issues. Games are also a good vehicle for the 'Better understanding' part, as they offer insights not only into what happens, but how actor decisions contribute to outcomes of the game.

**Hypothesis 1.** *The impact of serious games on 'agenda setting' and 'better understanding' parts of issue cycles is reflected in increased consensus about what is important to do and what changes can be expected to result from actions.*

However, more is needed to nudge decisions locally and/or globally into desirable transformative and re-imaginative (game-changing) directions: commitment to aspirational goals.

**Hypothesis 2.** *Embedding serious games in a stakeholder negotiation process, can contribute to commitment to aspirational goals for addressing underlying causes ('drivers'), while addressing immediate symptoms.*

Real progress depends, beyond ambitious policy language, on action [6]. Most 'serious games', however, will have limited precision and use an oversimplified problem description. Scenario analysis at 'implementation' level will often require more detailed spatial analysis of tradeoffs.

**Hypothesis 3.** *Games are not a safe basis for operational decisions due to their limited specificity and (spatial) precision.*

We expect that the SESAM studies will provide further evidence to judge the contexts in which 'boundary work' in the forest–water–people nexus can be supported through serious games to support game-changing transformations at 'driver' level.

## 8. Conclusions

Four challenges have been raised to the use of serious games in addressing issues such as those in the forest–water–people nexus: games so far (1) appear to be ad hoc, case dependent, with poorly defined extrapolation domains, (2) require heavy research investment, (3) have untested cultural limitations and (4) lack clarity on where and how they can be used in policy making. Reviewing the literature and considering a set of case study landscapes, we conclude that these challenges can be addressed at the design, testing and communication stages of games to be shared in a wider community.

The SESAM program of networked PhD research programs will be geared at credible, salient and legitimate action research designing, testing and using 'serious games', that are meant to (1) be systematic in their coverage of the pantropical forest–water–people nexus in its main manifestations and issues, using generic forest and tree cover transitions as continuum description rather than forest–agriculture dichotomies, supporting easy 'localization' of games to match local contexts, (2) use 'basic hydrological, ecological, social and economic concepts, (3) be cognizant of the main dimensions in which cultural contexts of inter-human and human–nature relations vary to guide responsible game use, and (4) be explicitly adapted (or adaptable) to different stages of local and global issue cycles. The 13 landscapes provide a wide diversity of forest–water–people nexus aspects, as they range from low (<1 $km^{-2}$) to very high (>1000 $km^{-2}$) human population densities, high (>90%) to low (<5%) forest cover, five ecoclimatic zones (from drylands, sub-humid, via 'water tower' to humid and per-humid), and have focal issues that range from concerns over groundwater and streamflow depletion, flood-and-drought flow regimes, erosion and water quality to atmospheric moisture recycling, all (supposedly) influenced by quantity, quality and spatial pattern of tree cover ('agroforestry'). Game prototypes are described for (1) a diagnostic phase where multiple explanations for identified local issues are explored, (2) a deeper understanding of individual and collective land-use decisions, (3) explicit consequences for a range of ecohydrological landscape functions of such decisions and (4) societal feedback to land users based on landscape-level consequences, through land and water use rights ('planning'), global trade and climate action. Games are expected (hypothesized) to help in agenda setting phases, in achieving a common understanding of what is at stake and in political commitment to solutions—but will need more specific information to guide decisions on actual solutions. These hypotheses will be tested in the coming years.

**Author Contributions:** Conceptualization, M.v.N., E.S. and G.J.H.; Writing—original draft, M.v.N., E.S., G.J.H., A.F., A.Y.A., A.M., A.L.H., C.N.W., E.L., F.A., G.K., G.G.C.A., L.B., L.T., M.G., P.R., R.R.S. and U.S.; Writing—review and editing, S.A., A.L., P.v.O., C.M., M.P.-C., E.P., D.R., D.S. and A.J.T. All authors have read and agreed to the published version of the manuscript.

**Acknowledgments:** The authors acknowledge assistance by Toha Zulkarnain for interpreting Forest Transition stages of the sites.

## Appendix A. Brief Descriptions of the SESAM Landscapes

### *Appendix A.1. Upstream Remote Forests of the Suriname River Basin*

Suriname, a country on the Guiana Shield in Latin America, is rich in rainforests and freshwater resources [78,79]. Its average annual rainfall in the area of 2700 mm [80] is an important source of evaporation that precipitates further inland in South America [41]. The Upper Suriname River Basin (USRB) is still mostly covered by primary and secondary forests [81]. The USRB is since several centuries inhabited by the Saamaka afro-descendent groups who, now around 20,000 people [82], live in 62 villages along the river. The Saamaka do not have legal land tenure rights, permanent electricity or running water, but live off ecosystem services provided by the forest (e.g., food, water, medicines, materials for housing and boats) [77]. The main income sources are rain-fed shifting cultivation and non-timber forest products, especially for the women. However, nature-based tourism has also become important. Some villages have community forest concessions, exploited by third parties. Competing interests and the community's increasing participation in market economies lead to land-use conflicts and sustainability challenges. There is a clear need for better coordination and improved capacities to manage the USRB in a sustainable way, benefitting rights and stakeholders. Government decentralization was only partly successful as district governments have limited capacity and budget, and coordination between stakeholders remains poor. There have been recent improvements in forestry policies, but enforcement is a challenge. Forest, water and people in the USRB are inextricably linked, yet knowledge and understanding are lacking about these interactions and how different land uses affect them. Sustainably managing forests for their vital role in watersheds, precipitation sheds [83] and global climate is crucial, especially for vulnerable communities such as the Saamaka. Our study aims to improve shared understanding on the multi-scale effects of land-use changes on the forest–water–people nexus and apply gaming as a tool to enhance coordinated management of the landscape.

*Appendix A.2. Upstream Forests of the Amazon in Madre de Dios, Peru*

The Madre de Dios region, in the western part of the Amazon basin, is located in the south-eastern part of Peru and shares borders both with Brazil and Bolivia. The region is almost entirely covered (95%) by forest [84] and its mean annual precipitation is 2200 mm [85]. The region's name originates from the Madre de Dios river that forms part of the vast Amazon River watershed [86]. The population of about 150,000 inhabitants is growing at a 2.6% yearly rate, partly due to immigration [87]. This migration originates mainly in the highlands and is driven by the economic perspectives offered to rural workers in the sectors of (illegal) gold mining and informal agriculture. Both these sectors have benefited of the recent pavement of the Inter-oceanic Highway, along which all recent deforestation hotspots are located [88]. In order to provide economic alternatives to mining, the regional government has started investing more resources in the agricultural sector. This strategy is mainly focused on cash crops, i.e., cacao (native to the Amazon) with a cultivated surface that has risen about 100% between 2010 and 2017 [87]. This agricultural intensification is likely to affect most smallholder farmers in the region and the high agrobiodiversity that is traditionally managed on their land. While small-scale farming remains little studied in the region, there is a growing need for transdisciplinary research that, by giving a voice to the farmers, allows them to become part of the conversation about land use. This case-study aims to provide a better understanding of smallholder farmer's decision-making processes and how it is influenced by their systemic perception, i.e., the forest–water–people nexus. In the longer run, this will provide NGOs and the regional government with tools to better adapt their development programs to the local farmer's needs and visions [89].

*Appendix A.3. Mangrove Coasts of Suriname*

The majority of the population of Suriname resides in the coastal area where also 90% of the economic activities is concentrated. The 370-km-long coastline harboring the largest and pristine mangrove forest of the Guianan Ecoregion serves the entire nation with its numerous ecosystem services. The Districts Nickerie and Paramaribo represent rural and urban parts of this coastal zone. The Suriname climate is tropical humid, with an average air humidity in the coastal area of 80-90% and a north-eastern (land inward) wind direction. Poor mangrove forest management and its ultimate destruction threaten the livelihoods of coastal communities, but also removes flood protection for the whole coastal zone. The root cause of the problem is lack of awareness about the mangrove ecosystem, its services and the effects of stakeholder activities on mangrove ecosystem services. Mangrove management can be achieved with the current legislation; however, the legislation is very much fragmented and sectoral in its orientation. In 2019, a national mangrove strategy has been developed for the ministry of Spatial Planning, Land and Forest Management. This study will apply serious games to increase awareness as well as to stimulate an effective decision-making process among stakeholders.

*Appendix A.4. Amazonian Agroforestry Mosaics in Para State, Brazil*

Pará State, which comprises a large portion of the Eastern Amazon basin in Brazil, has the highest level of deforestation of any subnational area in the tropical world [90] and is deemed a critical hotspot to contain the Amazon dieback [91] (breakdown of rainfall recycling). It thus has become a priority landscape for corporations to promote deforestation-free supply chains [92]. Nestled along the Acará river in the mesoregion of Northeastern Pará, the municipality of Tomé-Açu, has a population density of 10 $km^{-2}$ and an area of roughly 5000 $km^2$. It is currently dominated by oil palm, pastures and perennial crops, particularly cocoa, cupuaçu (Theobroma grandiflorum), black pepper, and açaí (Euterpe oleracea), a native palm, as well as slash and burn agriculture for cassava production, a mainstay of local livelihoods, and other annual crops [93]. Tomé Açu has seen 56% of its original forest cleared, mostly dating back to the 1970s and 1980s, whereas some neighboring municipalities still face high deforestation rates to this day. Since 2010, the municipality has been a hotspot of oil palm expansion against a backdrop of logging and extensive grazing juxtaposed by clusters of agroforestry innovation. Plummeting prices and fusarium disease that ravaged monocrop black pepper plantations in the 1960s and 1970s led Japanese settlers [94] to diversify their production systems to reduce such risks, developing what became known as the Tomé Açu Agroforestry Systems (SAFTAS) supported by a fruit-processing cooperative (CAMTA) [95]. These commercially oriented systems have become a key example widely followed by family farmers in Tomé Açu and far beyond in the Brazilian Amazon. This case study aims to shed light on the factors underlying agroecological intensification and agroforestry transition pathways in the Tomé Açu landscape, focusing on the constraints and levers for scaling agroforestry, with and without oil palm, and trade-offs between different agroforestry systems and other land-use options.

*Appendix A.5. Tropical Peatland Restoration in Indonesia*

Tropical peatlands form where drainage is limited, and organic matter decomposition cannot keep up with its production—often in lowland interfluvial locations. In the Ketapang Peatland in the coastal zone of West Kalimantan, (Indonesia) two Peatland Hydrological Unit (PHU): PHU Tolak river—Pawan river (948 km$^2$) and PHU Pawan river—Pesaguan river (1048 km$^2$) were selected to explore options for hydrological restoration. Human population density in these areas is low by Indonesian standards (20 people km$^{-2}$ in 2017) and is dominated by people of Malay background, along with migrants from Java and Bali. Average annual rainfall is 3168 mm [96], but the June—October period is relatively dry in most years and, especially in El Niño years serious droughts develop. The area is dominated by logged-over peatland forest and oil palm plantations, with some rubber plantations and paddy rice fields. Expansion of plantations and agricultural areas has led to massive canal construction [97]. Besides their functions for log transport and human access, these canals also drain water from the peatlands so that oil-palm, rubber and crops can grow. As a result, however, the peatland becomes drier during the dry season and more vulnerable to forest fires, with haze directly affecting human health and well-being. Drainage can also cause saline seawater intrusion into the coastal zone. Peatland restoration aims to restore the hydrological function in the peatland area by increasing and stabilizing the groundwater levels in the peat dome, especially in the dry season, to reduce fire risk. Understanding the relationship between the environment and people is the key to build commitment among all stakeholders, so they can engage and cooperate to address the environmental issues. The objective of this study is to develop communication tools that can use to increase the level of understanding and to facilitate the communication process among stakeholders.

*Appendix A.6. A Tropical Mountain Lake: Singkarak in Sumatra, Indonesia*

Lake Singkarak is one of the highland lakes in the Bukit Barisan mountain range that runs the length of Sumatra. Inflow to the lake derives from an area of 1135 km$^2$ through a number of streams and rivers, some with smaller natural lakes that provide some buffering. The watershed is home to 440,000 people with intensive horticulture in the upstream areas and extensive paddy cultivation before inflow to the lake. Located across two districts (Solok and Tanah Datar) the lake provides opportunities for fishing (an endemic fish, overfished due to high demand), year-round water supply for surrounding villages and tourism. The outflow of the lake provides irrigation water for downstream farmers and hydropower for West Sumatra Province. With annual rainfall ranging from 1700–3200 mm there is enough water for the hydropower plant in average years, but in years with long dry seasons there is a shortfall, as dropping the water level in the lake disturbs local livelihoods. The W side of the lake is dominated by forest, mixed gardens (agroforests) and agricultural fields, but the E Side is drier and dependent on the natural outflow from the lake that was disturbed by the construction in the 1990's of the Singkarak Hydro Electric Power Plant [98]. There have been a number of reforestation project implemented in the area to increase amount of forest cover [99] and the local village surrounding the lake tried to protect the water quality and endemic fish by setting up their village regulation. Understanding the water allocation management (between water flows in the original riverbed flowing East), supporting traditional rice farmers, and the use for hydropower and irrigation schemes west of the mountain range [96]. The objective of this study to have a tool to support and enhance social learning and action by actors involved in multi-level decision-making processes around the nexus, and to explore how participatory decision-making on water and (agro)forest landscape management can be improved as part of climate change adaptation.

*Appendix A.7. Water Tower for Adjacent Drylands: Mount Elgon, Uganda*

Overlapping the international boundary between Uganda and Kenya, Mount Elgon is the 7th highest mountain in Africa rising to 4320 m a.s.l. It is approximately 100 km North-east of Lake Victoria. The mean annual rainfall is 1600 mm [100]. Forests in the Mt. Elgon ecosystem have become restricted to the protected upper slopes (23% forest cover in 2016). With 1000 people km$^{-2}$ and a 3.4% annual increase the mid-slope is a densely populated agricultural landscape [101]. The region is a highly productive agricultural zone, growing arabica coffee and horticultural crops. The watershed contributes to Lake Victoria (and thus to its outflow, the Nile river), and lake Turkana) [102]. However, the ecosystem functionality and integrity has been compromised and impacted by climate change. The population increase has directly raised demand and competition for natural resources including land and water [103,104]. Due to land inheritance, land fragmentation is common and is expected to worsen with population increment mounting pressures on resources [105,106]. Besides, the declining land productivity has led to reduction in food produce thus deficit in food supplies to the continuously fast-growing human population. The resultant food gap has sparked encroachment into the national park and on riparian zones of riverbanks, swamps and steep slopes leading to soil erosion, siltation and flooding. There is overgrazing, destruction of forest for urban extension and high levels of conflict between the park authorities and locals. Landscape actors with divergent values/interests include resource users, local government, national conservation agencies, international donors and NGOs, local politicians, etc. To strengthen local natural resource governance, participatory scenario evaluation games for supporting and enhancing joint planning and social learning by actors involved in multi-level decision-making processes around the forest–water–people nexus will be developed and implemented.

*Appendix A.8. Water Tower for Adjacent Drylands: the Ewaso Ng'iro River NW of Mount Kenya*

Mount Kenya is located on the equator and rises to 5199 m a.s.l, 180 km north of Nairobi. The mountain has different zones of influence. The eastern climatic gradient is relatively humid (windward) and the western climatic gradient (leeward) stretches from humid (upstream) to semi-arid areas in the downstream. The mountain contributes about 50% of the entire flow of the Tana River, the largest river basin in Kenya providing water supply to 50% of the Kenyan population and 70% of the country's hydroelectric power. Both rainfed and irrigated agriculture are major economic activities in the upstream. Further downstream many others rely on the river flows for irrigation, energy generation, pastoralism and hotels/tourism. Problems are related to over-abstraction of water [107], farm encroachment to fragile lands and deforestation [108]. Over-abstraction of river water in upper reaches and unwillingness to be held accountable through metering has been a major threat to meeting downstream flows. Different water management institutions exist, but activities are uncoordinated, leading to gaps in water resources management. The national government reacts by closing down water intakes in desperate attempts to resolve downstream water crises. Such orders affect schools, hospitals and thousands of households in the downstream. The local politicians strongly oppose the national government and defend the community water projects (most of them are unmetered). Consequently, an endless cycle of water crises characterizes the Mt. Kenya region [109]. Climate change and land-use changes continue to exert pressure on limited water resources. There is a need for increased knowledge and understanding among stakeholders for sustainable solutions. This study will explore the influence of climate change and land-use changes on downstream hydrology and apply serious games to explore decision making processes among stakeholders. This will provide a better understanding and knowledge for sustainable development.

*Appendix A.9. Mountain Farming in the Andes, Peru*

The Peruvian region of Puno is located in the south-eastern Peru on the shore of Lake Titicaca close to the border with Bolivia. The region is poorly covered (0.1%) by forest [84] and the mean annual rainfall averages 700 mm [110]. The watershed of Lake Titicaca extends for 8490 km$^2$. The highlands—over 3000 m high—region of Lake Titicaca supports a population density of 5.1 km$^{-2}$ [87]. In the region of Puno of the Altiplano, in Peru, farmers with their knowledge and practices are growing the highest quinoa diversity hotspot in the world [111]. The potential of quinoa was promoted by the United Nations during the International Year of Quinoa (IYQ) in 2013 [112]. However, IYQ-2013 did not cover aspects of the worldwide spread of and commercial interests in quinoa, and the unbalanced competition between producers from the Andes and producers from North America and Europe [113]. Collective governance instruments as collective trademark (CT) are used to defend property rights on trading products and to recognize their anteriority and origin. Co-developing an Andean quinoa CT rise our research objective: explore the process of co-constructing a CT for recognize and promote quinoa in the global market. The region of Puno in Peru (the highest quinoa diversity hotspot in the world) is selected as case study. Participatory games and agent-based models will be developed and explored to assess the gap between local and regional farmers, and higher level including Andean quinoa farmers for developing an Andean CT for quinoa. The proposed research will explore the role CT can play to preserve the Andean generic hotspot of quinoa as well as to provide a potentially relevant governance tool for other neglected crops that suddenly get global consumer attention.

*Appendix A.10. Farming Drylands on the Mossi Plateau, Burkina Faso*

The Mossi Plateau in the central part of Burkina-Faso in West-Africa spans the Sudano–Sahelian climate gradient, with unimodal rainfall from June to September, with an annual average ranging from 400-700 mm [114]. Two villages (Yilou and Tansin) represent this gradient, with sandy clay loam soils, low in soil organic carbon (0.2 and 0.4, respectively) [115]. Agriculture in this area is challenged by low and unstable biomass production, limiting farmers' resilience. Water, nutrients, biomass, labor, information and money move in the landscape, while many studies focused on biomass production at farm scale only considering single households [116–118], without their interactions. Biomass management through farmers' decisions on biomass allocation [119], include use of crop residues as livestock feed [120], indirectly influencing manure availability for crop production. At village scale, biomass production is determined by farmers' organization, the spatial and temporal interactions amongst farmers, and between farmers and their biophysical environment (soil fertility), as farmers can organize themselves in "labor-groups" to till each other's fields and thus be able to cultivate more land. As rainfall tends to vary spatially, risk management depends on the collective use of space. A spatio-temporal modelling approach [121,122] needs to take the various spatial and social interactions into account. We aim to achieve this by co-designing tailored and realistic biomass management and organization options to improve farmers' production and livelihoods through participatory modelling.

*Appendix A.11. Upland Agroforestry Mosaics in Kali Konto, East Java, Indonesia*

The Kali Konto sub-watershed in East Java contributes to the Brantas river, and covers an area of approximately 240 km$^2$ with elevation ranging from 600-2800 m a.s.l. The Kali Konto river is approximately 40 km long and empties into Selorejo reservoir. The average daily temperature is 20–22 °C with a mean annual rainfall ranging from 2995–4422 mm [123]. Forest area has significantly decreased in the last two decades (from 30.5% of forest cover in 1990 to 20.4% in 2005), following the increased demand for timber, firewood, fodder and other agriculture (and dairy) products [124,125]. On the other hand, annual crops area increased by 26%. In the last three decades, the population increased by approximately 20% from 1990 adding more pressure to the system [126]. Intensive agricultural farming with minimum tree cover in hill slope leads to the increase of soil surface exposure, which is followed by higher soil erosion and sedimentation (top 'fertile' soil washed away) to the river [127] and impacts on crop production [128]. The polluted reservoir brings negative consequences for both the state water management corporation that is responsible for managing water quality and quantity in the region and downstream 'water' beneficiary (HIPAM–drinking water user association) [129]. Efforts have been made by PJT to encourage soil conservation practices. However, it has not yet accommodated farmers' choice on the type of land management in the program. It is essential to understand farmers' preferences and decision making along with the direct consequences for livelihoods and environmental impact. It benefits on exploring environmentally friendly management options based on their preference and increase the awareness on the importance of achieving sustainable watershed conservation through collaborative action among stakeholders.

*Appendix A.12. Water Tower for a Metropole: Rejoso, East Java, Indonesia*

The Rejoso watershed stretches from the Mount Bromo volcano crater (summit at 2329 m a.s.l.) to the Pasuruan coastal area (0 m a.s.l.) covering an area of 62,773 ha. Land in the watershed is used in a variety of ways: a national park (conservation forests), protected forests (natural jungle forests), monoculture production forests, agroforestry (production forest + agricultural + livestock), irrigated lowland agriculture and built-up areas. The watershed covers 14 sub-districts in Pasuruan Regency and three sub-districts in Pasuruan City with a total population in 2018 of 838,313 people [130,131], with 100,497 people living in the upstream part, 271,908 in the middle zone and 465,728 in the downstream area. Most of the population, except residents of Pasuruan City, depend for their livelihoods on their agricultural and agroforestry systems, supported by the ecosystem services provided by land and water resources, including ecotourism [132–134]. They worked collaboratively (but there are conflicts as well) with multi-stakeholder institutions in charge of managing natural resources. Since 2003, for example, residents in all parts who are members of the forest village community (LMDH) collaborate with Perhutani (a state-owned forestry company) in 'the Community Joint Forest Management Program (PHBM). They are allowed to plant (intercropping) various agricultural commodities (horticulture, food crops, and fruit trees) and fodder in between the stands of production forest trees. They also get wages and/or profit-sharing after contributing to maintaining the main trees (teak, pine, mahogany, eucalyptus) which are managed together with Perhutani. There are at least 114 village communities that use the Perhutani area of 11,713 ha [135]. Before the PHBM was implemented, conflicts often occurred between communities and Perhutani. In the upstream part, communities collaborate with national park managers (Ministry of Environment and Forestry (MoEF)), natural resource conservation offices (MoEF), and tourism agencies authorities. Another collaboration is 'the Rejoso-Kita Program' which implements an integrated watershed management model. Hydrological analysis suggests that the tree cover required for 'infiltration friendly' land uses depends on the altitudinal zone [136]. The program involved various stakeholders: communities, NGOs, private sectors, and multi-level governments [137–139]. Although, there have been collaborations, conflicts have also been reported, including communities and the private sector (mining industry in the middle and water industry in the downstream). The study to be carried out at this location aims to design participatory collective action games to strengthen sustainability community–private–government collaboration in agroforestry management.

*Appendix A.13. Rehabilitating a Water Tower Under Pressure: Brantas, East Java, Indonesia*

The Upper Brantas sub watershed is the source area of the Brantas river that starts from the southwestern mountainous slope of active volcanoes of the Arjuna-Anjasmara mountain complex (a protected forest area). Originating at the Sumberbrantas spring, the river flows southward through the cities of Batu and Malang, before bending to the West and then North to the Provincial capital Surabaya. Total area of the sub watershed is 180 km$^2$. Annual rainfall ranges from 875 to 3000 mm [140]. Its forest fraction is 23.8%. The Upper Brantas sub-watershed population size in 2018 was 207,490, with a density 1092 km$^{-2}$ and an annual population growth of 0.95 per year [141]. Forest encroachment was a serious environmental problem in the 1998–2000 era. Conflicts over state forest land became visible in 1998 during the political reformation, when many local farmers occupied state forest land [142]. In this period power was decentralized to provincial and district levels. This period meant the end of timber forest management. Over 2003–2007, the settlement area increased by 9%, plantation and farms increased by 7%, and the forest area decreased by 6%. The watershed response can be observed through the increasing of run-off coefficient from 0.59 to 0.67 [143] and an increase in peak discharge (at the outflow of the sub-watershed) from 96.8 $m^3 s^{-1}$ in 2003 to 189.2 $m^3 s^{-1}$ in 2007 [144]. Meanwhile the springs were affected (data to be further collected) and groundwater recharge was probably reduced. Policy level response to the forest encroachment was a number of new programs. In 2001 the state forest agency introduced a co-management program, specially focused on forest resource management. In 2007 the co-management concept became relabeled as community-based forest management. The objective of the case study is to develop games and simulations of land-use change of recharge area management for sustainable groundwater supply and demand. There are many volcanic slope watersheds in Indonesia similar to the upper Brantas sub-watershed.

## References


1. van Noordwijk, M.; Coe, R.; Sinclair, F.L. Agroforestry paradigms. In *Sustainable Development through Trees on Farms: Agroforestry in its Fifth Decade*; van Noordwijk, M., Ed.; World Agroforestry (ICRAF): Bogor, Indonesia, 2019; pp. 1–12. Available online: http://www.worldagroforestry.org/trees-on-farms (accessed on 20 July 2020).
2. Sustainable Developent Goals. Available online: https://sustainabledevelopment.un.org/sdgs (accessed on 21 June 2020).
3. Taleb, N.N.; Douady, R. Mathematical definition, mapping, and detection of (anti) fragility. *Quant. Financ.* **2013**, *13*, 1677–1689. [CrossRef]
4. Meadows, D. Leverage Points: Places to Intervene in a System. 1999. Available online: http://donellameadows.org/archives/leverage-points-places-to-intervene-in-a-system/ (accessed on 21 June 2020).
5. Minang, P.A.; van Noordwijk, M.; Freeman, O.E.; Mbow, C.; de Leeuw, J.; Catacutan, D. (Eds.) *Climate-Smart Landscapes: Multifunctionality in Practice*; World Agroforestry Centre (ICRAF): Nairobi, Kenya, 2015; p. 405.
6. Climate-Smart Landscapes: Multifunctionality in Practice. Available online: http://www.worldagroforestry.org/publication/climate-smart-landscapes-multifunctionality-practice (accessed on 20 July 2020).
7. van Noordwijk, M.; Lusiana, B.; Villamor, G.B.; Purnomo, H.; Dewi, S. Feedback loops added to four conceptual models linking land change with driving forces and actors. *Ecol. Soc.* **2011**, *16*. Available online: http://www.ecologyandsociety.org/vol16/iss1/resp1/ (accessed on 20 July 2020). [CrossRef]
8. van Noordwijk, M. Prophets, Profits, Prove It: Social Forestry under Pressure. *One Earth* **2020**, *2*, 394–397. [CrossRef]
9. Creed, I.F.; van Noordwijk, M. *Forest and Water on a Changing Planet: Vulnerability, Adaptation and Governance Opportunities: A Global Assessment Report (No. 38)*; International Union of Forest Research Organizations (IUFRO): Vienna, Austria, 2018; p. 188. Available online: https://www.iufro.org/fileadmin/material/publications/iufro-series/ws38/ws38.pdf (accessed on 20 July 2020).
10. Creed, I.F.; Jones, J.A.; Archer, E.; Claassen, M.; Ellison, D.; McNulty, S.G.; van Noordwijk, M.; Vira, B.; Wei, X.; Bishop, K.; et al. Managing Forests for both Downstream and Downwind Water. *Front. For. Glob. Chang.* **2019**. [CrossRef]
11. Nobre, C.A. Land-use and climate change risks in the Amazon and the need of a novel sustainable development paradigm. *Proc. Natl. Acad. Sci. USA* **2016**, *113*, 10759–10768. [CrossRef] [PubMed]
12. Page, S.E. *The Difference: How the Power of Diversity Creates Better Groups, Firms, Schools, and Societies (New Edition)*; Princeton University Press: Princeton, NJ, USA, 2008; p. 448.
13. Walker, B. Resilience management in social-ecological systems: A working hypothesis for a participatory approach. *Conserv. Ecol.* **2002**, *6*, 14. Available online: http://www.consecol.org/vol6/iss1/art14 (accessed on 20 July 2020). [CrossRef]
14. Reed, M.S. Stakeholder participation for environmental management: A literature review. *Biol. Conserv.* **2008**, *141*, 2417–2431. [CrossRef]
15. Scholz, G. An Analytical Framework of Social Learning Facilitated by Participatory Methods. *Syst. Prac. Action Res.* **2013**, *27*, 575–591. [CrossRef]

16. Angelstam, P.; Andersson, K.; Annerstedt, M.; Axelsson, R.; Elbakidze, M.; Garrido, P.; Grahn, P.; Jönsson, K.I.; Pedersen, S.; Schlyter, P.; et al. Solving problems in social–ecological systems: Definition, practice and barriers of transdisciplinary research. *Ambio* **2013**, *42*, 254–265. [CrossRef]
17. Tschakert, P. Anticipatory Learning for Climate Change Adaptation and Resilience. *Ecol. Soc.* **2010**, *15*, 11. Available online: http://www.ecologyandsociety.org/vol15/iss2/art11 (accessed on 20 July 2020). [CrossRef]
18. Rodela, R.; Ligtenberg, A.; Bosma, R. Conceptualizing Serious Games as a learning-Based Intervention in the Context of Natural Resources and Environmental Governance. *Water* **2019**, *11*, 245. [CrossRef]
19. García-Barrios, L.E.; Speelman, E.N.; Pim, M. An educational simulation tool for negotiating sustainable natural resource management strategies among stakeholders with conflicting interests. *Ecol. Model.* **2008**, *210*, 115–126. [CrossRef]
20. Villamor, G.B.; van Noordwijk, M. Social role-play games vs individual perceptions of conservation and PES agreements for maintaining rubber agroforests in Jambi (Sumatra), Indonesia. *Ecol. Soc.* **2011**, *16*, 27. [CrossRef]
21. Speelman, E.N.; García-Barrios, L.E.; Groot, J.C.J.; Tittonell, P. Gaming for smallholder participation in the design of more sustainable agricultural landscapes. *Agric. Syst.* **2014**, *126*, 62–75. [CrossRef]
22. Speelman, E.N.; van Noordwijk, M.; Garcia, C. Gaming to better manage complex natural resource landscapes. In *Co-investment in Ecosystem Services: Global Lessons from Payment and Incentive Schemes*; Namirembe, S., Leimona, B., van Noordwijk, M., Minang, P., Eds.; World Agroforestry Centre (ICRAF): Nairobi, Kenya, 2017; pp. 480–489.
23. García-Barrios, L.; Cruz-Morales, J.; Vandermeer, J.; Perfecto, I. The Azteca Chess experience: Learning how to share concepts of ecological complexity with small coffee farmers. *Ecol. Soc.* **2017**, 22. [CrossRef]
24. Mayer, I.; Bekebrede, G.; Harteveld, C.; Warmelink, H.; Zhou, Q.; Van Ruijven, T.; Lo, J.; Kortmann, R.; Wenzler, I. The research and evaluation of serious games: Toward a comprehensive methodology. *Br. J. Educ. Technol.* **2014**, *45*, 502–527. [CrossRef]
25. Hofstede, G.J. One game does not fit all cultures. In *Organizing and Leaning through Gaming and Simulation, Proceedings of ISAGA, Nijmegen, The Netherlands, 9–13 July 2007*; Mayer, I., Mastik, H., Eds.; Eburon Uitgeverij BV: Nijmegen, The Netherlands, 2007; pp. 103–110.
26. Jann, W.; Wegrich, K. Theories of the policy cycle. In *Handbook of Public Policy Analysis: Theory, Politics, and Methods*; CRC Press: Boca Raton, FL, USA, 2007; Volume 125, pp. 43–62.
27. Clark, W.C.; Tomich, T.P.; van Noordwijk, M.; Guston, D.; Catacutan, D.; Dickson, N.M.; McNie, E. Boundary work for sustainable development: Natural resource management at the Consultative Group on International Agricultural Research (CGIAR). *Proc. Natl. Acad. Sci. USA* **2016**, *113*, 4615–4622. [CrossRef]
28. Kristensen, P. The DPSIR framework. In *Workshop on a Comprehensive/Detailed Assessment of the Vulnerability of Water Resources to Environmental Change in Africa Using River Basin Approach*; UNEP Headquarters: Nairobi, Kenya, 2004; p. 10. Available online: http://enviro.lclark.edu:8002/rid%3D1145949501662_742777852_522/DPSIR%20Overview.pdf (accessed on 20 July 2020).
29. Svarstad, H.; Petersen, L.K.; Rothman, D.; Siepel, H.; Wätzold, F. Discursive biases of the environmental research framework DPSIR. *Land Use Policy* **2008**, *25*, 116–125. [CrossRef]
30. Diaz, S.; Settele, J.; Brondízio, E.; Ngo, H.; Guèze, M.; Agard, J.; Arneth, A.; Balvanera, P.; Brauman, K.; Butchart, S.; et al. *Summary for Policymakers of the Global Assessment Report on Biodiversity and Ecosystem Services of the Intergovernmental Science-Policy Platform on Biodiversity and Ecosystem Services*; IPBES: Bonn, Germany, 2020; p. 56.
31. van Noordwijk, M. Integrated natural resource management as pathway to poverty reduction: Innovating practices, institutions and policies. *Agric. Syst.* **2019**, *172*, 60–71. [CrossRef]
32. Mithöfer, D.; van Noordwijk, M.; Leimona, B.; Cerutti, P.O. Certify and shift blame, or resolve issues? Environmentally and socially responsible global trade and production of timber and tree crops. *Int. J. Biodivers. Sci. Ecosyst. Serv. Manag.* **2017**, *13*, 72–85. [CrossRef]
33. Hofstede, G.J. GRASP agents: Social first, intelligent later. *Ai Soc.* **2019**, *34*, 535–543. [CrossRef]
34. van Noordwijk, M.; Dewi, S.; Leimona, B.; Lusiana, B.; Wulandari, D. *Negotiation-Support Toolkit for Learning Landscapes*; World Agroforestry Centre: Bogor, Indonesia, 2013; p. 285. Available online: http://www.worldagroforestry.org/publication/negotiation-support-toolkit-learning-landscapes (accessed on 20 July 2020).

35. van Noordwijk, M.; Leimona, B.; Jindal, R.; Villamor, G.B.; Vardhan, M.; Namirembe, S.; Catacutan, D.; Kerr, J.; Minang, P.A.; Tomich, T.P. Payments for Environmental Services: Evolution towards efficient and fair incentives for multifunctional landscapes. *Annu. Rev. Environ. Resour.* **2012**, *37*, 389–420. [CrossRef]
36. Shyamsundar, P.; Ahlroth, S.; Kristjanson, P.; Onder, S. Supporting pathways to prosperity in forest landscapes–A PRIME framework. *World Dev.* **2020**, *125*, 104622. [CrossRef]
37. Dewi, S.; van Noordwijk, M.; Zulkarnain, M.T.; Dwiputra, A.; Hyman, G.; Prabhu, R.; Gitz, V.; Nasi, R. Tropical forest-transition landscapes: A portfolio for studying people, tree crops and agro-ecological change in context. *Int. J. Biodivers. Sci. Ecosyst. Serv. Manag.* **2017**, *13*, 312–329. [CrossRef]
38. van Noordwijk, M.; Bruijnzeel, S.; Ellison, D.; Sheil, D.; Morris, C.E.; Sands, D.; Gutierrez, V.; Cohen, J.; Sullivan, C.A.; Verbist, B.; et al. Ecological rainfall infrastructure: Investment in trees for sustainable development. In *ASB Policy Brief 47*; World Agroforestry Centre: Nairobi, Kenya, 2015; p. 6. Available online: http://www.asb.cgiar.org/Publications%202015/PBs/ASB_PB47.p (accessed on 20 July 2020).
39. Ellison, D.; Morris, C.E.; Locatelli, B.; Sheil, D.; Cohen, J.; Murdiyarso, D.; Gutierrez, V.; van Noordwijk, M.; Creed, I.F.; Pokorny, J.; et al. Trees, forests and water: Cool insights for a hot world. *Glob. Environ. Chang.* **2017**, *43*, 51–61. [CrossRef]
40. van Noordwijk, M.; Creed, I.F.; Jones, J.A.; Wei, X.; Gush, M.; Blanco, J.A.; Sullivan, C.A.; Bishop, K.; Murdiyarso, D.; Xu, J.; et al. Climate-forest-water-people relations: Seven system delineations. In *Forest and Water on a Changing Planet: Vulnerability, Adaptation and Governance Opportunities: A Global Assessment Report (No. 38)*; International Union of Forest Research Organizations (IUFRO): Vienna, Austria, 2018; pp. 27–58.
41. van Noordwijk, M.; Tanika, L.; Lusiana, B. Flood risk reduction and flow buffering as ecosystem services: I. Theory on a flow persistence indicator. *Hydrol. Earth Syst. Sci.* **2017**, *21*, 2321–2340. [CrossRef]
42. van Noordwijk, M.; Tanika, L.; Lusiana, B. Flood risk reduction and flow buffering as ecosystem services II. Land use and rainfall intensity effects in Southeast Asia. *Hydrol. Earth Syst. Sci.* **2017**, *21*, 2341–2352. [CrossRef]
43. van der Ent, R.J.; Savenije, H.H.G.; Schaefli, B.; Steele-Dunne, S.C. Origin and fate of atmospheric moisture over continents. *Water Resour. Res.* **2010**, *46*, W09525. [CrossRef]
44. Ellison, D.; Wang-Erlandsson, L.; van der Ent, R.; van Noordwijk, M. Upwind forests: Managing moisture recycling for nature-based resilience. *Unasylva* **2019**, *251*, 14–26.
45. Teuling, A.J.; Taylor, C.M.; Meirink, J.F.; Melsen, L.A.; Miralles, D.G.; Van Heerwaarden, C.C.; Vautard, R.; Stegehuis, A.I.; Nabuurs, G.J.; de Arellano, J.V.G. Observational evidence for cloud cover enhancement over western European forests. *Nat. Commun.* **2017**, *8*, 14065. [CrossRef]
46. Tuinenburg, O.A.; Staal, A. Tracking the global flows of atmospheric moisture and associated uncertainties. *Hydrol. Earth Syst. Sci.* **2020**, *24*, 2419–2435. Available online: https://www.hydrol-earth-syst-sci.net/24/2419/2020/ (accessed on 17 July 2020). [CrossRef]
47. van Noordwijk, M.; Bargues-Tobella, A.; Muthuri, C.W.; Gebrekirstos, A.; Maimbo, M.; Leimona, B.; Bayala, J.; Ma, X.; Lasco, R.; Xu, J.; et al. Agroforestry as part of nature-based water management. In *Sustainable Development through Trees on Farms: Agroforestry in its Fifth Decade*; van Noordwijk, M., Ed.; World Agroforestry (ICRAF): Bogor, Indonesia, 2019; pp. 261–287.
48. Guswa, A.J.; Tetzlaff, D.; Selker, J.S.; Carlyle-Moses, D.E.; Boyer, E.W.; Bruen, M.; Cayuela, C.; Creed, I.F.; van de Giesen, N.; Grasso, D.; et al. Advancing ecohydrology in the 21st century: A convergence of opportunities. *Ecohydrology* **2020**, e2208. [CrossRef]
49. Kadykalo, A.N.; López-Rodriguez, M.D.; Ainscough, J.; Droste, N.; Ryu, H.; Ávila-Flores, G.; Le Clec'h, S. Disentangling 'ecosystem services' and 'nature's contributions to people'. *Ecosyst. People* **2019**, *15*, 269–287. [CrossRef]
50. van Noordwijk, M. Did Mother Nature consent in the human 'appropriation' of 'gifts' so sweet, or would a #MeToo hashtag get flooded if she could tweet? *Sci. E-Lett.* **2019**, *359*, 270.
51. Beugelsdijk, S.; Welzel, C. Dimensions and dynamics of national culture: Synthesizing Hofstede with Inglehart. *J. Cross-Cult. Psychol.* **2018**, *49*, 1469–1505. [CrossRef] [PubMed]
52. Minkov, M. A revision of Hofstede's model of national culture: Old evidence and new data from 56 countries. *Cross Cult. Strateg. Manag.* **2018**, *25*, 231–256. [CrossRef]
53. Dimension Data Matrix. Available online: https://geerthofstede.com/research-and-vsm/dimension-data-matrix/ (accessed on 20 July 2020).

54. Hofstede, G.; Hofstede, G.J.; Minkov, M. *Cultures and Organizations: Software of the Mind*, 3rd ed.; McGraw-Hill: New York, NY, USA, 2010; p. 561.
55. Hofstede, G.J. Culture's causes: The next challenge (Distinguished scholar essay). *Cross Cult. Manag.* **2015**, *22*, 545–569. [CrossRef]
56. Hofstede, G.J.; Murff, E.J. Repurposing an Old Game for an International World. *Simul. Gaming* **2012**, *43*, 34–50. [CrossRef]
57. Hofstede, G.J.; de Caluwé, L.; Peters, V. Why Simulation Games Work-In Search of the Active Substance: A Synthesis. *Simul. Gaming* **2010**, *41*, 824–843. [CrossRef]
58. Villamor, G.B.; Badmos, B.K. Grazing game: A learning tool for adaptive management in response to climate variability in semiarid areas of Ghana. *Ecol. Soc.* **2016**, *21*. [CrossRef]
59. Barreteau, O. Our companion modelling approach. *J. Artif. Soc. Soc. Simul.* **2003**, *6*, 1. Available online: http://jasss.soc.surrey.ac.uk/6/2/1.html (accessed on 20 July 2020).
60. International Simulation and Gaming Association. Available online: https://www.isaga.com/ (accessed on 20 July 2020).
61. Reed, M.S.; Graves, A.; Dandy, N.; Posthumus, H.; Hubacek, K.; Morris, J.; Prell, C.; Quinn, C.H.; Stringer, L.C. Who's in and why? A typology of stakeholder analysis methods for natural resource management. *J. Environ. Manag.* **2009**, *90*, 1933–1949. [CrossRef] [PubMed]
62. Etienne, M.; Du Toit, D.R.; Pollard, S. ARDI: A co-construction method for participatory modeling in natural resources management. *Ecol. Soc.* **2011**, *16*. Available online: http://www.ecologyandsociety.org/vol16/iss1/art44/ (accessed on 20 July 2020). [CrossRef]
63. Vanwindekens, F.M.; Baret, P.V.; Stilmant, D. A new approach for comparing and categorizing farmers' systems of practice based on cognitive mapping and graph theory indicators. *Ecol. Model.* **2014**, *274*, 1–11. [CrossRef]
64. Langston, J.; McIntyre, R.; Falconer, K.; Sunderland, T.; Van Noordwijk, M.; Boedhihartono, A.K. Discourses mapped by Q-method show governance constraints motivate landscape approaches in Indonesia. *PLoS ONE* **2019**, *14*, e0211221. [CrossRef] [PubMed]
65. Amaruzaman, S.; Leimona, B.; van Noordwijk, M.; Lusiana, B. Discourses on the performance gap of agriculture in a green economy: A Q-methodology study in Indonesia. *Int. J. Biodivers. Sci. Ecosyst. Serv. Manag.* **2017**, *13*, 233–247. [CrossRef]
66. Page, C.L.; Dray, A.; Perez, P.; Garcia, C. Exploring how knowledge and communication influence natural resources management with ReHab. *Simul. Gaming* **2016**, *47*, 257–284. [CrossRef]
67. Andreotti, F.; Speelman, E.N.; Van den Meersche, K.; Allinne, C. Combining participatory games and backcasting to support collective scenario evaluation: An action research approach for sustainable agroforestry landscape management. *Sustain. Sci.* **2020**. [CrossRef]
68. Ligtenberg, A.; van Lammeren, R.J.A.; Bregt, A.K.; Beulens, A.J.M. Validation of an agent-based model for spatial planning: A role-playing approach. *Comput. Environ. Urban Syst.* **2010**, *34*, 424–434. [CrossRef]
69. Van Oel, P.R.; Mulatu, D.W.; Odongo, V.O.; Willy, D.K.; Van der Veen, A. Using data on social influence and collective action for parameterizing a geographically-explicit agent-based model for the diffusion of soil conservation efforts. *Environ. Model. Assess.* **2019**, *24*, 1–19. [CrossRef]
70. Aubert, A.H.; Bauer, R.; Lienert, J. A review of water-related serious games to specify use in environmental Multi-Criteria Decision Analysis. *Environ. Model. Softw.* **2018**, *105*, 64–78. [CrossRef]
71. Meinzen-Dick, R.; Janssen, M.A.; Kandikuppa, S.; Chaturvedi, R.; Rao, K.; Theis, S. Playing games to save water: Collective action games for groundwater management in Andhra Pradesh, India. *World Dev.* **2018**, *107*, 40–53. [CrossRef]
72. Marini, D.; Medema, W.; Adamowski, J.; Veissière, S.P.; Mayer, I.; Wals, A.E. Socio-psychological perspectives on the potential for serious games to promote transcendental values in IWRM decision-making. *Water* **2018**, *10*, 1097. [CrossRef]
73. Buyinza, J.; Nuberg, I.K.; Muthuri, C.W.; Denton, M.D. Psychological Factors Influencing Farmers' Intention to Adopt Agroforestry: A Structural Equation Modeling Approach. *J. Sustain. For.* **2020**. [CrossRef]
74. Guyot, P.; Honiden, S. Agent-based participatory simulations: Merging multi-agent systems and role-playing games. *J. Artif. Soc. Soc. Simul.* **2006**, *9*, 8. Available online: http://jasss.soc.surrey.ac.uk/9/4/8.html (accessed on 20 July 2020).

75. Speelman, E.N.; Rodela, R.; Doddema, M.; Ligtenberg, A. Serious gaming as a tool to facilitate inclusive business; a review of untapped potential. *Curr. Opin. Environ. Sustain.* **2019**, *41*, 31–37. [CrossRef]
76. Zafra-Calvo, N.; Balvanera, P.; Pascual, U.; Merçon, J.; Martin-Lopez, B.; van Noordwijk, M.; Heita Mwampamba, T.; Lele, S.; Ifejika Speranza, C.; Arias-Arévalok, P.; et al. Plural valuation of nature for equity and sustainability: Insights from the Global South. *Glob. Environ. Chang.* **2020**, *63*, 102115. [CrossRef]
77. Leimona, B.; Lusiana, B.; van Noordwijk, M.; Mulyoutami, E.; Ekadinata, A.; Amaruzaman, S. Boundary work: Knowledge co-production for negotiating payment for watershed services in Indonesia. *Ecosyst. Serv.* **2015**, *15*, 45–62. [CrossRef]
78. Ouboter, P.E. (Ed.) *The Fresh Water Ecosystems of Suriname*; Kluwer Academic Publishers: Dordrecht, The Netherlands, 1993; p. 315.
79. Misiedjan, D.J.E. Towards a Sustainable Human Right to Water: Supporting Vulnerable People and Protecting Water Resources with Suriname as a Case Study. Ph.D. Thesis, Utrecht University, Utrecht, The Netherlands, 2017; p. 220.
80. Nurmohamed, R.; Naipal, S.; Becker, C. Changes and variation in the discharge regime of the Upper Suriname River Basin and its relationship with the tropical Pacific and Atlantic SST anomalies. *Hydrol. Process.* **2008**, *22*, 1650–1659. [CrossRef]
81. Ramirez, S.O.I.; Verweij, P.; Best, L.; van Kanten, R.; Rambaldi, G.; Zagt, R. P3DM as a socially engaging and user-useful approach in ecosystem service assessment among marginalized communities. *Appl. Geogr.* **2017**, *83*, 63–77. [CrossRef]
82. *Census Statistics: Population by Resort*; Statistics Bureau Suriname: Paramaribo, Suriname, 2012.
83. Keys, P.W.; van der Ent, R.J.; Gordon, L.J.; Hoff, H.; Nikoli, R.; Savenije, H.H.G. Analyzing precipitationsheds to understand the vulnerability of rainfall dependent regions. *Biogeosciences* **2012**, *9*, 733–746. [CrossRef]
84. Global Forest Watch. Dashboard by Country and Region. Available online: https://www.globalforestwatch.org/dashboards/country/PER (accessed on 20 July 2020).
85. Climate-data, Puerto Maldonado Climate. Available online: https://en.climate-data.org/south-america/peru/madre-de-dios/puerto-maldonado-27856/ (accessed on 20 July 2020).
86. Madre de Dios Web Portal, Ríos y Quebradas. Available online: https://madrededios.com.pe/rios-y-quebradas.html (accessed on 20 June 2020).
87. *Madre de Dios, Compendio Estadistico*; Instituto Nacional de Estadística e Informática (INEI): Puerto Maldonado, Peru, 2018.
88. Finer, M.; Mamani, N.; García, R.; Novoa, S. MAAP: 78: Deforestation Hotspots in the Peruvian Amazon. 2017. Available online: http://maaproject.org/2018/hotspots-peru2017/ (accessed on 20 July 2020).
89. Rojas, R. *Vulnerabilidad al cambio climático en el sector agropecuario de la región Madre de Dios*; Proyecto Especial Madre de Dios; Consorcio Madre de Dios: La Molina, Peru, 2014.
90. Stickler, C.; Duchelle, A.E.; Nepstad, D.; Ardila, J.P. Subnational jurisdictional approaches. In *Transforming REDD*; CIFOR: Bogor, Indonesia, 2018; pp. 145–159.
91. Lovejoy, T.E.; Nobre, C. Amazon tipping point. *Sci. Adv.* **2018**, *4*, eaat2340. [CrossRef] [PubMed]
92. Tropical Forest Alliance (TFA). Available online: https://www.tropicalforestalliance.org/ (accessed on 20 July 2020).
93. *Censo Demográfico*; Instituto Brasileiro de Geografia e Estatística (IBGE): Brasilia, Brasil, 2010.
94. Smith, N.J.; Falesi, I.C.; Alvim, P.D.T.; Serrão, E.A.S. Agroforestry trajectories among smallholders in the Brazilian Amazon: Innovation and resiliency in pioneer and older settled areas. *Ecol. Econ.* **1996**, *18*, 15–27. [CrossRef]
95. Saes, M.S.M.; Silva, V.L.; Nunes, R.; Gomes, T.M. Partnerships, learning, and adaptation: A cooperative founded by Japanese immigrants in the Amazon rainforest. *Int. J. Bus. Soc. Sci.* **2014**, *5*, 131–141.
96. Salafsky, N. Drought in the rain forest: Effects of the 1991 El Niño-Southern Oscillation event on a rural economy in West Kalimantan, Indonesia. *Clim. Chang.* **1994**, *27*, 373–396. [CrossRef]
97. Carlson, K.M.; Curran, L.M.; Ratnasari, D.; Pittman, A.M.; Soares-Filho, B.S.; Asner, G.P.; Trigg, S.N.; Gaveau, D.A.; Lawrence, D.; Rodrigues, H.O. Committed carbon emissions, deforestation, and community land conversion from oil palm plantation expansion in West Kalimantan, Indonesia. *Proc. Natl. Acad. Sci. USA* **2012**, *109*, 7559–7564. [CrossRef] [PubMed]

98. Farida, A.; Jeanes, K.; Kurniasari, D.; Widayati, A.; Ekadinata, A.; Hadi, D.P.; Joshi, L.; Suyamto, D.; van Noordwijk, M. *Rapid Hydrological Appraisal (RHA) of Singkarak Lake in the Context of Rewarding Upland Poor for Environmental Services (RUPES)*; World Agroforestry Centre: Bogor, Indonesia, 2005; p. 119.
99. Burgers, P.; Farida, A. Community management for agro-reforestation under a voluntary carbon market scheme in West Sumatra. In *Co-Investment in Ecosystem Services: Global Lessons from Payment and Incentive Schemes*; Namirembe, S., Leimona, B., van Noordwijk, M., Minang, P., Eds.; World Agroforestry Centre (ICRAF): Nairobi, Kenya, 2017; Available online: http://www.worldagroforestry.org/sites/default/files/chapters/Ch29%20Collective%20action_ebookB-DONE2.pdf (accessed on 20 July 2020).
100. Musau, J.; Sang, J.; Gathenya, J.; Luedeling, E. Hydrological responses to climate change in Mt. Elgon watersheds. *J. Hydrol. Reg. Stud.* **2015**, *3*, 233–246. [CrossRef]
101. Shames, S.; Heiner, K.; Kapukha, M.; Kiguli, L.; Masiga, M.; Kalunda, P.N.; Ssempala, A.; Recha, J.; Wekesa, A. Building local institutional capacity to implement agricultural carbon projects: Participatory action research with Vi Agroforestry in Kenya and ECOTRUST in Uganda. *Agric. Food Secur.* **2016**, *5*, 13. [CrossRef]
102. Muhweezi, A.B.; Sikoyo, G.M.; Chemonges, M. Introducing a Transboundary Ecosystem Management Approach in the Mount Elgon Region. *Mt. Res. Dev.* **2007**, *27*, 215–219. [CrossRef]
103. Mugagga, F. The Effect of Land Use on Carbon Stocks and Implications for Climate Variability on the Slopes of Mount Elgon, Eastern Uganda. *Int. J. Reg. Dev.* **2015**, *2*, 58–75. [CrossRef]
104. Nsubuga, F.N.W.; Namutebi, E.N.; Nsubuga-Ssenfuma, M. Water Resources of Uganda: An Assessment and Review. *J. Water Resour. Prot.* **2014**, *6*, 1297–1315. [CrossRef]
105. Mugagga, F.; Buyinza, M. Land tenure and soil conservation practices on the slopes of Mt Elgon National Park, Eastern Uganda. *J. Geogr. Reg. Plan.* **2013**, *6*, 255–262. [CrossRef]
106. Sassen, M. Conservation in a Crowded Place: Forest and people on Mount Elgon Uganda. Ph.D. Thesis, Wageningen University, Wageningen, The Netherlands, 2014. Available online: https://edepot.wur.nl/293853 (accessed on 20 July 2020).
107. Mutiga, J.K.; Mavengano, S.T.; Zhongbo, S.; Woldai, T.; Becht, R. Water allocation as a planning tool to minimise water use conflicts in the Upper Ewaso Ng'iro North Basin, Kenya. *Water Resour. Manag.* **2010**, *24*, 3939–3959. [CrossRef]
108. Ngigi, S.N.; Savenije, H.H.; Gichuki, F.N. Land use changes and hydrological impacts related to up-scaling of rainwater harvesting and management in upper Ewaso Ng'iro river basin, Kenya. *Land Use Policy* **2007**, *24*, 129–140. [CrossRef]
109. Chepyegon, C.; Kamiya, D. Challenges faced by the Kenya water sector management in improving water supply coverage. *J. Water Resour. Prot.* **2018**, *10*, 85–105. [CrossRef]
110. Climate-Data, Puno Climate. Available online: https://en.climate-data.org/search/?q=puno (accessed on 20 July 2020).
111. Fagandini, R.F. Distribution des parents sauvages du quinoa cultivé en lien avec les pratiques et usages des communautés andines dans la région de Puno au Pérou. Ph.D. Thesis, AgroParisTec, Paris, France, 2019.
112. International Year of Quinoa 2013. Available online: http://www.fao.org/quinoa-2013/iyq/en/?no_mobile=1 (accessed on 25 June 2020).
113. Winkel, T.; Álvarez-Flores, R.; Bertero, D.; Cruz, P.; Castillo, C.D.; Joffre, R.; Parada, S.P.; Tonacca, L.S. Calling for a reappraisal of the impact of quinoa expansion on agricultural sustainability in the Andean highlands. *Idesia* **2014**, *32*, 95–100. [CrossRef]
114. Diarisso, T.; Corbeels, M.; Andrieu, N.; Djamen, P.; Douzet, J.-M.; Tittonell, P. Soil variability and crop yield gaps in two village landscapes of Burkina Faso. *Nutr. Cycl. Agroecosyst.* **2016**, *105*, 199–216. [CrossRef]
115. Hengl, T.; de Jesus, J.M.; Heuvelink, G.B.; Gonzalez, M.R.; Kilibarda, M.; Blagotić, A.; Shangguan, W.; Wright, M.N.; Geng, X.; Bauer-Marschallinger, B. SoilGrids250m: Global gridded soil information based on machine learning. *PLoS ONE* **2017**, *12*, e0169748. [CrossRef]
116. Chikowo, R.; Zingore, S.; Snapp, S.; Johnston, A. Farm typologies, soil fertility variability and nutrient management in smallholder farming in Sub-Saharan Africa. *Nutr. Cycl. Agroecosyst.* **2014**, *100*, 1–18. [CrossRef]
117. Tittonell, P.; van Wijk, M.T.; Herrero, M.; Rufino, M.C.; de Ridder, N.; Giller, K.E. Beyond resource constraints–Exploring the biophysical feasibility of options for the intensification of smallholder crop-livestock systems in Vihiga district, Kenya. *Agric. Syst.* **2009**, *101*, 1–19. [CrossRef]

118. van Wijk, M.T.; Tittonell, P.; Rufino, M.C.; Herrero, M.; Pacini, C.; de Ridder, N.; Giller, K.E. Identifying key entry-points for strategic management of smallholder farming systems in sub-Saharan Africa using the dynamic farm-scale simulation model NUANCES-FARMSIM. *Agric. Syst.* **2009**, *102*, 89–101. [CrossRef]
119. Grillot, M.; Guerrin, F.; Gaudou, B.; Masse, D.; Vayssières, J. Multi-level analysis of nutrient cycling within agro-sylvo-pastoral landscapes in West Africa using an agent-based model. *Environ. Model. Softw.* **2018**, *107*, 267–280. [CrossRef]
120. Rufino, M.C.; Dury, J.; Tittonell, P.; van Wijk, M.T.; Herrero, M.; Zingore, S.; Mapfumo, P.; Giller, K.E. Competing use of organic resources, village-level interactions between farm types and climate variability in a communal area of NE Zimbabwe. *Agric. Syst.* **2011**, *104*, 175–190. [CrossRef]
121. Andrieu, N.; Vayssières, J.; Corbeels, M.; Blanchard, M.; Vall, E.; Tittonell, P. From farm scale synergies to village scale trade-offs: Cereal crop residues use in an agro-pastoral system of the Sudanian zone of Burkina Faso. *Agric. Syst.* **2015**, *134*, 84–96. [CrossRef]
122. Baudron, F.; Delmotte, S.; Corbeels, M.; Herrera, J.M.; Tittonell, P. Multi-scale trade-off analysis of cereal residue use for livestock feeding vs. soil mulching in the Mid-Zambezi Valley, Zimbabwe. *Agric. Syst.* **2015**, *134*, 97–106. [CrossRef]
123. Meteorological, Climatological, and Geophysical Agency of Karangploso. *Rainfall Data of Karangploso Station Year 2013–2017*; BMKG Karangploso: Malang, East Java, Indonesia, 2018.
124. Kurniawan, S.; Prayogo, C.; Widianto, M.; Zulkarnain, M.T.; Lestari, N.D.; Aini, F.K.; Hairiah, K. *Estimasi Karbon Tersimpan di Lahan-lahan Pertanian di DAS Konto, Jawa Timur: RACSA (Rapid Carbon Stock Appraisal). Working Paper no.120*; World Agroforestry Centre: Bogor, Indonesia, 2010; p. 59.
125. Lusiana, B.; van Noordwijk, M.; Cadisch, G. Land sparing or sharing? Exploring livestock fodder options in combination with land use zoning and consequences for livelihoods and net carbon stocks using the FALLOW model. *Agric. Ecosyst. Environ.* **2012**, *159*, 145–160. [CrossRef]
126. BPS (Central Bureau of Statistics). *Malang Regency in Figures 2018*; BPS-Statistic of Malang Regency: Malang, East Java, Indonesia, 2018.
127. Andriyanto, C.; Sudarto, S.; Suprayogo, D. Estimation of soil erosion for a sustainable land use planning: RUSLE model validation by remote sensing data utilization in the Kalikonto watershed. *J. Degrad. Min. Lands Manag.* **2015**, *3*, 459–468. [CrossRef]
128. Jackson, L.; Bawa, K.; Pascual, U.; Perrings, C. *Agro-Biodiversity–A New Science Agenda for Biodiversity in Support of Sustainable Agroecosystems*; Diversitas: Paris, France, 2005; p. 40. Available online: http://www.diversitasinternational.org/cross_agriculture.html (accessed on 20 July 2020).
129. Rijsdijk, A.; Bruijnzeel, L.S.; Sutoto, C.K. Runoff and sediment yield from rural roads, trails and settlements in the upper Konto catchment, East Java, Indonesia. *Geomorphology* **2007**, *87*, 28–37. [CrossRef]
130. *Pasuruan Regency in Figures*; BPS-Statistics of Pasuruan Regency: Pasuruan, Indonesia, 2019.
131. *Pasuruan City in Figures*; BPS-Statistics of Pasuruan City: Pasuruan, Indonesia, 2019.
132. Ross, H.; Adhuri, D.S.; Abdurrahim, A.Y.; Phelan, A. Opportunities in community-government cooperation to maintain marine ecosystem services in the Asia-Pacific and Oceania. *Ecosyst. Serv.* **2019**, *38*, 100969. [CrossRef]
133. Charles, A.; Loucks, L.; Berkes, F.; Armitage, D. Community science: A typology and its implications for governance of social-ecological systems. *Environ. Sci. Policy* **2020**, *106*, 77–86. [CrossRef]
134. van Noordwijk, M.; Tomich, T.P.; Verbist, B. Negotiation support models for integrated natural resource management in tropical forest margins. *Conserv. Ecol.* **2001**, *5*, 21. Available online: http://www.consecol.org/vol5/iss2/art21 (accessed on 17 July 2020). [CrossRef]
135. Perhutani. *KPH Pasuruan*; Jakarta, Indonesia. Available online: https://perhutani.co.id/tentang-kami/struktur-organisasi-perum-perhutani/divisi-regional/jatim/kph-pasuruan/ (accessed on 20 July 2020).
136. Suprayogo, D.; van Noordwijk, M.; Widianto; Hairiah, K.; Meilasari, N.; Rabbani, A.L.; Ishaq, R.M. Infiltration-Friendly Land Uses for Climate Resilience on Volcanic Slopes in the Rejoso Watershed, East Java, Indonesia. *Land* **2020**, *9*, 240. [CrossRef]
137. Amaruzaman, S.; Khasanah, N.; Tanika, L.; Dwiyanti, E.; Lusiana, B.; Leimona, B. *Landscape Characteristics of Rejoso Watershed: Land Cover Dynamics, Farming Systems, and Community Strategies*; World Agroforestry Centre (ICRAF) Southeast Asia Regional Program: Bogor, Indonesia, 2018.

138. Leimona, B.; Khasanah, N.; Lusiana, B.; Amaruzaman, S.; Tanika, L.; Hairiah, K.; Suprayogo, D.; Pambudi, S.; Negoro, F.S. *A Business Case: Co-Investing for Ecosystem Service provIsions and Local Livelihoods in Rejoso Watershed*; World Agroforestry Centre (ICRAF) Southeast Asia Regional Program: Bogor, Indonesia, 2018.
139. Tanika, L.; Khasanah, N.; Leimona, B. *Simulasi Dampak Perubahan Tutupan Lahan Terhadap Neraca Air di DAS dan Sub-DAS Rejoso Menggunakan Model GenRiver*; World Agroforestry Centre (ICRAF) Southeast Asia Regional Program: Bogor, Indonesia, 2018.
140. Mahzum, M.M. *Analisis Ketersediaan Sumber Daya Air dan Upaya Konservasi SUb DAS Brantas Hulu Wilayah Kota Batu*; Civil Engineering and Planning Facultry, Institut Tehnologi Sepuluh November: Magister, Surabaya, 2015; p. 187.
141. *Batu Municipality in Figures*; BPS_Statistical Bureau: Jakarta, Indonesia, 2020; p. 151.
142. Maryudi, A.; Citraningtyas, E.R.; Purwanto, R.H.; Sadono, R.; Suryanto, P.; Riyanto, S.; Siswoko, B.D. The emerging power of peasant farmers in the tenurial conflicts over the uses of state forestland in Central Java, Indonesia. *For. Policy Econ.* **2016**, *67*, 70–75. [CrossRef]
143. Witjaksono, A.; Surjono, S.; Suharyanto, A.; Muhammad, B. Spatial analysis of land use in Bumiaji subdistrict, Batu city, East Java, Indonesia. *Int. J. GEOMATE* **2018**, *15*, 139–144. [CrossRef]
144. Nurrizqi, E.; Suyono, S. Pengaruh Perubahan Penggunaan Lahan Terhadap Perubahan Debit Puncak Banjir Di Sub DAS Brantas Hulu. *Jurnal Bumi Indonesia* **2012**, *1*, 363.

# 8

# A Discounted Cash Flow and Capital Budgeting Analysis of Silvopastoral Systems

**Stephanie Chizmar [1,*], Miguel Castillo [1], Dante Pizarro [2], Hector Vasquez [3], Wilmer Bernal [3], Raul Rivera [1], Erin Sills [1], Robert Abt [1], Rajan Parajuli [1] and Frederick Cubbage [1]**

1. Department of Forestry and Environmental Resources, North Carolina State University, 2800 Faucette Dr., Raleigh, NC 27695, USA; mscastil@ncsu.edu (M.C.); rrivera4@ncsu.edu (R.R.); sills@ncsu.edu (E.S.); bobabt@ncsu.edu (R.A.); rparaju@ncsu.edu (R.P.); cubbage@ncsu.edu (F.C.)
2. Livestock and Climate Change Center, Universidad Nacional Agraria La Molina, Lima 15000, Peru; dpizarro@lamolina.edu.pe
3. Departamento de Zootecnia, Universidad Nacional Toribio Rodriguez de Mendoza de Amazonas, Chachapoyas 477002, Peru; hvasquez@untrm.edu.pe (H.V.); wilmer.bernal@untrm.edu.pe (W.B.)

* Correspondence: sjchizma@ncsu.edu

**Abstract:** Silvopasture is a type of agroforestry that could deliver ecosystem services and support local livelihoods by integrating trees into pasture-based livestock systems. This study modeled the financial returns from silvopastures, planted forests, and conventional cattle-pasture systems in Amazonas, Peru using capital budgeting techniques. Forests had a lower land expectation value (USD 845 per hectare) than conventional cattle systems (USD 1275 per hectare) at a 4% discount rate. "Typical" model silvopastures, based on prior landowner surveys in the Amazonas region, were most competitive at low discount rates. The four actual silvopastoral systems we visited and examined had higher returns (4%: USD 1588 to USD 9524 per hectare) than either alternative pure crop or tree system, more than likely through strategies for generating value-added such as on-site retail stands. Silvopasture also offers animal health and environmental benefits, and could receive governmental or market payments to encourage these practices.

**Keywords:** silvopasture; economics; financial analysis; carbon payment; Peru

---

## 1. Introduction

Expanding forest cover remains a global priority to combating trends in deforestation and global climate change. Countries in Latin America, where the livestock sector generates 58–70% of overall agricultural emissions, are particularly interested in mitigating climate change through integrating forest cover into agricultural production systems [1]. Agroforestry operations combine forest or horticultural species and pasture or cropland to make mixed land-use systems that produce commercial benefits to landowners [2]. Silvopasture, a branch of agroforestry, is a strategic and managed agroecosystem in which livestock, forage, and trees or shrubs are integrated to help improve individual components [3,4]. Silvopastoral systems (SPS) diversify earnings to landowners by generating products on various harvest schedules, from daily as in the case of milk and cheese to multi-year for forest products such as fuelwood, posts, and boards [5].

As such, silvopasture has been identified as a key integrated landscape approach that could potentially increase returns to landowners by generating timber and non-timber products, while also improving conditions for forage production and cattle [6]. This most likely occurs when the components of the system have complementary relationships such that tree and cattle–forage production are mutually beneficial, which has been found true at low tree densities in the USA [7]. However, this may not be the case in the absence of established markets and technical knowledge networks. In other words, without access to knowledge and markets, silvopastoral practices may not achieve the social and environmental benefits they have been documented to produce [8].

Recently, the Peruvian government defined its Nationally Determined Contributions, which contemplate reducing 30% of the greenhouse gas emissions projected for the year 2030. The government recommended considering strategies such as the recovery of degraded soils with SPS in the Peruvian Amazon. Incorporating SPS in the Peruvian Amazon has the potential to mitigate 1344 metric tons of $CO_2e$ through intervention on 102,000 hectares [9]. Peru also committed to restore 3.2 million hectares as part of the Bonn Challenge's mission to restore 150 million hectares of the world's degraded and deforested lands by 2020 [10]. Silvopasture is one strategy aimed at increasing forest cover in agricultural systems, augmenting a region's carbon sequestration potential, especially in deforested and degraded ecosystems [1].

Amazonas, a jurisdiction of Northeastern Peru, has an economically disadvantaged rural population that depends on agriculture and animal husbandry as primary income sources. Poverty rates exceed 50% and malnutrition levels are 30% or more in parts of Amazonas [11], suggesting a need for governmental intervention aimed at improving the livelihoods of the lower income populace. The rural population of Amazonas, encompassing over half of the total population [12], has opportunities to improve the welfare of landowners, farmers, and associated laborers. Rural landowners, both onsite and absentee, may generate additional income flows from timber and non-timber products by incorporating trees into their agricultural systems. Rural households may also become eligible for payments for ecosystem services through helping to expand tree cover and thereby provide habitat, sequester carbon, or provide other benefits.

The main goal of this paper is to evaluate the financial returns of Peruvian production systems, including silvopasture, typical cattle-pasture operations, and planted tree monocultures, through the application of discounted cash flow and capital budgeting analyses. The main hypothesis of SPS states that mixed-use systems can diversify farm income, reduce biophysical and financial risks, and perhaps, increase total farm returns by providing timber and non-timber products in addition to forage for livestock production. We postulated that the degree of tree and forage competition determines the profitability of SPS. In other words, at a modest tree density, trees may complement forage systems; however, at high tree densities, the production of forage will decline due to competition for resources, especially light. The collaborative team determined productivity data as well as prices of inputs and outputs as reported by local silvopasture practitioners and industry experts to model whole-system cash flows. Finally, we assessed the potential impact of incentive payments on a range of actual and typical landowner income for both absentee and onsite landowners. The rest of this paper is structured in the following five sections: (i) a review of the literature on silvopasture, particularly in tropical and subtropical regions; (ii) the methods we utilized to determine cash flows for multiple land-use systems; (iii) the results of the discounted cash flow and capital budgeting analyses; (iv) a discussion on the profitability of different land-uses; and (v) concluding remarks.

## 2. Literature Review

Loker [13] modeled low-input agroforests to determine their suitability as alternative, sustainable production systems for small-to-medium farmers in the Amazon Basin, as landowners in the region are attracted to cattle management for its multiple economic benefits. Landowners in Uruguay recognized that shelter from silvopasture benefits cattle, fostering increased calving rates [14]. In Argentina, small-scale farmers perceived cash flow diversification through the sale of livestock and forest products as the main advantage of silvopasture [15]. Silvopastures also likely provide better microenvironments, reducing climatic-induced stress, such that grazing animals are "happier" and gain weight at a faster rate than when grazing in traditional open pastures [3]. For many species of livestock, deviations in core body temperature more than 2 to 3 °C negatively impact performance, productivity, and fertility, potentially leading to decreased successful pregnancies [16]. Faster and increased weight gain from improved microenvironments as well as augmented crop growth may lead to higher profits for landowners, as well as better animal welfare in Amazonas.

Agroforests potentially sequester more carbon than pastures or field crops growing under similar ecological conditions [17]. Oliva et al. [18] measured 337.2 tons of carbon stored in *Pinus patula* silvopastures with 8- to 10-year-old trees in Amazonas. Dube et al. [19] reported that silvopastures in Patagonia, Chile, as well as in Minas Gerais, Brazil [20] required less time than a plantation monoculture to reach similar carbon gains above and below ground. The researchers attributed gains in carbon to the positive interactions between cattle, tree, and pasture components, including increased tree growth. Cubbage et al. [21] reviewed global SPS in eight regions, including areas in Argentina, Uruguay, Chile, and Brazil, and reported that silvopasture is financially competitive with alternative land-use systems and offers biophysical and financial diversity and resilience, attributes critical with the increased occurrences of extreme weather events due to climate change.

Payments for ecosystem (environmental) services (PES) compensate landowners for managing their property to sequester carbon, protect biodiversity, or provide watershed services. Some scientists have argued that in tropical systems the greatest potential for carbon sequestration is through the establishment of tree-based systems on degraded pastures [22]. Programs that provide incentives for forest management on degraded productive lands and supply additional sources of revenue may aid profitability as well as conservation of land-use systems in Amazonas. Orefice et al. [23] found that silvopastures in the U.S. North were financially superior to open pasture or thinned forests. Bruck et al. [24] observed that SPS are less financially profitable than pasture in the southern and western U.S. The study, however, did not account for any animal welfare or joint output effects.

Silvopastures can also shift forage production for livestock and soil quality, resulting in an expanded grazing season and livestock diets higher in protein [3,23]. Livestock provides essential elements such as nitrogen, phosphorus, and potassium through nutrient cycling to fertilize forages and trees, which may reduce dependence on additional external inputs [25,26]. Pent and Fike [25] suggested there is a complementary biophysical relationship between forage production for livestock and trees, assuming a modest tree density. Silvopastoral systems with relatively low tree densities required reduced weeding, provided increased available nitrogen, improved the microclimate, reduced erosion control costs, and fostered better animal health such as increased pregnancy success rates [26].

## 3. Methods

The Molinopampa district, located in the northeastern section of the Chachapoyas Province, in the southern portion of the Amazonas Region, has an oceanic climate (dry forest lower montane tropical according to Holdridge's Life Zone) [27]. Approximately 75% of the population in Molinopampa make a living in agriculture, ranching, hunting, and silviculture [28]. The Huayabamba Valley, in the Rodríguez de Mendoza Province, is characterized as having a tropical savanna and warm-humid climate (moisture forest premontane tropical according to Holdridge's Life Zone) [27]. Past surveys of the study area [28–31] found that 61% and 43% of the silvopasture producers in the Molinopampa district and the Huayabamba Valley, respectively, own farms that are on less than 10 hectares (ha). Conversely, 10-ha to 30-ha (30-ha+) farms represented 28% and 42% (11% and 16%) of the surveyed systems in the Molinopampa district and the Huayabamba Valley, respectively. The field sites for this study, which participated in the past surveys, are located on privately-owned farms, are managed as part of cooperatives with multiple institutions, and range from 10-ha to 65-ha in size [28–31]. Therefore, our sample farms are among the larger size class ownerships for these similar regions, but still within a representative range.

Universidad Nacional Agraria La Molina (UNALM), El Porvenir Research Station (INIA), the Universidad Nacional Toribio Rodríguez de Mendoza de Amazonas (UNTRM), and North Carolina State University (NCSU) have been collaborating with private landowners over the past five years to better understand the dynamics of applied agroforestry systems [28]. In July 2017, the collaborative team interviewed three landowners and one land manager well-known locally for practicing silvopasture and who have established relationships with the local university UNTRM. We selected these farms since they have established connections to the researchers and are most likely local leaders in their farm practices. Analyzing examples of successful programs may help to illuminate cost-effective characteristics and strategies that could be adapted throughout Amazonas. Cash flow estimations based on the four landowner responses, as well as average costs and benefits from discussions among co-authors, were predicted for each site in addition to "typical" model systems representing average farms in the region documented in the previously mentioned surveys [28–31].

The study collaborators gathered production function data for the case studies and typical model systems, organized in a database, to estimate values from soils, crops, and grazing. We estimated the associated economic impacts of inputs and outputs over a 25-year time horizon, both to calculate Net Present Value and Land Expectation Value. We then performed analyses in Microsoft Excel for Microsoft 365 using discounted cash flows associated with SPS, conventional grazing, and planted forests. The spreadsheets included necessary measures of productivity, product prices and costs, and management schedules, among other site-specific information, to calculate the net returns of annual activities. The operations were modeled as whole land-use regimes due to their complexity and the available information.

Tools in capital budgeting analysis such as Net Present Value (NPV), Land Expectation Value (LEV), Annual Equivalent Income (AEI), and Benefit-Cost Ratio (BCR) allow for comparison of land-use systems. Real discount rates of 4%, 8%, and 12% were used in all financial formulas to represent a range of the opportunity cost of the next best investment. Discount rates account for the future value of income that would equate to income earned in the present through representing preferences for incurring costs and benefits now or in the future, i.e., they capture the opportunity cost of the investment. Since discount rates are often unknown to landowners, most analyses use the internal rate of return of the next best option for comparisons, such as a bank investment [32]. The Internal Rate of Return (IRR) was not relevant and inapplicable for this analysis for ongoing farms since annual benefits always exceeded costs. We excluded land values from NPV and LEV calculations. As such, each metric represents use-values, or income-based values, of the land. We then compared these metrics to the alternative values, market prices for land in the region, to facilitate a discussion on the financial competitiveness of different land-use systems.

*NPV* measures the amount of capital that an investment returns at a given discount rate through summing the total expenditures and subtracting them from total income [33]. Formula (1) demonstrates how NPVs are calculated in terms of farm and forest costs and revenues.

$$NPV = \sum_{i=0}^{t} \frac{(B - C)}{(1 + i)^t} \tag{1}$$

where $B$ and $C$ represent the annual total benefits and costs, respectively, of the land-use system, $i$ signifies the interest or discount rate, and $t$ is the year of the cash flow.

Similar to NPV, LEV utilizes expenditures, income, and a discount rate to measure the expected cash flow of a land-use in perpetuity. LEV has four assumptions to be viable: (1) identical costs and revenues in all rotations; (2) land-use will be maintained in perpetuity; (3) land requires identical regeneration costs at the beginning of each rotation; (4) the land value does not enter the calculation. In other words, landowners must replicate the following activities, including identical costs of inputs, prices of outputs, and cash flow schedules, into the future: purchase inputs, produce goods and services, extract products, and use or sell products. LEV allows researchers to compare systems of different rotation ages as well as how much an investor is willing to pay for the land at a given discount rate by assuming land-use will be continued indefinitely [26]. Formula (2) describes the method of estimating a production system's periodic and perpetual net returns in the present value [34].

$$LEV = NPV + \frac{NPV}{(1+i)^T - 1}$$
$$\text{Or } LEV = \frac{V_n}{[(1+i)^n - 1]} \tag{2}$$

where $NPV$ equates to the net present value of the system, $T$ is the final year of the system's rotation, $V_n$ equates to the net future value of the system at $n$ number of years per period, and $i$ again signifies the interest or discount rate.

AEI expresses NPV or LEV in annual payments equally distributed over the life of the investment (Formula (3)). AEI allows comparison of long-term timber investments with seasonal returns from agriculture by expressing the income of each alternative in annual payments [32].

$$AEI = LEV * i \tag{3}$$

where $LEV$ represents the system's land expectation value and $i$ signifies the interest or discount rate.

BCR relates the total discounted benefits to the total discounted costs. The relation describes present value benefits and costs as a unitless proportion rather than as a difference such as in the case of NPV. The proportion (Formula (4)) reveals the return landowners receive per dollar invested [26].

$$BCR = \frac{\sum_{i=0}^{t} B_p}{\sum_{i=0}^{t} C_p} \tag{4}$$

where $B_p$ and $C_p$ represent the benefits or revenues and costs or expenditures, respectively, in present value terms.

For price transformations of Peruvian wood products, the co-authors converted from Nuevo Sol per product to USD per cubic meter of wood product. Derived residual stumpage prices, equating to 50% of the roadside value, were calculated to account for logging and transportation when not given labor amounts for transport and/or sale. As a result, the income generated by the primary landowner as well as additional laborers is captured. For this reason, primary landowner labor was not included in the analysis. In other words, the determined income includes earned income from self-employment of the primary landowner. However, for farms with more than 15 heads of cattle, average annual income data were used to proportionately increase income to outside laborers, additional landowners, and/or land managers, bore as a cost in the modeled scenarios. The average annual income of a farm laborer in Amazonas at the time of the study was USD 11,111.11 [28–31]. The most common land-uses in Amazonas, Peru are dual-production systems of cattle and dairy and cultivation of horticultural and forest products such as coffee, guava, citrus fruits, eucalyptus, and cedar [29,30]. Agroforestry systems in Amazonas are increasingly incorporating multi-strata systems and improved fallows [29,35,36]. The study collaborators represented this trend through the selection of field sites of various production complexity.

## 4. Results

Table 1 summarizes the operations based on actual farms and survey observations. The first three rows review typical model systems defined by our project cooperators and previously completed surveys [28–31]. The last four rows describe systems observed in field visits. Both typical and actual farms included a variety of predominantly exotic cattle breeds including Brown Swiss and Holstein and planted non-native Eucalypts and pines.

**Table 1.** Description of evaluated scenarios in Amazonas, Peru.

| Scenario | Management Intensity * | Tree Species | Tree Growth ($m^3ha^{-1}$ $year^{-1}$) | Cattle Breeds | Number of Cattle (Lactating Cows) |
|---|---|---|---|---|---|
| **Typical Model Systems (based on prior surveys)** | | | | | |
| 10-ha Planted Forest Only | Low | *Pinus patula* | 5 | - | 0 (0) |
| 10-ha Cattle-Pasture Only | Medium | - | 0 | Brown Swiss, Simmental, Holstein | 16 (5) |
| 10-ha SPS | Medium | *Alnus acuminata, Eucalyptus globulus, P. patula* | 5 | Brown Swiss, Simmental, Holstein | 16 (5) |
| **Actual Systems (field sites)** | | | | | |
| 10-ha Cattle + Trees + Fruit | Medium | *E. globulus* | 5 | Holstein | 21 (6) |
| 25-ha Cattle + Trees + Fruit + Store | High | *P. patula, Cupressus macrocarpa, E. globulus, A. acuminate* | 5 | Holstein, Simmental, Brown Swiss | 38 (8) |
| 30-ha Cattle + Trees | High | *E. globulus, A. acuminate, P. patula* | 5 | Brown Swiss | 49 (24) |
| 65-ha Cattle + Trees + Restaurant | Medium | *P. patula, C. macrocarpa* | 5 | Brown Swiss | 38 (11) |

* In terms of inputs including labor.

The income from the typical and actual SPS predominately comes from the production of agricultural and horticultural products. Table 2 separately identifies the present value income (at a 4% discount rate) from wood products as well as agricultural and horticultural products for the one SPS scenario based on the past surveys and the four actual systems. Table 3 displays the capital budgeting results for the typical and actual regimes. Returns for the modeled systems varied from positive to negative net profits depending on management choices, local markets, and the discount rate.

**Table 2.** Total present value income by product type (discounted at 4%).

| SPS Scenario | Total Income at 4% | | | |
|---|---|---|---|---|
| | Wood Products | | Ag. and Hort. Products | |
| Typical SPS | USD | 3369 | USD | 148,060 |
| 10-ha C + T + F | USD | 2616 | USD | 322,077 |
| 25-ha C + T + F + S | USD | 4657 | USD | 715,614 |
| 30-ha C + T | USD | 2616 | USD | 217,352 |
| 65-ha C + T + S | USD | 2322 | USD | 67,951 |

C—Cattle; T—Trees; F—Fruit; S—Store (25-ha farm) or Restaurant (65-ha farm).

**Table 3.** Net present values (NPV) and land expectation values (LEV) of the evaluated scenarios.

| Scenario | NPV (USD/ha) | | | LEV (USD/ha) | | |
|---|---|---|---|---|---|---|
| | 4% | 8% | 12% | 4% | 8% | 12% |
| Typical Forest | 527.87 | −210.61 | −583.94 | 844.74 | −246.62 | −620.44 |
| Typical Cattle-Pasture | 796.88 | 318.64 | 69.53 | 1275.25 | 373.12 | 73.87 |
| Typical SPS | 992.52 | 321.91 | 2.52 | 1588.33 | 376.96 | 2.68 |
| 10-ha C + T + F | 5794.18 | 4045.62 | 3040.01 | 9272.43 | 4737.36 | 3230.01 |
| 25-ha C + T + F + S | 3626.78 | 1433.91 | 276.45 | 5803.93 | 1679.08 | 293.73 |
| 30-ha C + T | 5951.49 | 1840.67 | −261.03 | 9524.17 | 2155.39 | −277.35 |
| 65-ha C + T + S | 4603.27 | 2615.55 | 1555.65 | 7366.60 | 3062.77 | 1652.88 |

For the three regimes modeled based on prior farm owner surveys [28–31], pure timber investments provided the lowest returns to landowners (LEV: 4% = USD 844.74; 8% = −USD 246.62; 12% = −USD 620.44 per ha). Dual-purpose cattle grazing systems earned about 50% more than the returns of plantation forests at the lowest discount rate (LEV: 4% = USD 1275.25; 8% = USD 373.12; 12% = USD 73.87 per ha). The "typical" silvopasture scenario based on survey findings generated the highest landowner profitability of the hypothetical regimes (LEV: 4% = USD 1588.33; 8% = USD 376.96; 12% = USD 2.68 per ha). At a 12% discount rate, however, the wait for timber income was penalized more. Consequently, the ranking of systems in respect to landowner profitability changed, signaling conventional cattle–forage regimes to earn more net income than conventional SPS.

The four actual agroforestry systems that we surveyed in Amazonas provided landowners high returns. Again, it is worth noting that these farms were selected by the project co-investigators, and probably are local leaders in their farm practices. The 10-ha farm including planted trees for fruit and timber production and cattle grazing earned relatively high net returns (LEV: 4% = USD 9272.43; 8% = USD 4737.36; 12% = USD 3230.01 per ha). The second smallest farm visited, a 25-ha system with planted timber and fruit trees, cattle grazing, and a store on-site for sale of final products, provided acceptable, positive returns at low discount rates (LEV: 4% = USD 5803.93; 8% = USD 1679.08; 12% = USD 293.73 per ha). The second largest agroforest, a management-intensive 30-ha cattle grazing regime with a "living-fence" of planted trees, generated positive returns at low discount rates but failed to return positive earnings at a 12% discount rate (LEV 4% = USD 9524.17; 8% = USD 2155.39; 12% = −USD 277.35 per ha). The largest property analyzed, a 65-ha farm with multiple cattle grazing paddocks and tree plantations, produced acceptable incomes at all discount rates (LEV: 4% = USD 7366.60; 8% = USD 3062.77; 12% = USD 1652.88 per ha).

Figure 1 demonstrates the AEI of the conventional and monoculture regimes. At 4%, the values of all typical systems in Peru ranged from USD 33.79 to USD 51.01/ha/year. Forest monocultures experienced negative returns at higher discount rates. Figure 2 shows the AEI values for mixed land-use regimes. Typical silvopastures experienced positive annual returns at all discount rates. The calculated annual net revenue of actual systems ranged between USD 134.33 and USD 378.99/ha/year at 8%. The AEI may be multiplied by the number of acres to estimate the system's annual income, including earned income for the self-employed primary landowner. Only the two largest actual operations, the 30-ha (4% discount rate) and 65-ha systems (all discount rates), earned more than the average annual income for farmers in Amazonas when accounting for total net system returns. Thus, the two larger systems were the only scenarios that support sustainable livelihoods for the landowner(s).

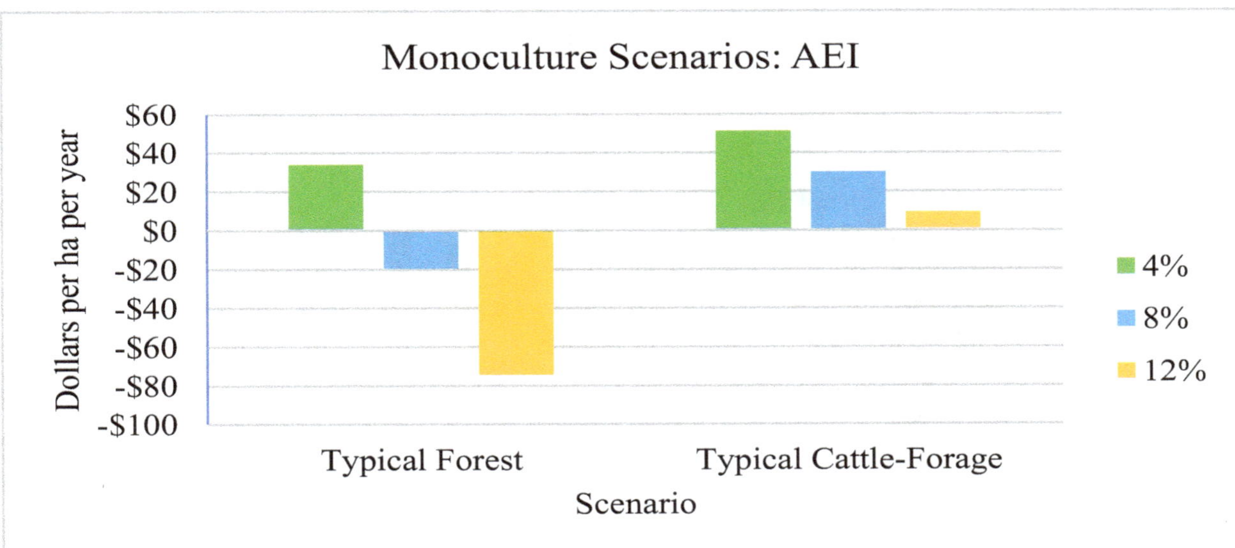


**Figure 1.** Annual equivalent income (AEI) of traditional forest and cattle systems.

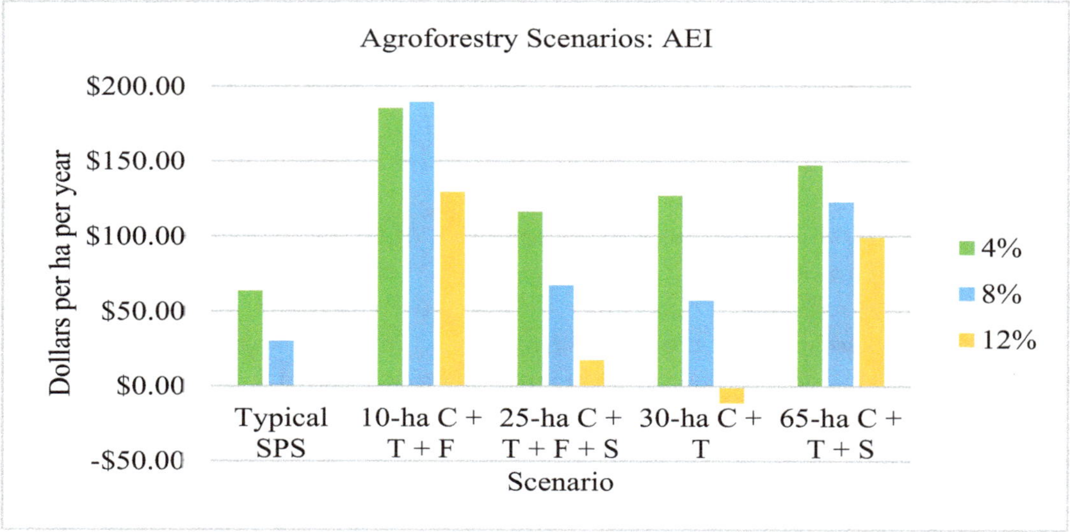


**Figure 2.** Annual equivalent income (AEI) of agroforestry systems.

Figure 3 displays the BCRs, a measure of the financial returns received per dollar spent in an investment, of all systems analyzed in the study. Tree plantations offered profitable returns per dollar invested at the lowest discount rate, as observed in its BCR, 1.27. At 8% and 12%, the BCR of forest monocultures was less than 1, signifying a loss per dollar invested (0.87, 0.61), due to the rapid diminution of future values. Dual-purpose cattle–forage regimes were cost-intensive and generated approximately USD 1.11 per dollar invested at the lowest discount rate (BCR: 4%= 1.11, 8% = 1.07, 12% = 1.02). The returns per dollar invested in silvopastures included a combination of costs and benefits from each industry (BCR: 4% = 1.16, 8% = 1.08, 12% = 1.00). The returns per dollar invested in the 10-ha actual farm increased as the discount rate increased (BCR: 4% = 1.62, 8% = 1.66, 12% = 1.71). The 25-ha mixed-use farm required high management intensity and generated just over a dollar per dollar spent at 12% (BCR: 4% = 1.27, 8% = 1.15, 12% = 1.04). The 30-ha silvopasture with an absentee landowner lost USD 0.02 per dollar invested at the highest discount rate (BCR: 4% = 1.24, 8% = 1.11, 12% = 0.98). The 65-ha silvopastoral system with a restaurant and timber sale on site generated over USD 2 per dollar invested at all discount rates (BCR: 4% = 2.54, 8% = 2.33, 12% = 2.12).

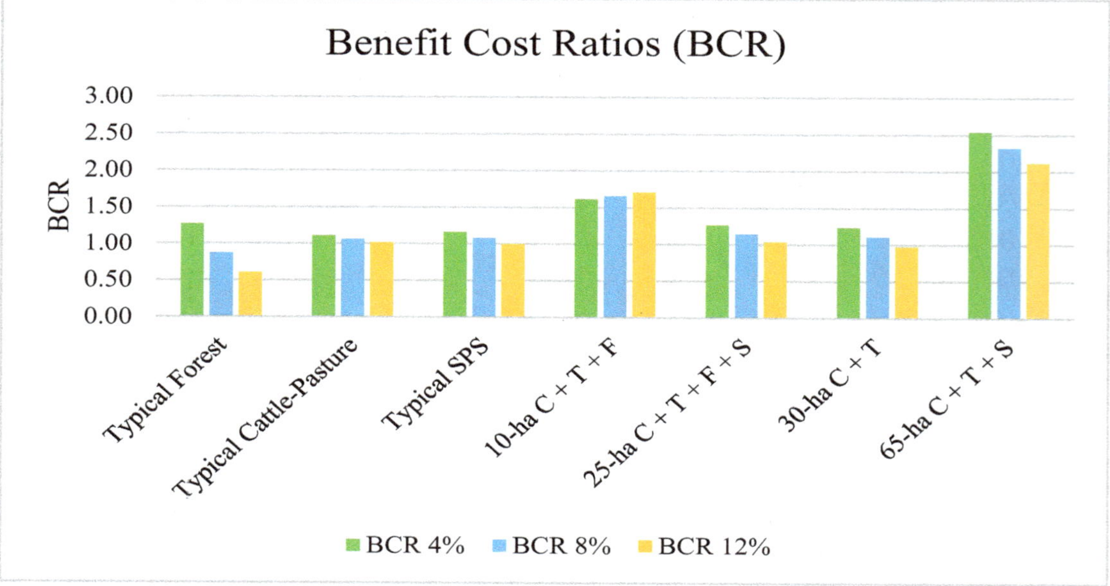


**Figure 3.** Benefit-cost ratios (BCR) for the evaluated scenarios.

The production systems analyzed required varying degrees of establishment intensity (Table 4). Of the three model scenarios, pure forest investments demanded the least in establishment inputs (USD 1130.50/ha). However, due to the longer turn-around time before generating a positive net income, forest monocultures also were associated with the longest payback period, in non-present value terms (8 years). While traditional cattle-pasture regimes required higher cost inputs for establishment (USD 1197.53/ha), they produced positive net returns sooner than planted forests (3 years). Establishing the typical cattle-pasture operation and the typical SPS required similar costs, differentiated only by the inputs needed to plant trees in the SPS. Cattle purchases were not included in this metric as they occur after pasture and trees are established. Establishment costs of hypothetical and actual silvopastures ranged from USD 623.99 to USD 1183.10 per ha and required 1 to 5 years to pay back, in non-present value terms.

**Table 4.** Establishment costs (not discounted) and payback periods for all scenarios.

| Scenario | Establishment Cost (USD/ha) | Payback Period (Years) |
|---|---|---|
| Typical Forest | USD 1130.50 | 8 |
| Typical Cattle + Pasture | USD 1197.53 | 3 |
| Typical SPS | USD 1203.35 | 4 |
| 10-Ha C + T + F | USD 915.56 | 1 |
| 25-Ha C + T + F + S | USD 1183.10 | 1 |
| 30-Ha C + T | USD 1082.35 | 5 |
| 65-Ha C + T + S | USD 623.99 | 5 |

Cost-share payments at various rates pay landowners part of the establishment costs necessary to develop a tree plantation or other conservation practices. These are common in the United States now and were in Latin America in the past, and increase the net present value exactly by the amount paid in year 0, as well shorten the payback period needed to generate positive net earnings. An assumed one-time 50% cost-share payment during the first years of USD 87.61 per hectare increased the LEV of the typical silvopasture scenario to USD 90.29 per hectare at a 12% discount rate. The cost-share payments in the initial years increased the NPV or LEV by the amount of the payment, *ceteris paribus*. Payments for environmental services (PES), such as carbon sequestration and water quality protection, can help increase the returns of agroforestry systems and provide landowners an equal choice between land uses. Mixed systems, both hypothetical and actual, experienced higher returns than the monoculture alternatives at discount rates of 4% and 8%. At 12%, typical silvopastures earned USD 71.19 per ha less in present value terms than conventional cattle–forage systems and required an annuity of USD 12.60 per ha for 10 years to breakeven.

Using a growth rate of 5 $m^3$/ha/year and an approximate biomass volume of 42.7% dry wood volume [18], we estimated the net carbon flow of each tree-based land-use system. The baseline comparison to show additional carbon storage was traditional cattle–pasture systems. The derived aboveground carbon accumulation rate for pure planted forests in Amazonas was 2.56 tons/ha/year (9.4 metric tons $CO_2$e/ha/year). Typical silvopastures with 15–20% trees store 0.38–0.51 tons of carbon/ha/year (1.39–1.88 metric tons $CO_2$e/ha/year). In a 25-year rotation, landowners may sequester 34.75–47 metric tons $CO_2$e/ha.

## 5. Discussion

In Amazonas, Peru, we modeled three representative crop, forest, and silvopasture systems (SPS) based on previous farm owner surveys, interviewed four landowners or land managers as empirical case studies, and estimated their financial returns using capital budgeting techniques. The case study systems included dual-purpose dairy cattle with some forests, and often with on-farm retail stands. The span of systems analyzed in this study provided a mix of inputs and returns. The increased profitability of complex agroforestry scenarios to private landowners may have resulted from the reduced management costs, compared to monoculture systems, and the resulting increased system productivity through complementary biophysical characteristics. Some owners also made and sold cheese, yogurt, and jams at their farms—adding value and a handy outlet for their raw products. The "typical" model systems identified by the co-authors and based on past surveys [28–31] were moderately profitable. The actual operations visited demonstrated that with increased system complexity, and some home stores with added value for the products, returns could increase as well, at least by the best landowners. With a limited sample of farms, more replications are required to better determine impacts from returns to scale and added value products.

Planted forests appeared to have the most opportunity to be cost-effective for landowners with limited capital, generating the highest ratio of returns in proportion to total expenditures, as seen through the system's BCR. Low-input land-use methods may be appealing to those interested in systems with moderate profitability and limited costs. However, tree plantations earned relatively low returns at all discount rates, compared to cattle or mixed-use systems, required longer payback periods to cover establishment costs, and had less certain and established markets. PES incentives such as cost-share payments encourage landowners to invest in land-use systems by reducing initial costs and increasing income sooner. The calculated cost-share payment necessary to breakeven with conventional cattle systems was less than the U.S. national average rental payment, USD 133.86 per ha, offered in the Conservation Reserve Program (CRP) [37]. With PES, the payment amount needs to be at least the difference in returns between incorporating forest cover through silvopasture and the conventional more profitable use to provide an equal choice to landowners.

The returns for the 10-ha dairy–fruit SPS were the greatest in our small sample in Amazonas. However, we did not include on-farm home labor of the primary landowner as a cost, whether active onsite or absentee. The profits essentially represented the landowner income, so their returns per hectare were high, but the total area was relatively small. Nevertheless, only the 30-ha and the 65-ha farms earned a greater annual income than the average annual income of farmers in the region. Generally, systems utilized family labor when possible. As system size increases, the need for additional labor increases. The 25-ha and 65-ha systems, including a store and restaurant, respectively, did not earn as high of returns as the 10-ha and 30-ha systems at the lowest discount rate. The 10-ha and 65-ha systems used mostly family labor, while the other two systems employed more outside laborers. Nonetheless, salaries were deducted from the returns to represent the opportunity cost of all necessary laborers, both familial and hired, outside of the primary landowner in the analysis.

The varying results suggested that different mixes of production systems along with management intensities play a role in determining landowner profitability. Both the 10-ha and 65-ha regimes were associated with relatively low-intensity management, in terms of inputs, including but not limited to their self-sufficiency in labor. However, the 10-ha farm, both smaller in size and less profitable per dollar invested, generated higher net returns in present value terms than the 65-ha system at all discount rates. This smaller farm along with the larger 25-ha farm sold fruit and home-made cheese, which increased returns. This corroborated that larger investments with more value—added can increase discounted present values, but they required more technical and managerial time and skill. The relationship between the resource variables and the capital budgeting results requires more analysis to determine their marginal impacts on returns per unit of land.

Small landowners often have limited capital and need frequent streams of income, which limits long-term investments. In addition, all systems other than the two largest actual farms earned less than the average annual income for farm laborers. Likewise, it took more time to cover establishment costs in forest systems when compared to cattle and agroforestry systems. Tremblay et al. [38] found fruit-bearing agroforests generated positive profits after 7 years, reinforcing this study's estimates. The difference in returns as well as the turn-around time for landowners to generate positive profits represent challenges that governmental policies can alleviate in order to encourage investment in expanding forest cover in agricultural systems [39].

For example, PES such as offsets for forest carbon storage have the potential to increase and diversify returns as well as make less profitable land uses more attractive. The average price of offsets for carbon stored in forests in 2016 was USD 5.10 per ton $CO_2e$ [40]. This means that with favorable carbon markets, silvopasture practitioners in Amazonas may be able to earn USD 177–USD 240/ha per 25-year rotation. Governmental intervention may be necessary through PES programs such as payments for carbon sequestration to make less profitable land-use systems competitive with conventional cattle-forage systems [39]. This would be consistent with the ambitions announced by the Peruvian authorities, in their plans to reduce net carbon emissions in the country. PES may also supplement annual income to landowners to make their profits comparable to income earned by workers in the farming industry, thus, ensuring an adequate income to landowners of rural lands.

Moreover, since the cost of land purchase was not considered in the analysis, we were able to compare LEV values with the average sale price of land to determine if the expected returns from the systems correlated with market prices for agricultural land. The average price of productive land for sale in Northern Peru, based on three lots on the market at the time of the study, was quite expensive, at USD 9618 per hectare [41]. At a 4% discount rate, our typical systems based on prior farm surveys and expert opinion earned less, but the four actual systems met or exceeded that level. None of the LEVs were greater than USD 9618 at the 12% discount rate.

Accordingly, using discount rates of 4%, 8%, and 12% provided a range of potential returns from land-use systems with various opportunity costs. Without the established literature and known landowner discount rates, using multiple discount rates allowed us to estimate returns with multiple preferences in mind. For instance, lower discount rates give more weight to long-term returns and less weight to short-term costs and benefits. Meanwhile, higher discount rates represent higher opportunity costs of alternative investments and/or involve riskier investments. For the scope of the study, we were able to see which land-use practices were more sensitive to high discount rates. Forests, which returned periodic, long-term revenues and required high establishment costs, experienced negative returns at 8% and 12% discount rates. Conventional cattle–forage systems in Amazonas appeared more resilient to high discount rates with relatively high returns at each discount rate. The lower-input case studies, the 10-ha and 65-ha regimes, maintained high returns at all discount rates, unlike the input—intensive 25-ha and 30-ha systems.

To our knowledge, no other projects have analyzed the economic returns of SPS compared to planted trees monocultures and traditional cattle grazing regimes in Amazonas, Peru. Our findings from Amazonas compare favorably with those of other global SPS research. In the Amazon lowlands of Ecuador, SPS featuring traditional and improved species of forage earned negative NPVs at 8%, −USD 356.17 per hectare and −USD 206.58 per hectare, respectively, but higher returns than traditional cattle ranching [42]. Other studies also highlighted the potential of added value from fruit–tree-based systems. For example, Tremblay et al. [38] recognized the economic advantages of silvopasture as a sustainable alternative land-use for small-scale agriculture in the Tapajos region of the Brazilian Amazon. At a 10% discount rate, the net present values of the agroforests returned USD 17,800 for the medium-sized orchard and USD 21,844 for the larger orchard. In addition, Hoch et al. [43] estimated positive NPVs at 3% and 12% discount rates for successful intercropping studies in Brazil, Bolivia, Peru, and Ecuador as part of a non-governmental agency's initiative.

Silvopasture systems are posited to provide significant risk reduction benefits because they produce at least two different commodities, which would have different, uncorrelated market price cycles given their different product life-cycle time spans [21]. In financial investment portfolio theory, any two or more uncorrelated investments will lead to larger overall investment returns. There is not much empirical literature on such financial risk benefits of silvopasture systems, and the complexity of finding historical returns and distributions and using them to estimate risk was beyond the scope of this study. We could have varied input costs and output prices for sensitivity or risk analyses, but given the many different factors of production and the lack of empirical cost distributions for any of our inputs, varying some arbitrary combination of these would not be supportable for this research.

Frey et al. [15] surveyed small-scale farmers in northeast Argentina, and found they believed that diversified cash flows, and thus, less risk of single commodity price fluctuations, were significant advantages of silvopasture. The mixed-use systems analyzed by Dube et al. [19,20] in Chile and Brazil showed positive potential for benefits by risk reduction from multiple products and carbon storage benefits. Similar approaches could be facilitated in Peru with financial assistance. Comparable studies acknowledged establishment costs to adopt silvopasture as well as increased labor as barriers for landowners with limited income [38,43,44]. A trade-off exists between the benefits of diverse mutually compatible species and its increased demand of specialized labor [13]. However, they also recognized the potential of timber sales to cover upfront and additional management costs [44], in addition to increased social and conservation benefits that are not always financially captured from silvopastures [38,45].

## 6. Conclusions

We estimated the financial returns for silvopasture systems in Amazonas, Peru based on previous farm surveys and representative models and four selected case studies of successful SPS farms based on local contacts with the co-authors. Pure cattle-pasture systems had larger returns in the synthetic calculations, but for the four integrated case studies that had pasture, forests, and some processing or retail parts of the value chain, silvopasture systems all had the greatest returns. However, the markets for wood are local and informal, so prices and quantity purchased depend on farm labor and individual negotiations, not established timber markets.

We made the analyses using the best available market prices or the estimated market equivalent of factor costs and output prices. We did not estimate social welfare or shadow prices for factors of production or output. Per strict economics jargon, we performed a financial analysis using discounted cash flows and capital budgeting criteria. We have referred to the research as an economic analysis as well, using the terms interchangeably per common usage [32]. Purists suggest that comprehensive economic analyses should include all the social (e.g., shadow prices, social welfare, non-market values) estimates of goods and services, but we used the most common convention and term of an economic or financial analysis for market prices, as is applied almost universally.

The sampled landowners captured excellent economic returns through SPS systems and diverse farm product sales, but they were exceptional opinion and business leaders. Achieving similar success for silvopasture with smaller typical landowners will require more research into the best practices for limited income farmers, extension advice, and implementation of good farm management practices as well. The opportunity cost of alternative investments will determine whether diversifying land-use through agroforestry will increase landowner returns.

Government support through direct payments could increase the profitability of silvopastures in order to gain ecosystem service benefits such as carbon storage or improved water quality. Cost-share payments may make land uses that expand forest cover more desirable to landowners by reducing establishment costs. Cost-share payments also have the potential of shortening payback periods so limited income landowners can reach a net positive cash flow sooner. When cost-share payments are not enough to make silvopasture more profitable, other PES, such as carbon storage, may be employed to provide a stream of annual benefits over some time period, e.g., the first ten years when revenue is at its lowest. Payments to landowners for carbon sequestration represent a promising solution to the low profits of silvopasture regimes at high discount rates. However, with low carbon prices, the absence of mature markets, and competitive resource relationships, carbon payments may not be sufficient to make silvopasture competitive with conventional cattle-pasture systems. In these cases, multiple forms of aid may be necessary to encourage landowners to adopt land-uses such as agroforestry. Overall, mixed-use systems in Amazonas represent essential models of land-use regimes which can be altered and/or replicated to produce supportable livelihoods for landowners.

**Author Contributions:** Conceptualization, M.C. and F.C.; methodology, F.C. and R.R.; formal analysis, S.C.; interviews and investigation, S.C., F.C., D.P., H.V., W.B.; resources, F.C.; data curation, S.C., F.C., D.P., W.B. and H.V.; writing—original draft preparation, S.C.; writing—review and editing, S.C., F.C., M.C., D.P., W.B., R.R., H.V., and R.P.; supervision, F.C., R.A., R.P. and E.S.; funding acquisition, S.C., R.A. and F.C. All authors have read and agreed to the published version of the manuscript.

**Acknowledgments:** We would like to thank the landowners that graciously allowed us to visit and use their farms in this analysis and the Peruvian collaborators that provided their guidance and expertise. We appreciate the suggestions and comments from both our own and *Land*'s reviewers and editors.

## References

1. Aynekulu, E.; Suber, M.; van Noordwijk, M.; Arango, J.; Roshetko, J.M.; Rosenstock, T.S. Carbon Storage Potential of Silvopastoral Systems of Colombia. *Land* **2020**, *9*, 309. [CrossRef]
2. Zomer, R.J.; Neufeldt, H.; Xu, J.; Ahrends, A.; Bossio, D.; Trabucco, A.; Wang, M. Global Tree Cover and Biomass Carbon on Agricultural Land: The Contribution of Agroforestry to Global and National Carbon Budgets. Scientific Reports, 6. 2016. Available online: http://search.proquest.com.prox.lib.ncsu.edu/docview/1817890421?pq-origsite=summon (accessed on 9 January 2018).
3. Orefice, J.N.; Carroll, J. Silvopasture—It's not a Load of Manure: Differentiating Between Silvopasture and Wooded Livestock Paddocks in the Northeastern United States. *J. For.* **2017**, *115*, 71. Available online: http://search.proquest.com.prox.lib.ncsu.edu/docview/1863558982?pq-origsite=summon&accountid=12725 (accessed on 7 January 2018). [CrossRef]
4. USDA National Agroforestry Center. Silvopasture: An Agroforestry Practice [PowerPoint Slides]. 2012. Available online: https://nac.unl.edu/practices/silvopasture.htm (accessed on 1 April 2017).
5. Cotta, J.N. Revisiting Bora fallow agroforestry in the Peruvian Amazon: Enriching ethnobotanical appraisals of non-timber products through household income quantification. *Agroforest. Syst.* **2017**, *91*, 17–36. Available online: https://link-springer-com.prox.lib.ncsu.edu/article/10.1007%2Fs10457-016-9892-4 (accessed on 30 March 2017). [CrossRef]
6. Lacerda, A.E.B.; Hanisch, A.L.; Nimmo, E.R. Leveraging Traditional Agroforestry Practices to Support Sustainable and Agrobiodiverse Landscapes in Southern Brazil. *Land* **2020**, *9*, 176. [CrossRef]
7. Chizmar, S.; Sills, E.; Cubbage, F.; Castillo, M.; Abt, R. A Comparative Economic Assessment of Silvopasture Systems in the Amazonas Region of Peru and in North Carolina. Master's Thesis, North Carolina State University, Raleigh, NC, USA, 2018. Available online: http://www.lib.ncsu.edu/resolver/1840.20/35059 (accessed on 5 May 2018).
8. Van Noordwijk, M.; Gitz, V.; Minang, P.A.; Dewi, S.; Leimona, B.; Duguma, L.; Pingault, N.; Meybeck, A. People-Centric Nature-Based Land Restoration through Agroforestry: A Typology. *Land* **2020**, *9*, 251. [CrossRef]
9. Ministry of Environment [MINAM]. Intended Nationally Determined Contribution from the Republic of Peru. 2015. Available online: http://www4.unfccc.int/ndcregistry/PublishedDocuments/Peru%20First/iNDC%20Per%C3%BA%20english.pdf (accessed on 18 May 2018).
10. Bonn Challenge. n.d. Commitments: Peru. Available online: http://www.bonnchallenge.org/content/peru (accessed on 1 May 2017).
11. Sistema Nacional de Evaluation Acreditacion y Certificacion de la Calidad Educativa [SINEACE]. Caracterización de la Región Amazonas. 2017. Available online: https://www.sineace.gob.pe/wp-content/uploads/2017/08/PERFIL-AMAZONAS.pdf (accessed on 18 May 2018).
12. City Population. Amazonas. 2 October 2018. Available online: https://www.citypopulation.de/php/peru-admin.php?adm1id=01 (accessed on 1 February 2018).
13. Loker, W.M. Where's the beef? Incorporating Cattle into Sustainable Agroforestry Systems in the Amazon Basin. *Agroforest. Syst.* **1994**, *25*, 227–241. Available online: https://link.springer.com/article/10.1007/BF00707462 (accessed on 21 May 2018). [CrossRef]
14. Bussoni, A.; Juan, C.; Fernandez, E.; Boscana, M.; Cubbage, F.; Bentancur, O. Integrated beef and wood production in Uruguay: Potential and limitations. *Agroforest. Syst.* **2015**, *89*, 1107–1118. Available online: https://link-springer-com.prox.lib.ncsu.edu/article/10.1007%2Fs10457-015-9839-1 (accessed on 4 May 2017). [CrossRef]
15. Frey, G.E.; Fassola, H.E.; Pachas, A.N.A.; Colcombet, L.; Lacorte, S.M.; Pérez, O.; Renkow, M.; Warren, S.T.; Cubbage, F.W. Perceptions of silvopasture systems among adopters in northeast Argentina. *Agric. Syst.* **2012**, *105*, 21–32. Available online: https://www-sciencedirect-com.prox.lib.ncsu.edu/science/article/pii/S0308521$\times$11001272 (accessed on 1 February 2018). [CrossRef]

16. Walthall, C.L.; Hatfield, J.; Backlund, P.; Lengnick, L.; Marshall, E.; Walsh, M.; Adkins, S.; Aillery, M.; Ainsworth, E.A.; Ammann, C.; et al. Climate Change and Agriculture in the United States: Effects and Adaptation. In *USDA Technical Bulletin 1935*; USDA: Washington, DC, USA, 2012; pp. 1–193. Available online: https://www.usda.gov/oce/climate_change/effects_2012/CC%20and%20Agriculture%20Report%20(02-04-2013)b.pdf (accessed on 1 May 2017).
17. Nair, P.K.R. Methodological challenges in estimating carbon sequestration potential of agroforestry systems. In *Carbon Sequestration Potential of Agroforestry Systems: Opportunities and Challenges*; Kumar, B.M., Nair, P.K.R., Eds.; Springer: Dordrecht, The Netherlands, 2011; pp. 3–16. Available online: https://link-springer-com.prox.lib.ncsu.edu/chapter/10.1007%2F978-94-007-1630-8_1 (accessed on 8 April 2017).
18. Oliva, M.; Mirano, L.C.; Leiva, S.; Collazos, R.; Salas, R.; Vásquez, H.V.; Quintana, J.L.M. Reserva de carbono en un sistema silvopastoril compuesto de Pinus patula y herbáceas nativas. *Sci. Agropecu.* **2017**, *8*, 149–157. [CrossRef]
19. Dube, F.; Thevathasan, N.V.; Zagal, E.; Gordon, A.M.; Stolpe, N.B.; Espinosa, M. Carbon sequestration potential of silvopastoral and other land use systems in the Chilean Patagonia. In *Carbon Sequestration Potential of Agroforestry Systems: Opportunities and Challenges*; Kumar, B.M., Nair, P.K.R., Eds.; Springer: Dordrecht, The Netherlands, 2011; pp. 101–127. Available online: https://link-springer-com.prox.lib.ncsu.edu/chapter/10.1007%2F978-94-007-1630-8_6 (accessed on 15 June 2017).
20. Dube, F.; Couto, L.; Silva, M.L.; Leite, H.G.; Garcia, R.; Araujo, G.A. A simulation model for evaluating technical and economic aspects of an industrial eucalyptus-based agroforestry system in Minas Gerais, Brazil. *Agroforest. Syst.* **2002**, *55*, 73–80. Available online: https://link-springer-com.prox.lib.ncsu.edu/article/10.1023%2FA%3A1020240107370 (accessed on 15 June 2017). [CrossRef]
21. Cubbage, F.W.; Balmelli, G.; Bussoni, A.; Noellemeyer, E.; Pachas, A.N.; Fassola, H.; Colcombet, L.; Rossner, B.; Frey, G.; Dube, F.; et al. Comparing silvopastoral systems and prospects in eight regions of the world. *Agroforest. Syst.* **2012**, *86*, 303–314. [CrossRef]
22. Montagnini, F.; Nair, P.K.R. Carbon sequestration: An underexploited environmental benefit of agroforestry systems. *Agroforest. Syst.* **2004**, *61*, 281–295. Available online: https://link-springer-com.prox.lib.ncsu.edu/content/pdf/10.1023%2FB%3AAGFO.0000029005.92691.79.pdf (accessed on 1 February 2018).
23. Orefice, J.N.; Smith, R.G.; Carroll, J.; Asbjornsen, H.; Howard, T. Forage productivity and profitability in newly-established open pasture, silvopasture, and thinned forest production systems. *Agroforest. Syst.* **2019**, *93*, 51–65. [CrossRef]
24. Bruck, S.R.; Bishaw, B.; Cushing, T.L.; Cubbage, F.W. Modeling the financial potential of silvopasture agroforestry in eastern North Carolina and northeastern Oregon. *J. For.* **2019**, *177*, 13–20. [CrossRef]
25. Pent, G.J.; Fike, J.H. Winter Stockpiled Forages, Honeylocust Pods, and Lamb Performance in Hardwood Silvopastures & Sheep Performance and Behavior in Silvopasture Systems. In Proceedings of the Agroforestry for a Vibrant Future: Connecting People, Creating Livelihoods, Sustaining Places 15th NAAC on the Campus of Virginia Tech in Blacksburg, Blackburg, VA, USA, 27–29 June 2017.
26. Mercer, D.E.; Frey, G.E.; Cubbage, F.W. Economics of agroforestry. In *Handbook of Forest Economics*; Kant, S., Alavalapati, J.R.R., Eds.; Routledge: New York, NY, USA, 2014; pp. 188–209. Available online: https://www.srs.fs.usda.gov/pubs/ja/2014/ja_2014_mercer_001.pdf (accessed on 9 October 2017).
27. Holdridge, L.R. *Life Zone Ecology*; Tropical Science Center: San Jose, CA, USA, 1967.
28. Gómez, C.A. Sistemas Silvopastoriles: Investigación e Innovación en el caso de Amazonas y San Martín. In *I Curso Internacional Ganadería y Agroecosistemas Tropicales Sostenibles Para Enfrentar el Cambio Climático: Sistemas Silvopastoriles*; Print: Pucallpa, Peru, 2017.
29. Pizarro, D.; Vasquez, H.; Bernal, W.; Fuentes, E.; Alegre, J.; Castillo, M.S.; Gomez, C. Assessment of silvopasture systems in the northern Peruvian Amazon. *Agroforest. Syst.* **2020**, *94*, 173–183. [CrossRef]
30. Alegre, J.C.; Sánchez, Y.; Pizarro, D.M.; Gómez, C.A. *Manejo de los Suelos con Sistemas Silvopastoriles en las Regiones de Amazonas y San Martín*; Print: Lima, Peru, 2019.
31. Echevarría, M.G.; Pizarro, D.M.; Gómez, C.A. *Alimentación de Ganadería en Sistemas Silvopastoriles de la Amazonia Peruana*; Print: Lima, Peru, 2019.
32. Cubbage, F.W.; Davis, R.R.; Frey, G.E.; Behr, D.C.; Sills, E.O. Financial and economic evaluation guidelines for international forestry projects. In *Tropical Forestry Handbook*; Pancel, L., Kohl, M., Eds.; Springer Publishing: Berlin, Germany, 2016. [CrossRef]

33. Sharrow, H. Natural Resource Economics: Considering the Time Element of Investments. 2008. Available online: http://www.doctorrange.com/PDF/TimeinNRInvest.pdf (accessed on 1 April 2017).
34. Bullard, S.H.; Straka, T.J. *Basic Concepts in Forest Valuation and Investment Analysis*, 3rd ed.; Faculty Publications: Nacogdoches, TX, USA, 2011; p. 460. Available online: http://scholarworks.sfasu.edu/forestry/460 (accessed on 19 September 2020).
35. Porro, R.; Miller, R.P.; Tito, M.R.; Donovan, J.A.; Vivan, J.L.; Trancoso, R.; Van Kanten, R.F.; Grijalva, J.E.; Ramirez, B.L.; Gonçalves, A.L. Agroforestry in the Amazon Region: A pathway for balancing conservation and development. In *Agroforestry—The Future of Global Land Use*; Nair, P.K.R., Garrity, D., Eds.; Springer: Dordrecht, The Netherlands, 2012; pp. 391–427. Available online: https://link-springer-com.prox.lib.ncsu.edu/book/10.1007%2F978-94-007-4676-3 (accessed on 1 April 2017).
36. Alegre, J.C. Sistema agroforestal multiestrato. Recuperación de suelos degradados en la amazonía. *Rev. Leisa* **2015**, *31*, 28–30. Available online: http://www.leisa-al.org/web/index.php/volumen-31-numero-1/1091-sistema-agroforestal-multiestrato-recuperacion-de-suelos-degradados-en-la-amazonia (accessed on 1 April 2017).
37. USDA Farm Service Agency. The Conservation Reserve Program: 49th Signup Results. USDA. 2016. Available online: https://www.fsa.usda.gov/Assets/USDA-FSA-Public/usdafiles/Conservation/PDF/SU49Book_State_final1.pdf (accessed on 1 April 2017).
38. Tremblay, S.; Lucotte, M.; Reveret, J.; Davidson, R.; Mertens, F.; Sousas Passos, C.J.; Romaña, C.A. Agroforestry systems as a profitable alternative to slash and burn practices in small-scale agriculture of the Brazilian Amazon. *Agroforest. Syst.* **2015**, *89*, 193–204. [CrossRef]
39. Vosti, S.A.; Braz, E.M.; Carpentier, C.L.; d'Oliveira, M.V.; Witcover, J. Rights to forest products, deforestation and smallholder income: Evidence from the western Brazilian Amazon. *World Dev.* **2003**, *31*, 1889–1901. [CrossRef]
40. Hamrick, K.; Gallant, M. Unlocking Potential State of the Voluntary Carbon Markets 2017. Forest Trends' Ecosystem Marketplace. 2017. Available online: https://www.cbd.int/financial/2017docs/carbonmarket2017.pdf (accessed on 4 January 2018).
41. Realigro Real Estate. Peru. 2015. Available online: https://peru.realigro.com/for-sale/farmland/#annunci (accessed on 9 October 2017).
42. Ramírez, A.; Seré, C.; Uquillas, J. An economic analysis of improved agroforestry practices in the Amazon lowlands of Ecuador. *Agroforest. Syst.* **1992**, *17*, 65–86. Available online: https://link.springer.com/article/10.1007/BF00122928 (accessed on 1 April 2017).
43. Hoch, L.; Pokorny, B.; de Jong, W. Financial attractiveness of smallholder tree plantations in the Amazon: Bridging external expectations and local realities. *Agroforest. Syst.* **2012**, *84*, 361–375. [CrossRef]
44. Adams, M.L. Pasture extensification in the Southern Ecuadorian Andes: Appraisal and recommendations. *J. Sustain. For.* **2009**, *28*. [CrossRef]
45. McDermott, M.E.; Rodewald, A.D. Conservation value of silvopastures to Neotropical migrants in Andean forest flocks. *Biol. Conserv.* **2014**, *175*. [CrossRef]

9

# Soil Organic Matter, Mitigation of and Adaptation to Climate Change in Cocoa–Based Agroforestry Systems

**Sikstus Gusli [1,2,*], Sri Sumeni [1], Riyami Sabodin [1], Ikram Hadi Muqfi [1], Mustakim Nur [1], Kurniatun Hairiah [3], Daniel Useng [2,4] and Meine van Noordwijk [5,6]**

1 Department of Soil Science, Faculty of Agriculture, Hasanuddin University, Makassar 9245, Indonesia; srisumeni31@gmail.com (S.S.); riyamial@gmail.com (R.S.); ikramhadimuqfi@gmail.com (I.H.M.); takimkure.20@gmail.com (M.N.)

2 Natural Resource Research and Development Center, Hasanuddin University, Makassar 9245, Indonesia; daniel.useng@agri.unhas.ac.id

3 Department of Soil Science, Faculty of Agriculture, Brawijaya University, Malang 65415, Indonesia; kurniatun_h@ub.ac.id

4 Department of Agricultural Engineering, Hasanuddin University, Makassar 9245, Indonesia

5 World Agroforestry Centre (ICRAF), Bogor 16001, Indonesia; m.vannoordwijk@cgiar.org

6 Plant Production Systems, Wageningen University, 6708 PB Wageningen, The Netherlands

* Correspondence: sikstusgusli@unhas.ac.id

**Abstract:** Belowground roles of agroforestry in climate change mitigation (C storage) and adaptation (reduced vulnerability to drought) are less obvious than easy-to-measure aspects aboveground. Documentation on these roles is lacking. We quantified the organic C concentration ($C_{org}$) and soil physical properties in a mountainous landscape in Sulawesi (Indonesia) for five land cover types: secondary forest (SF), multistrata cocoa–based agroforestry (CAF) aged 4–5 years (CAF4), 10–12 years (CAF10), 17–34 years (CAF17), and multistrata (mixed fruit and timber) agroforest (MAF45) aged 45–68 years. With four replicate plots per cover type, we measured five pools of C-stock according to IPCC guidelines, soil bulk density (BD), macro porosity (MP), hydraulic conductivity ($K_s$), and available water capacity of the soil (AWC). The highest C-stock, in SF, was around 320 Mg ha$^{-1}$, the lowest, 74 Mg ha$^{-1}$, was in CAF4, with the older agroforestry systems being intermediate with 120 to 150 Mg ha$^{-1}$. Soil compaction after forest conversion led to increased BD and reduced MP, $K_s$, and AWC. Older agroforestry partly recovered buffering: AWC per m of rooted soil profile increased by 5.7 mm per unit (g kg$^{-1}$) increase of $C_{org}$. The restored AWC can support about a week's worth of evapotranspiration without rain, assisting in climate change adaptation.

**Keywords:** cocoa agroforestry; climate adaptation; soil restoration; soil organic carbon; soil macro-porosity; soil water availability; inceptisols

---

## 1. Introduction

Agroforestry, integrating trees in farms, implies reforestation (or at least re–treeing) within an agricultural framework of land tenure and practice [1,2]. Growing cocoa between various other trees as shade or companion trees to increase overall income from farming is one of the many forms of agroforestry [3]. Agroforestry can connect the climate change (CC) policy imperatives on mitigation —reducing net emissions of greenhouse gasses—and adaptation: reducing human vulnerability to increased climate variability and global warming trends [4,5]. These connections are made aboveground as increased tree cover in agricultural landscapes stores carbon [6,7] and modifies microclimates [8].

Moreover, belowground soil carbon storage will increase in agroforestry [9] while the soil water buffer function can be enhanced [10]. Yet, the belowground effects are slower (both in their degradation and their restoration phases [11]) and less visible than those aboveground and need to be teased out from a substantial background variability of soil diversity [12,13]. In accordance with the 'cascade' perspective on ecological structure and function, which shapes ecosystem services that represent the benefits people derive from functioning (agro)ecosystems as a basis for valuation and explicit decision making (Figure 1A, [14]), the chain of required evidence for belowground relevance of agroforestry for CC adaptation can be considered in three steps (Figure 1B):

A. Effects of trees as part of land cover change on soil organic matter (and soil carbon storage),
B. Relations between soil organic matter and soil physical properties relevant to crop vulnerability to climate variability, and
C. Deliberate farmer use of trees to reduce their own vulnerability (the adaptation line), potentially supported by global climate policies (if incentives will reach the farmer, then the dotted line).

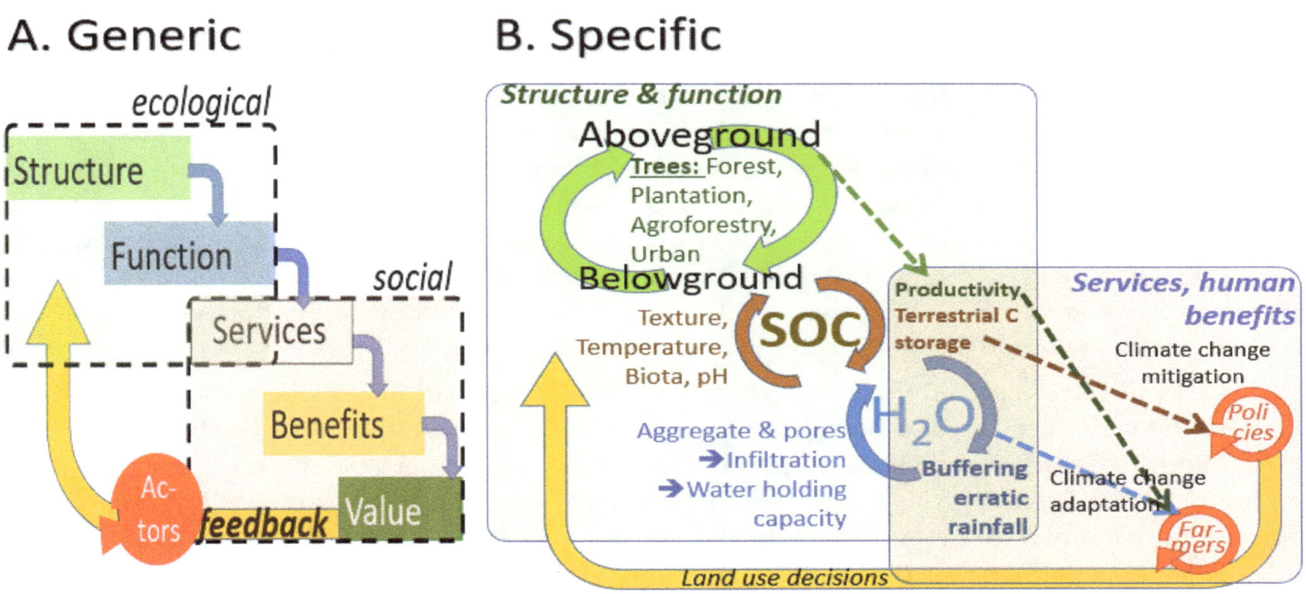


**Figure 1.** (**A**) Cascade representation of ecosystem structure and function, interacting with social systems via 'ecosystem services' [14]; (**B**) Relationship between the agroecosystem structure (observable), function (inferred), and human benefits (policy focus) in relation to the growth cycle of trees, the formation, and the decay of soil organic carbon (SOC), climate change, and the water ($H_2O$) cycle.

Soil organic matter needs to be understood in its central role in belowground C storage (mitigation) and its influence on soil water balance and hence on climate vulnerability (adaptation). Obtaining a full account of all inputs, both aboveground and belowground, and their associated rates of turnover and interaction with mineral soil particles in shaping soil physical properties is no simple task for even relatively simple forms of cocoa agroforestry. Stock changes over time can provide an indication of the net effects of interacting processes, as both plants [15] and farmers adapt to climate change [16]. Co-adaptation of people and trees is needed for resilience to climate change in the medium term [17].

Cocoa (*Theobroma cacao* L.) is grown across the humid tropics and is a major driver of tropical forest transitions [18]. Whether or not cocoa agroforestry increases or reduces vulnerability to uncertain rainfall is contested [19] and may depend on the tree species used, the soil, and climatic conditions. Within cocoa production, comparisons between more open and more shaded systems have revealed ecological, social, and economic variables [20–22], with farmer choices depending on location, markets, and evolving knowhow. According to global agriculture database data [23], Indonesia still is the third-largest global cocoa producer, while international cocoa organization data [24] suggests that it is the fifth-largest global cocoa producer. Indonesian smallholder cocoa production confronts multiple challenges on how to increase productivity as well as smallholder's prosperity, reduce pressure on forest conversion, and be adaptive and resilient to climate change [25]. Nearly a million hectares of forest was converted to cocoa farmland in Indonesia, especially in the island of Sulawesi [26], most notably in the 1980s [27] during the period of cocoa booming. Such conversion was intended to increase smallholder income, however it led to declining soil health by reducing soil biodiversity and the soil microbiome [28,29], compacting soils, reducing infiltration and soil water storage [30], depleting soil fertility [31], and reducing environmental services [32,33] such as carbon storage [34,35].

In the last twenty years, cocoa farmland management in Sulawesi has been dynamic, very much influenced by field experience and information gathered by returning migrant farmers from Malaysia [24]. In the early 1970s to 1980s, cocoa was grown as mixed cropping with food crops (mostly in flat land); much of the land was converted to monoculture (full sun) cocoa in the 1990s. The monoculture practice expanded rapidly including on sloping lands, in response to the information from other places that this system produces higher yields than cocoa grown under shade, especially in the short term [36]. Yet, farmers eventually learnt that, in the longer term, cocoa grown as a single crop suffered from pests and diseases, including the damaging vascular strike dieback, which can kill cocoa trees under heavy infestation [37]. Farmers also claimed that monoculture cocoa suffers more from extreme weather conditions (heavy rain or long dry periods), while monoculture cocoa only allows farmers to earn money in the months during the harvest periods. Learning from this experience, many cocoa farmers at present also grow other plants between cocoa trees, including horticultural crops, as shade trees for the cocoa and as extra income sources for the farmers. While the structure and functions of cocoa-based agroforestry practices are easily observed aboveground, their belowground roles are less obvious and remain poorly documented, including carbon storage dynamics in terms of the various agroforestry types and ages and its effects on climate change mitigation and adaptation, as well as changes of soil properties affecting vulnerability to drought.

In this study, we explored the relationships between A) Agroforestry land cover, aboveground and belowground carbon stocks, and B) Soil carbon and soil physical conditions, as steps towards assessing contributions to CC adaptation have occurred in the context of southern Sulawesi where farmers in the Polewali-Mandar District grow cocoa with three major systems: monoculture with minimum shade trees and both simple- and complex- (multilayers) agroforestry systems of varying ages. We tested the hypothesis that in cacao production systems, the farmer-level benefits from agroforestry through increased water storage, contributing to climate change adaptation, are parallel to global benefits of climate change mitigation through increased C storage. If such a synergy exists between local and global climate change agendas, then it is easier for climate policy to provide appropriate incentives.

## 2. Methods

### *2.1. Sampling Sites*

We conducted the study in a mountainous landscape of Sulawesi, Indonesia with sample sites ranging from 93 to 575 m above sea level (Figure 2), on five land cover types (all were directly converted from primary forest according to local land use history) consisting of young smallholder cocoa agroforestry (CAF4, 4 to 5 years old), medium-aged smallholder cocoa agroforestry (CAF10, 10 to 12 years old), old smallholder cocoa agroforestry (CAF17, 17 to 34 years), multistrata agroforest (MAF45, 45 to 68 years after forest conversion) owned by smallholders, and remnant secondary tropical forest (SF). Information on the year planted or converted from forest was obtained directly from the farm owners who lived on and managed the farms. The sites of these agroforestry systems (CAF4, CAF10, CAF1,7 and MAF45) are scattered at elevations between 100 to 300 m above the sea level, with an 8 to 30% slope. The three sites of SFs are at >400 m above the sea level and 30 to over 40% slope. The average annual rainfall of the area is 2113 mm.

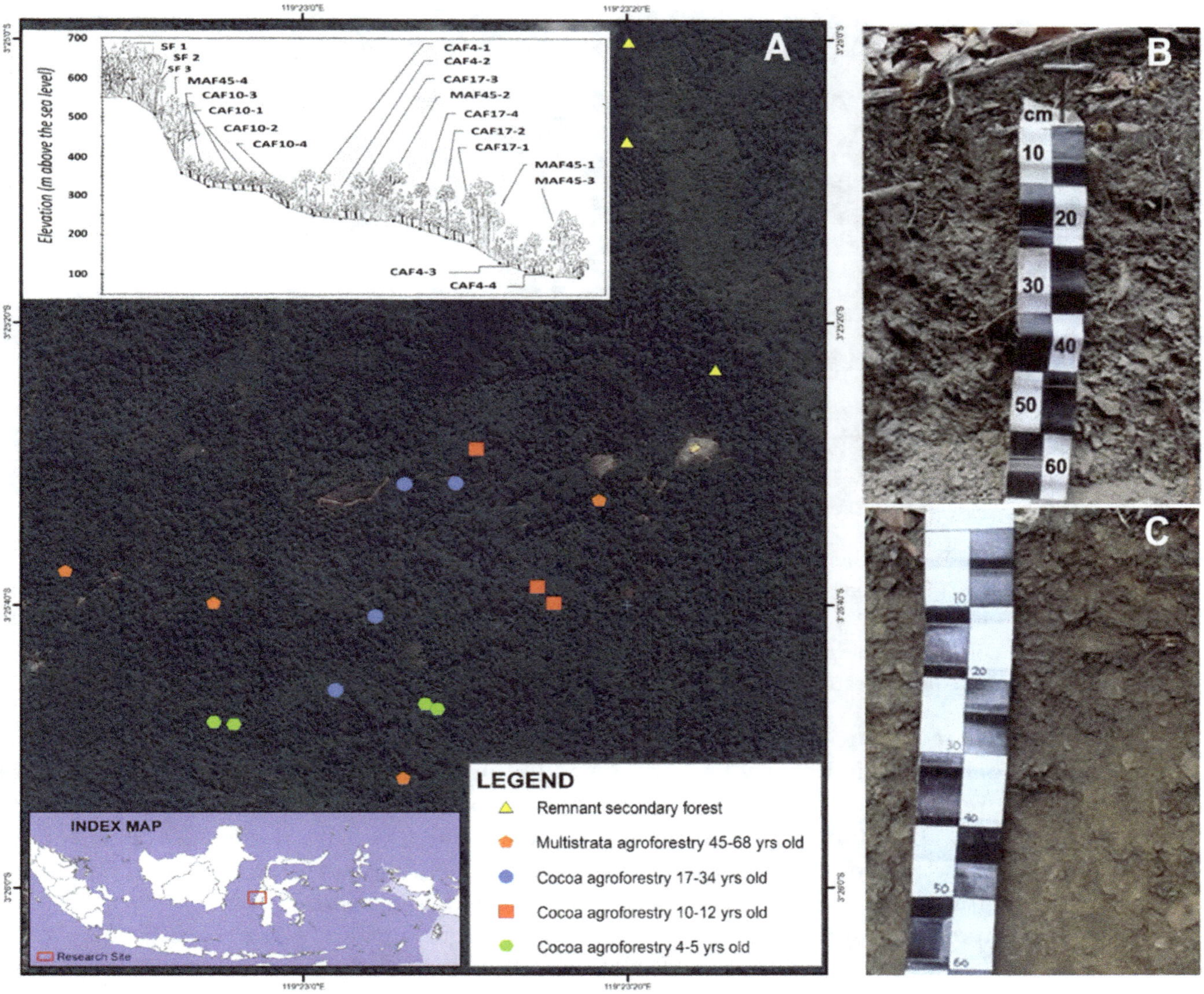


**Figure 2.** (**A**) The site distribution (satellite image) of remnant secondary forest (SF), multistrata agroforest that is 45–68 years old (MAF45), cocoa agroforestry that is 17–34 years old (CAF17), cocoa agroforestry that is 10–12 years old (CAF10), and cocoa agroforestry that is 4–5 years old (CAF4). The topo-sequence positions are shown to indicate slopes and elevations of the systems. The cocoa agroforestry sites are scattered randomly within 100 to 300 m above sea level (a.s.l.), generally 8 to 30% in slope, but the SFs are at >400 m a.s.l. with a slope from 30 to over 40%. CAF (of all ages) refers to cocoa farming practices where at least five other tree species (with shade and/or direct use functions) are used alongside cocoa trees as the main crop. MAF denotes a farming practice of multiple tree species, dominated by horticultural trees, cocoa is not the main crop. (**B**) A typical (common) soil profile of the study area for all land uses showing abundance of rock fragments, especially below a 20 or 30 cm depth. (**C**) A profile of less stony soil (uncommon).

The plots differed in vegetation structure and stratification, but all had permanent litter layers (Figure 3). We did not find remnants of natural (primary) forest in the neighborhood. The surface soil of all land-uses was well covered by mulch (litter layer). For each land-cover type, we selected four sites (replicates), managed by separate households; however, due to forest area limitations, only three sites were sampled in SF, in a single landscape location. Hence, across these five land-uses we evaluated 19 plots.

**Figure 3.** Typical look of the five land-use systems, namely remnant secondary forest (SF), multistrata agroforest that is 45–68 years old (MAF45), cocoa agroforestry that is 17–34 years old (CAF17), cocoa agroforestry that is 10–12 years old (CAF10), and cocoa agroforestry that is 4–5 years old (CAF4). Bottom right corner: SF being land–cleared using a chain saw and machete operated by the owner for a new mixed cocoa agroforestry (CAF), without burning.

At the soil surface, the rock fragments covered 10 to 20 percent of the soil surface, but beyond a 30 cm depth, rocks occupied nearly half of the soil volume. Soils in each site varied from coarse– to medium–textures (sandy loam, sandy clay loam, and loam), with an increase of clay content as the depth increases. To minimize confounding effects of soil texture and elevation on C-organic concentrations, we used reference values based on a pedotransfer function [9] as a basis of comparisons. The pedotransfer function we used accounts for texture (microaggregate C protection) and elevation (as proxy for temperature) effects on expected C-organic concentrations (with coefficients for silt, clay, pH, elevation, and specific soil types) [13]; this was reconfirmed in recent analysis [9], with an additional correction for sampling depths. The $C_{ref}$ values can help in judging the validity of the 'chronosequence' interpretation of observed differences between land uses as being 'caused' by land use in an otherwise homogeneous background situation. Variations in texture (Table 1) stayed within the 'Loam' and 'Sandy Loam' classes, with the lowest $C_{ref}$ value 9% below and the highest 14% above the mean for layer–specific results. This can be compared to $C_{org}$ with a lowest value 25% below and a highest one 59% above the average. Variation in texture can thus not be ignored in the interpretation of plot–level $C_{org}$ results in relation to land use. For part of the analysis, we relied on the $C_{org}/C_{ref}$ ratio as a potentially more sensitive indicator of land use effects [9], as it has considered that the 'chronosequence' interpretation assumes that there is no variation in texture.

**Table 1.** Particle–size distribution (USDA classification), pH and derived reference organic carbon concentration (C-ref, responding to texture and pH) of the soil to a depth of 30 cm, as test of homogeneity of the sample sites used for the five land uses; values in brackets are standard errors of the means of four measurements; for C-ref and measured C-org, the values are also provided after normalization on the mean across land uses per soil layer.

| Land Use * | Soil Depth (cm) | Particle-Size Distribution (g kg$^{-1}$) | | | Soil Texture | pH (1N KCl) | C-Ref (g kg$^{-1}$) | Normalized C-Ref Per Layer | C-org (g kg$^{-1}$) | Normalized C-Org Per Layer |
|---|---|---|---|---|---|---|---|---|---|---|
| | | Clay | Silt | Sand | | | | | | |
| SF | 0–10 | 128 (49) | 176 (25) | 674 (54) | Sandy loam | 3.8 (0.2) | 22.1 (1.5) | 1.13 | 21.8 (0.3) | 1.32 |
| | 10–20 | 192 (48) | 175 (41) | 610 (67) | Sandy loam | 3.6 (0.3) | 15.3 (1.3) | 1.14 | 21.6 (0.5) | 1.59 |
| | 20–30 | 118 (52) | 294 (14) | 571 (65) | Sandy loam | 3.8 (0.5) | 13.0 (1.5) | 1.09 | 17.7 (3.2) | 1.47 |
| CAF4 | 0–10 | 82 (13) | 309 (52) | 606 (42) | Sandy loam | 4.4 (0.3) | 17.0 (1.0) | 0.87 | 16.2 (0.6) | 0.98 |
| | 10–20 | 96 (46) | 323 (33) | 567 (32) | Sandy loam | 4.1 (0.3) | 12.4 (0.7) | 0.92 | 12.4 (1.3) | 0.91 |
| | 20–30 | 86 (22) | 375 (18) | 527 (24) | Loam | 3.7 (0.3) | 11.4 (0.5) | 0.96 | 10.4 (1.5) | 0.87 |
| CAF10 | 0–10 | 94 (19) | 288 (57) | 602 (43) | Sandy loam | 4.1 (0.3) | 19.8 (1.1) | 1.01 | 15.1 (1.2) | 0.92 |
| | 10–20 | 80 (25) | 325 (27) | 578 (39) | Sandy loam | 3.9 (0.2) | 13.2 (0.8) | 0.98 | 11.4 (1.3) | 0.84 |
| | 20–30 | 67 (19) | 390 (37) | 530 (34) | Loam | 3.8 (0.2) | 11.8 (0.4) | 0.99 | 10.4 (1.4) | 0.87 |
| CAF17 | 0–10 | 93 (17) | 341 (43) | 554 (47) | Sandy loam | 3.9 (0.4) | 20.7 (1.6) | 1.06 | 13.5 (1.1) | 0.82 |
| | 10–20 | 138 (42) | 319 (26) | 525 (28) | Loam | 3.4 (0.4) | 14.2 (0.7) | 1.05 | 10.2 (0.3) | 0.75 |
| | 20–30 | 138 (31) | 347 (44) | 504 (39) | Loam | 3.5 (0.4) | 12.4 (0.9) | 1.04 | 10.0 (0.6) | 0.83 |
| MAF45 | 0–10 | 119 (33) | 254 (28) | 607 (36) | Sandy loam | 4.3 (0.4) | 18.5 (1.1) | 0.94 | 15.7 (2.0) | 0.95 |
| | 10–20 | 131 (45) | 267 (15) | 583 (32) | Sandy loam | 4.2 (0.3) | 12.2 (0.5) | 0.91 | 12.2 (2.2) | 0.90 |
| | 20–30 | 124 (50) | 313 (45) | 540 (32) | Sandy loam | 4.1 (0.3) | 10.8 (0.4) | 0.91 | 11.6 (2.6) | 0.97 |

* SF = secondary forest, CAF = cocoa agroforestry, MAF—multistrata agroforest, with years since the plot establishment indicated.

## 2.2. Land Cover

Beyond cocoa, a range of other fruit tree species was part of local agroforestry systems, including langsat (*Lansium domesticum* Corr.), durian (*Durio zibethinus* Murr.), *rambutan* (*Nephelium lappaceum* L.), jackfruit (*Artocarpus heterophyllus* Lam.), cempedak (*Artocarpus integer* (Thunb.) Merr.), star fruit (*Averrhoa carambola* L.), and candlenut trees (*Aleurites moluccanus* L.). Timber trees include white teak (*Gmelina arborea* Roxb.), sandalwood (*Santalum album* L.), cananga (*Cananga odorata* (Lam.) Hook.), and *bitti* (*Vitex cofassus Reinw.*). Various bamboo species (*Bambusoideae sp.*) are also a source of construction material, with young shoots as a food source. Bananas (*Musa spp.*), papaya (*Carica papaya* L.), sugar palm (*Arenga pinnata* Merr.), and chili (*Capsicum annuum* L.) are present as readily marketable food sources. As the 'mother of cocoa', gliricidia (*Gliricidia sepium* Jack.) was common as nitrogen-fixing shade trees, as was leucaena (*Leucaena leucocephala* (Lam.) de Wit). Most of these trees were planted, although some sufficiently naturalized to spread spontaneously. Farmer management is based on selective retention of trees with use value, similar to what was documented for Southeast Sulawesi [38].

The vegetation found in the remnant forest (SF) was dominated by *Castanopsis acuminatissima* (Blume) A., *Symplocos polyandra* (Blanco) Brand., rattan (*Calamus deerratus* G.Mann & H.Wendl.), forest/wild pandan (*Pandanus amaryllifolius* Roxb.), bamboo (*Bambusa* spp), and some hardwood tree species (locally named *dao, bubun, nato*) including kluwak (*Pangium edule* Reinw.), trees that are locally called forest teak, *gattungan*, ebony (*Diospyros celebica* Bakh.), trees that are locally named *belating*, candlenut, coffee (*Coffea sp*), cananga trees, and various shrubs. All these species are part of Sulawesi's native flora.

### 2.3. C–stock Measurements

Within the selected sites, based on the field vegetation conditions that represent the land-use type, we set up four C-stock observation plots according to the RaCSA (Rapid Carbon Stock Appraisal) procedure outlined by [39] and used in a cocoa landscape context by [38]. While other C sampling protocols prefer circular plots to maximize plot homogeneity, the basic plot size in the RACSA protocol is a transect–like 40 m by 5 m to include variation within a plot; some adjustments to plot dimensions were made where 40 m × 5 m was not possible. Where we found trees with DBH greater than 30 cm as in the SF, plot size was expanded for such trees to 100 m by 20 m, as they tend to occur at lower frequency but have considerable impact on total biomass estimates [39]. On the other hand, in cocoa farms (CAF17, CAF4 and CAF10) with regular cocoa planting distance of 3 m × 4 m, the plot size was increased to be 20 m by 20 m. In these plots, we measured all trees with a stem diameter at breast height (DBH) of 5 to 30 cm. In every basic plot, we randomly selected six sub-plots of 50 cm by 50 cm, where we measured C–stock of understory (destructing sampling), non-woody necromass, and soil organic carbon (depth of 0–30 cm). For woody necromass, where present, we followed the protocol of [39].

### 2.4. Aboveground and Belowground Biomass

Tree biomass was estimated allometrically from the stem diameters, using allometric equations for Aboveground biomass (AGB in kg per tree) (Table 2), on the basis of stem diameter (D, in cm) at 1.3 m above the soil surface, with wood density ρ (in g $cm^{-3}$, obtained from the World Agroforestry database [http://db.worldagroforestry.org/wd, April 2014]. Wood density is used in allometric equations that are derived across many different tree species; as these equations are calibrated on a wider range of tree diameter data, equations that deviate from the simple power laws used for the other equations account for a larger share of the variation in the calibration datasets [40].

**Table 2.** Allometric equations used for aboveground biomass ($AGB_{est}$) based on stem diameter D.

| Component | Allometric Equation for $AGB_{est}$ | Reference |
|---|---|---|
| Cocoa | $0.1208\ D^{1.98}$ | [41] |
| Generic trees, humid tropics (rainfall 1500–4000 mm $yr^{-1}$) | $\rho \times \exp(-1.499 + 2.148\ \ln(D) + 0.207\ (\ln(D))^2 - 0.0281\ (\ln(D))^3)$ | [40] |
| Banana | $0.030\ D^{2.13}$ | [42] |
| Palm | $0.118\ D^{2.53}$ | [43] |

Tree basal area ($m^2\ ha^{-1}$) was calculated by summing their cross–sectional area $\pi\ D^2/4$ of all sampled trees and scaling up from the sampled area, while converting from $cm^2$ to $m^2$. Formulas used to calculate woody necromass depend on the wood decomposition stage [39]. The decomposition stage was assessed in the field based on ease of inserting a metal rod into woody necromass. On recent tree falls, where the decomposition stage was low, the equations of Table 2 were used. However, where the necromass was highly decomposed, Equation (1) was used, and wood density was assumed to be 0.4 g $cm^{-3}$ [39]:

$$N_D = \rho\ H\ D^2/40 \qquad (1)$$

where $N_D$ is dry necromass (kg).

Measurement of understory biomass was done on six randomly selected quadrants of 50 cm by 50 cm in every land–use. Understory vegetation in the quadrants was collected, weighed, then taken to laboratory for oven–drying at 80 °C for 48 h. Due to massive and bulky understory biomass samples, approximately 300 g sub–samples were taken from the field samples for dry weights. If the samples were less than 300 g, the whole biomass was oven–dried. The understory dry weight (UD) was calculated as the ratio of oven–dry weight to the sub–sample fresh weight multiplied by the total fresh weight of understory biomass sampled in the field. Understory dry biomass of each land-use is the average of the six replicates of the respective land–use. The litter collected was carefully cleaned to remove adhering soil particles, before being taken to a laboratory for oven–drying at 80 °C, for 48 h. Calculation of litter biomass follows the same procedure as that for understory biomass.

Belowground biomass was estimated from the aboveground–tree root biomass relationship. A default value of the canopy/root ratio of 4/1 for wet tropical forest trees was applied [44]. However, for cocoa trees, we estimated the value based on the 6.7/1 ratio developed by [45].

### 2.5. *Soil Properties*

On 19 sub–plots (four sub–plots each in CAF4, CAF10, CAF17 and MAF45, and 3 sub–plots in SF) we collected 57 of both disturbed and undisturbed soil samples at 0–10, 10–20, and 20–30 cm depths. In each sub–plot, we collected about 1 kg of disturbed soil samples per layer, mixed composite from six random spots within the same sub–plot for the measurements of soil organic carbon (C–org), soil texture, and soil carbon pools. The undisturbed soil samples (also collected from three soil depths at six random spots in each sub–plot) were used for bulk density, saturated hydraulic conductivity, water retention, soil porosity, macro–porosity and aeration, available water capacity (AWC), and soil structural quality measurements. C–org was determined according to the Walkley and Black procedure outlined by [46]; bulk density was determined in each land use using six large (20 cm long × 20 cm wide × 10 cm deep) core samples as well as six standard small (70 mm inner diameter, 50 mm length) ring samplers according to [47]; soil texture was determined using the pipet method outlined by [48] saturated hydraulic conductivity was measured in a laboratory at a constant head [49] using undisturbed samples with six replicates for each form of land use. Using the same undisturbed samples for saturated hydraulic conductivity and water retention at 0 and 5 kPa, matric suctions were determined using a hanging water column system [50,51]. Soil porosity was calculated from bulk density, assuming a particle density of mineral soils of 2.65 Mg $m^{-3}$ [52]. Macro–porosity (effective diameter >60 μm) was calculated according to the soil pore (capillarity radius)—matric suction relationship formula [53] at 5 kPa (0.50 m). Available water capacity was calculated as the difference between water content at field capacity (−5 kPa matric potential) and at the wilting point (−1500 kPa), measured using a pressure chamber [54]. Soil carbon stocks were calculated from bulk density and C-org data [9], pH was measured in 1N KCl.

## 3. Results

### 3.1. *Density and Basal Area*

Land occupancy by trees, expressed in terms of tree basal area, of secondary forest was the most common form of land use (over 14 $m^2$ $ha^{-1}$); it was significantly reduced ($p < 0.05$) when land was converted to other land uses (Table 3). Ten–to–twelve years old cocoa agroforestry (CAF10) had the lowest tree occupancy (just over 5 $m^2$ $ha^{-1}$), but its difference with younger cocoa agroforestry systems (CAF4) was not statistically significant. Generally, the basal area increased as the cocoa–based agroforestry system aged.

**Table 3.** Tree basal area and plant density of remnant secondary forest (SF), multistrata agroforest that is 45–68 years old (MAF45), cocoa agroforestry that is 17–34 years old (CAF17), cocoa agroforestry that is 10–12 years old (CAF10), and cocoa agroforestry that is 4–5 years old (CAF4). The high plant density for CAF4 was mainly due to a high population of cocoa and shade trees introduced into the system. Shade trees are trees mainly planted to provide shade for cocoa, but several types also fix atmospheric nitrogen and thus improve soil fertility.

| Type of Tree or Plant | SF | MAF45 | CAF17 | CAF10 | CAF4 |
|---|---|---|---|---|---|
| Basal area | | | ($m^2$ $ha^{-1}$) | | |
| Timber trees | 12.42 | 4.50 | 0.19 | 2.02 | |
| Fruit trees | | 4.02 | 3.11 | 0.50 | 2.73 |
| Cocoa | | 0.70 | 3.71 | 2.13 | 2.71 |
| Shade trees | | | 2.88 | 0.76 | 1.75 |
| Palms | | 0.90 | | | |
| Other trees | 1.91 | 0.67 | | | 0.84 |
| Total | 14.33 [a] | 10.78 [b] | 9.88 [b] | 5.41 [c] | 8.04 [bc] |
| Plant density | | | (trees $ha^{-1}$) | | |
| Timber trees | 498 | 79 | 5 | 5 | |
| Fruit trees | | 251 | 64 | 28 | 175 |
| Cocoa | | 44 | 263 | 188 | 650 |
| Shade trees | | | 225 | 87 | 469 |
| Palms | | 5 | | | |
| Bananas | | | | | 75 |
| Other trees | 550 | 183 | | | 25 |
| Total | 1048 [a] | 563 [b] | 55 6 [b] | 307 [b] | 1394 [a] |

a,b,c: Different letters along rows indicate significant differences ($p < 0.05$) according to Tukey's test.

Young cocoa agroforestry systems had similar tree density to secondary forests (over 1000 trees per hectare), which was two to four times that of other land uses (Table 3). The young cocoa–based agroforestry had a significantly higher tree density compared with old cocoa–agroforestry and multipurpose tree species agroforestry plots. While a 3 m × 3 m planting grid yields 1111 stems $ha^{-1}$, the tree population reduces as trees grow older.

*3.2. Soil Properties*

Converting natural forest to any land use under cocoa agroforestry systems, including multistrata agroforestry in Sulawesi, markedly reduced topsoil (0 to 30 cm depth) organic matter concentration. In the remnant secondary forest, soil organic carbon concentration at the 0–10 cm depth was 22 g $kg^{-1}$, categorized as 'high' according to [55] compared with only around 15 g $kg^{-1}$ (moderate) for non-forest systems (Figure 4a). Variations among the agroforestry land uses (non–forest) were relatively small, within a 4 g $kg^{-1}$ (0.4%) range, compared to the decline from forest to non–forest.

Surface soil bulk density values (Figure 4b) changed in the opposite direction from the pattern of soil organic matter (Figure 4a). Having the highest C-organic concentration, secondary forest was the least compacted. Consequently, inferred changes in the soil C stock in the top 30 cm of soil were small.

Converting forest to cocoa agroforestry led to soil compaction and therefore decreased total porosity (data not shown as it is the mirror of bulk density data); multistrata agroforestry mimics the forest condition closest (it is the least compacted).

Increased bulk density (reduced total porosity) resulted in the disappearance of part of the macropores when they are >60 μm (Figure 5a), which consequently reduced the soil's capacity to conduct water (Figure 5b) and affected soil water availability (Figure 5c). The reduction in macro-porosity caused by shifting a forest to CAF resulted in a decreased saturated hydraulic conductivity (Figure 5b). Cocoa farms which were recently converted from forest had the lowest hydraulic conductivity; apparently the hydraulic conductivity improved and became closer to that for forest as the land use was transformed to cocoa agroforestry, especially for old cocoa agroforestry and multistrata agroforest.

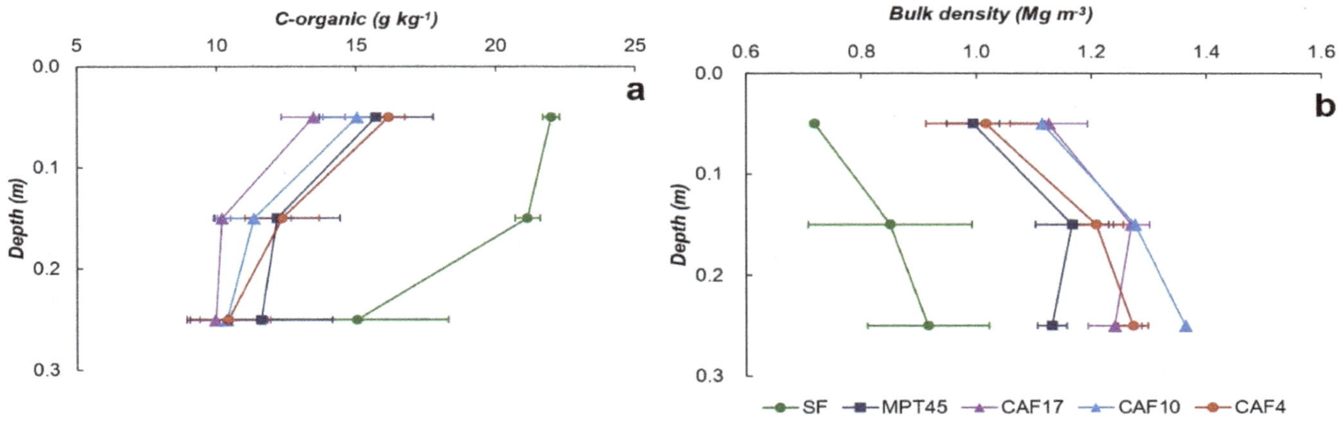


**Figure 4.** Surface soil (0–30 cm depth) soil organic carbon status (**a**) and bulk density (**b**) of secondary tropical forest (SF), multistrata agroforestry that is 45–68 years old (MAF45), cocoa agroforestry that is 17–34 years old (CAF17), cocoa agroforestry that is 10 years old (CAF10), and cocoa agroforestry that is 4–5 years old (CAF4). Bars are standard errors of the means.

Increased bulk density (reduced total porosity) resulted in the disappearance of part of the macropores when they are >60 µm (Figure 5a), which consequently reduced the soil's capacity to conduct water (Figure 5b) and affected soil water availability (Figure 5c). The reduction in macroporosity caused by shifting a forest to CAF resulted in a decreased saturated hydraulic conductivity (Figure 5b). Cocoa farms which were recently converted from forest had the lowest hydraulic conductivity; apparently the hydraulic conductivity improved and became closer to that for forest as the land use was transformed to cocoa agroforestry, especially for old cocoa agroforestry and multistrata agroforest.

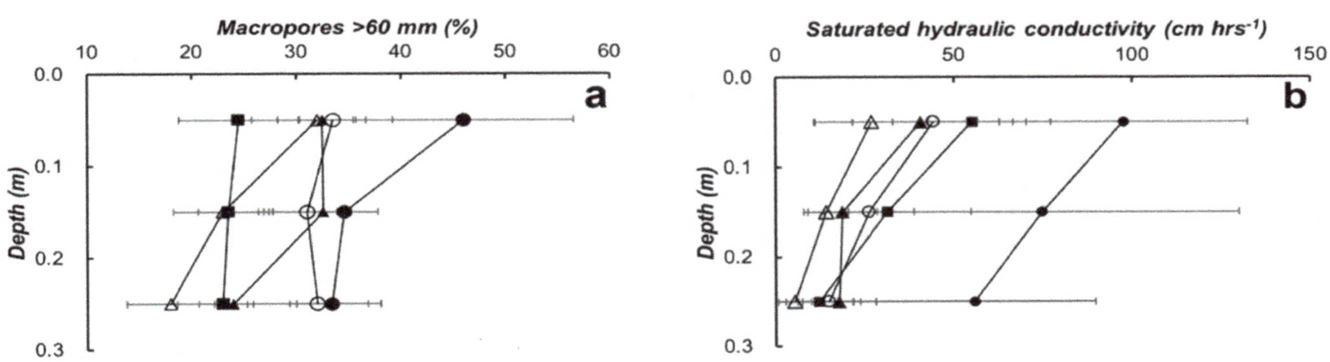


**Figure 5.** Changes of soil surface (0–30 cm depth) macro–porosity (**a**), saturated hydraulic conductivity (**b**) and available water capacity (**c**), resulted from shifting forest to cocoa farms and its restoration through aging cocoa agroforestry systems that are 4–5 years old (CAF4), 10–12 years old (CAF10), and 17–34 years old (CAF17), and through multistrata agroforest that is 45–68 years old (MAF45). Bars are standard errors of the means.

Converting forest to other land–uses also led to a change of soil water retention, which specifically reduces the available water capacity, AWC (Figure 5c). Young cocoa agroforestry led to the lowest hydraulic conductivity, but the figure improved closer toward that for forest as the land use was managed using aged cocoa agroforestry and multipurpose tree species.

### 3.3. C–stock

Converting tropical forest to cocoa–based agroforestry led to a substantial total C–stock decrease, from in total around 320 Mg ha$^{-1}$ of secondary forest to about 75 to 150 Mg ha$^{-1}$ of cocoa–based agroforestry (Figure 6). However, C–stock can be restored to approximately half of that found in the forest through aging cocoa agroforestry or practicing long–term multistrata agroforestry farming. While there was a large effect of land–use on the aboveground C–stock, no influence on the soil C–stock was evident (Figure 6), due to the inclusion of a higher soil dry weight in the 0–30 cm soil layers [9].

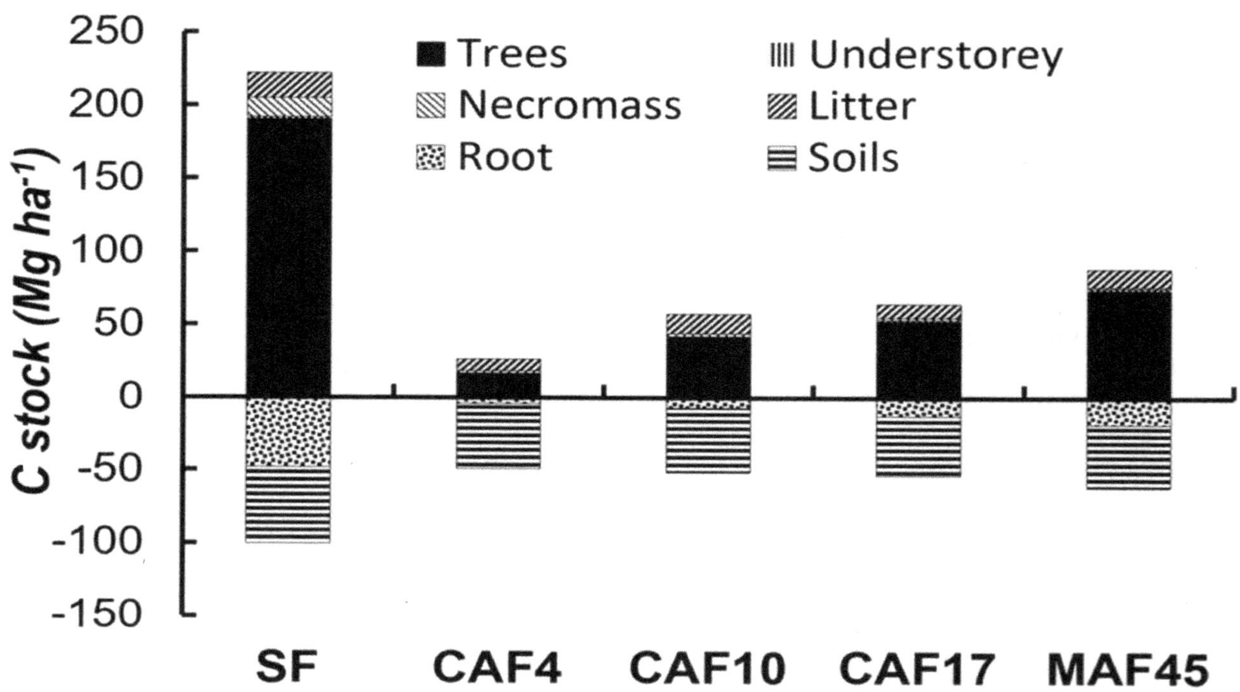


**Figure 6.** Total above– and below–ground carbon stock of remnant secondary forest (SF) compared with cocoa agroforestry that is 4–5 years old (CAF4), cocoa agroforestry that is 10–12 years old (CAF10), cocoa agroforestry that is 17–34 years old (CAF17), and multistrata agroforest that is 45–68 years old (MAF45). The figure demonstrates the total C–stock restoration from 4–5 (CAF4) to 17–34 years old cocoa agroforestry (CAF17) and to 45–68 years old multistrata agroforest (MAF45).

### 3.4. Available Water Capacity

Variable soil organic matter concentrations from different CAFs are associated with soil bulk density (Figure 7a), and hence with soil macro–pore reduction (Figure 7b) and soil hydraulic conductivity (Figure 7c). The decline of soil organic matter concentrations was correlated with soil macro–pore collapse and modified pore–size distribution, eventually affecting available water capacity (Figure 7d). Forest, having the highest organic matter concentrations, had the lowest bulk density, and highest macro–porosity, hydraulic conductivity, and plant water availability (Figure 8).

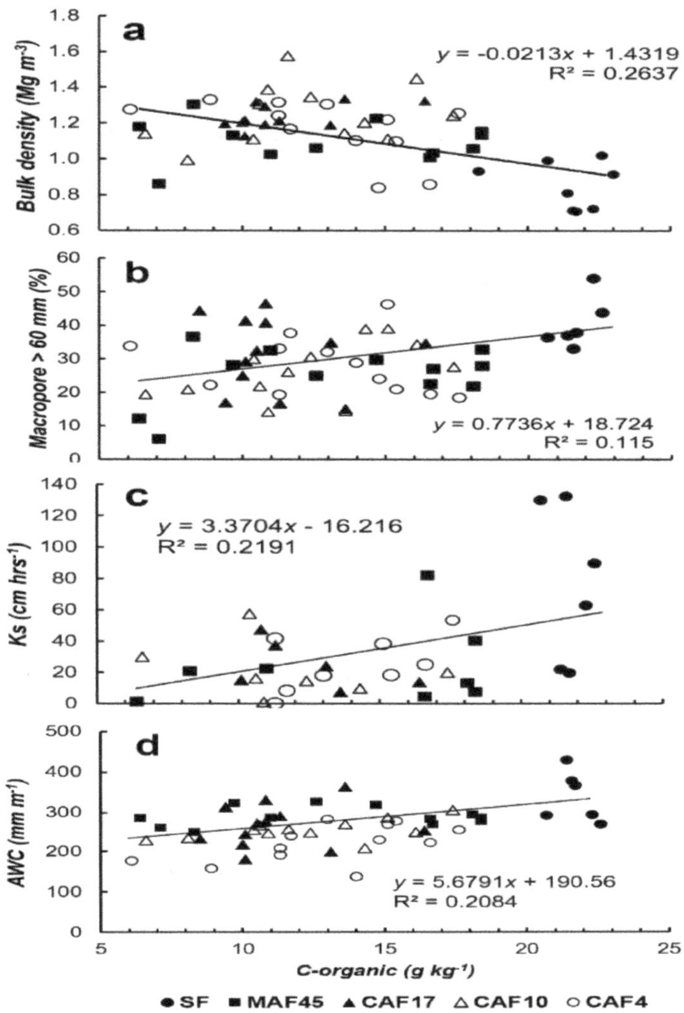


**Figure 7.** Surface soil bulk density (**a**), macro-porosity as percentage of total soil pores (**b**), saturated hydraulic conductivity – Ks (**c**) and available water capacity – AWC, expressed in mm of water per meter of soil (**d**) as related to soil organic matter measured at different cocoa-based agroforestry systems, i.e., secondary forest (SF), cocoa agroforestry that is 4–5 years old (CAF4), 10–12 years old (CAF10), and 17–34 years old (CAF17) and multistrata agroforest that is 45–68 years old (MAF45).

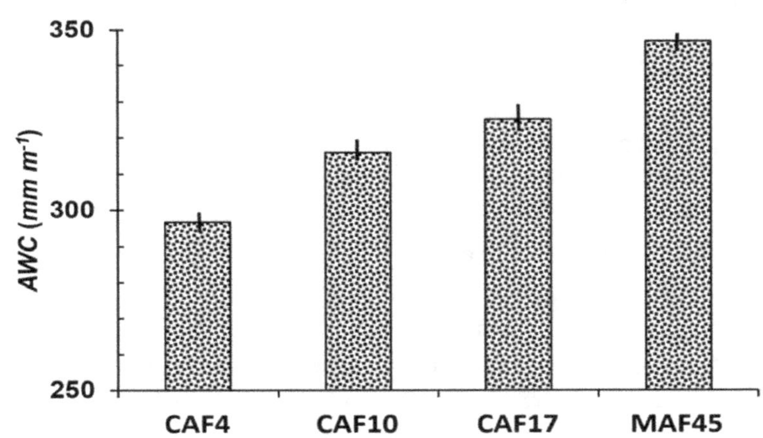


**Figure 8.** Available water capacity (AWC), expressed as mm of water per meter of soil of young (4–5 yrs.) cocoa agroforestry (CAF4) compared to aging (17 to 34 yrs.) cocoa-based agroforestry, and old (45–68 yrs.) multistrata agroforest (MAF45). Bars are standard errors of the means.

## 4. Discussion

Our results provided partial support for the hypothesis that in cacao production systems, the farmer–level benefits of agroforestry through increased water storage, contributing to climate change adaptation, are parallel to global benefits of climate change mitigation through increased C storage. However, the changes in soil properties are slow and gradual and only a partial recovery towards the conditions found in secondary forest was recorded, even for 45–year old mixed agroforest. This slow belowground change relative to aboveground appearance is in–line with restoration experience elsewhere [56].

Three steps are needed to assert a belowground contribution to climate change adaptation by farmers (Figure 1): A. an increase in soil organic matter content, B. a functional role of changed soil properties in the water balance reducing crop vulnerability to rainfall variability, and C. Deliberate use by farmers to reduce their vulnerability. Our data referred to steps A and B.

In the tradition of 'chronosequence' research where surveys of existing plot–level land cover are interpreted as a pseudo time–sequence, such as we did here, it is not easy, especially for soil properties, to separate cause–effect relations from background variability. While aboveground C stock data clearly differentiate the land cover types (Table 3), with only a partial recovery in older mixed agroforest of the C stocks lost when local forests were converted, soil changes are more subtle. As noted by [9] increased soil bulk density partly conceals reductions in $C_{org}$ (expressed per unit dry soil) when carbon stocks in the top 30 cm of soil are calculated. Thus, specific effects of land cover on $C_{org}$ and soil C stock are more clearly observed when $C_{org}/C_{ref}$ ratios are considered rather than direct $C_{org}$ observations.

As second step in the argument we similarly obtained clearer evidence when the plot–level $C_{org}$ parameter, rather than land cover classes, was used as the basis for correlation with other soil parameters. At around 20% of the variance accounted for (Figure 7), other sources of variation clearly dominate. According to the regression equation of Figure 7d, increasing soil organic carbon by 1 g per kilogram of soil (0.1% C-organic improvement) would give an increase of available soil water of about 6 mm per meter of soil. Our data (Figures 4 and 7) reveal that an increase of $C_{org}$ of 4 g $kg^{-1}$ (0.4%) is feasible when land cover is transformed to the MAF 45 conditions in multistrata agroforestry available plant water, which can be around 9.6 mm per 0.4 m of soil (as most of the cocoa roots on these stony soils are concentrated in this layer). An increase of 0.4% in 45 years from a base of around 2% corresponds with a soil C increase of 4 ‰ $y^{-1}$ that is desirable as part of global climate change policy [57], as the relative increase from 2.0 to 2.4 is 20% and $1.2^{1/45} = 1.004$, corresponding with a relative rate of increase of 4 ‰ $y^{-1}$.

As the average evapotranspiration rate for the humid tropics is 3–4 mm $day^{-1}$ [58] and measurements in the dry seasons in Central Sulawesi indicated around 3.6 mm $day^{-1}$ [59], this extra available water can support 50% of cocoa demand for water for any week without any rain in the growing season. In the study area, the average rainfall in the driest months is only 50 mm [60]. The additional AWC that can be expected with higher soil organic matter content, combines with the reduction of soil evaporation where a permanent litter layer is present [61,62]. Such effects are not dramatic but can at least partly offset increased demand for water by trees with active leaf canopies [63]. However, if shade trees are as shallowly rooted as those in a recent West African case study [64] were, then negative effects on cocoa survival are logical outcomes. Only with deep–rooted trees that have access to deeper soil layers and share up to 1 mm $day^{-1}$ to the topsoil by nighttime hydraulic equilibration [65,66] can we expect substantive increases in dry season tolerance in agroforestry systems. A number of the cocoa companion trees may have the deep rooting pattern necessary for such effects, but location–specific observations will be needed to test that [67]. Shade trees in cocoa agroforestry not only improved soil carbon concentrations in Southeast Sulawesi, but also soil structure and aggregation [68], possibly facilitating cocoa root development, in line with our results (Figure 7b) on macroporosity, that will also increase infiltration into the profile and reduction of surface runoff [69]. The water–related functionality of $C_{org}$ had coefficients of determination of around 20%, suggesting that other factors, such as details of soil structure, play a major role as well.

What matters to farmers is the overall effects on yield and economic performance of the plots (e.g., in terms of return to labor). Intercropped shade–tree species with cocoa had little or no effects on cocoa yield in earlier studies in Sulawesi [59,66]. Elsewhere the distance between shade trees and cocoa was found to be important [70]. Although cocoa yields maybe higher under monocultures system [71,72], returns derived from other agroforestry products compensated the difference in economic terms [73]. Part of the tree diversity on cocoa agroforestry in SE Sulawesi was recently attributed [39] to the fact that they harbor tree species prioritized by male farmers as well as those favored by female farmers. At the interface of aboveground and belowground effects and contributions to both mitigation and adaptation climate change agendas, cocoa agroforestry can indeed be 'climate–smart', contributing to resilience and policy synergy [74]. As detailed in a recent study in Southeast Sulawesi [75], the improvements of cocoa agroforestry relative to cocoa monoculture are modest steps in the direction of physically restoring forest soil conditions.

## 5. Conclusions

In terms of total system, C stock cocoa agroforestry could recover up to half of the stocks of secondary forest in the landscape, while conversion only left 23% of the original stock, essentially the soil component. Belowground losses after forest conversion but also restoration are slower than the changes aboveground.

Trees in agroforestry are not only relevant for C storage, but also as provider of farmer income, as modifier of microclimate and in support of soil C storage. In contrast to aboveground C storage, belowground increases in stock have functional relevance. Soil conditions connected to the capacity for CC adaptation, especially minimum soil compaction, greater macro–porosity, and hydraulic conductivity to allow high rain infiltration and higher available water capacity could be directly linked to higher $C_{org}$ concentrations across the land covers sampled. An increase of 1 g $kg^{-1}$ (0.1%) of soil organic carbon increased available soil water capacity by 6% (vol/vol). Our data confirmed an increase in available water capacity, relevant for survival of week–long dry spells. As such, a contribution can be made to human adaptation to increased rainfall variability.

The water–related functionality of $C_{org}$ had coefficients of determination of around 20%. Belowground consequences of tree cover in agroforestry are only quantifiable in large data sets but are likely as relevant as the more visible effects aboveground.

Our results confirmed that a shift from cocoa monoculture to forms of agroforestry has benefits for the farmer in terms of an increased soil water buffering and soil structure that, even though modest in effect size, contributes to tolerance of dry season conditions. These systems also contribute, in modest amounts, to increased storage of carbon in aboveground and belowground pools that ought to be reflected in the country's national C accounting system and be included in the Nationally Determined Contributions to the Paris Agreement on combatting global climate change. A shift from a focus on cocoa to be intensively managed as monoculture crop to one that is best grown in mixed agroforestry systems, along with other tree commodities will be appropriate.

**Author Contributions:** Conceptualization and methodology (S.G., K.H. and M.v.N.); investigation (S.S., R.S., I.H.M., M.N., D.U.); data curation (S.S., K.H., S.G., M.v.N.); writing—original draft preparation (S.G., K.H., M.v.N.); writing—review and editing (S.G., M.v.N.). All authors have read and agreed to the published version of the manuscript.

**Acknowledgments:** We are grateful to all farmers for fieldwork supports during the research, especially Annas, Sahabuddin, and Abdullah, and their team coordinated through the farmer group. The land clearing photo in Figure 3 was kindly provided by Anas and Abdullah. We also thank Samsuar for preparing the site map. The publication was prepared with the support of Ministry of Research and Technology, High Education, Republic Indonesia through WCU collaborative research scheme between Brawijaya, Hasanuddin, and Sebelas Maret Universities.

## References

1. Atangana, A.; Khasa, D.; Chang, S.; Degrande, A. *Tropical Agroforestry*; Springer Science: London, UK, 2013.
2. Van Noordwijk, M. *Sustainable Development through Trees on Farms: Agroforestry in its Fifth Decade*; World Agroforestry (ICRAF): Bogor, Indonesia, 2019.
3. Tscharntke, T.; Clough, Y.; Bhagwat, S.A.; Buchori, D.; Faust, H.; Hertel, D.; Hölscher, D.; Juhrbandt, J.; Kessler, M.; Perfecto, I.; et al. Multifunctional shade-tree management in tropical agroforestry landscapes—A review. *J. Appl. Ecol.* **2011**, *48*, 619–629. [CrossRef]
4. Duguma, L.A.; Minang, P.A.; Van Noordwijk, M. Climate change mitigation and adaptation in the land use sector: From complementarity to synergy. *Environ. Manag.* **2014**, *54*, 420–432. [CrossRef] [PubMed]
5. Harvey, C.A.; Chac, M.; Donatti, C.I.; Garen, E.; Hannah, L.; Andrade, A.; Bede, L.; Brown, D.; Calle, A.; Char, J. Climate-Smart Landscapes: Opportunities and Challenges for Integrating Adaptation and Mitigation in Tropical Agriculture. *Conserv. Lett.* **2014**, *7*, 77–90. [CrossRef]
6. Rosenstock, T.S.; Wilkes, A.; Jallo, C.; Namoi, N.; Bulusu, M.; Suber, M.; Mboi, D.; Mulia, R.; Simelton, E.; Richards, M.; et al. Making trees count: Measurement and reporting of agroforestry in UNFCCC national communications of non-Annex I countries. *Agric. Ecosyst. Environ.* **2019**, *284*, 106569. [CrossRef]
7. Zomer, R.J.; Neufeldt, H.; Xu, J.; Ahrends, A.; Bossio, D.; Trabucco, A.; Van Noordwijk, M.; Wang, M. Global Tree Cover and Biomass Carbon on Agricultural Land: The contribution of agroforestry to global and national carbon budgets. *Sci. Rep.* **2016**, *6*, 29987. [CrossRef]
8. Van Noordwijk, M.; Bayala, J.; Hairiah, K.; Lusiana, B.; Muthuri, C.; Khasanah, N.; Mulia, R. Agroforestry solutions for buffering climate variability and adapting to change. In *Climate Change Impact and Adaptation in Agricultural Systems*; CAB-International: Wallingford, UK, 2014; pp. 216–232.
9. Hairiah, K.; van Noordwijk, M.; Sasi, R.R.; Dwi, D.; Suprayogo, D.; Kurniawan, S.; Prayogo, C.; Gusli, S. Soil carbon stocks in Indonesian (agro) forest transitions: Compaction conceals lower carbon concentrations in standard accounting. *Agric. Ecosyst. Environ.* **2020**, *294*, 106879. [CrossRef]
10. Sanchez, P.A. *Properties and Management of Soils in the Tropics*; Cambridge University Press: Cambridge, UK, 2019. [CrossRef]
11. Wang, F.; Li, Z.; Ding, Y.; Sayer, E.J.; Li, Q.; Zou, B.; Mo, Q.; Li, Y.; Lu, X.; Tang, J.; et al. Tropical forest restoration: Fast resilience of plant biomass contrasts with slow recovery of stable soil C stocks. *Funct. Ecol.* **2017**, *31*, 2344–2355. [CrossRef]
12. Dollinger, J.; Jose, S. Agroforestry for soil health. *Agrofor. Syst.* **2018**, *92*, 213–219. [CrossRef]
13. Van Noordwijk, M.; Cerri, C.; Woomer, P.L.; Nugroho, K.; Bernoux, M. Soil carbon dynamics in the humid tropical forest zone. *Geoderma* **1997**, *79*, 187–225. [CrossRef]
14. Lusiana, B.; Kuyah, S.; Öborn, I.; van Noordwijk, M. Typology and metrics of ecosystem services and functions as the basis for payments, rewards and co-investment. In *Co-Investment in Ecosystem Services: Global Lessons from Payment and Incentive Schemes*; Namirembe, S., Leimona, B., van Noordwijk, M., Minang, P., Eds.; World Agroforestry Centre (ICRAF): Nairobi, Kenya, 2017.
15. Van Noordwijk, M.; Martikainen, P.; Bottner, P.; Cuevas, E.; Rouland, C.; Dhillion, S.S. Global change and root function. *Glob. Chang. Biol.* **1998**, *4*, 759–772. [CrossRef]
16. Van Noordwijk, M.; Bizard, V.; Wangpakapattanawong, P.; Tata, H.L.; Villamor, G.B.; Leimona, B. Tree cover transitions and food security in Southeast Asia. *Glob. Food Secur.* **2014**, *3*, 200–208. [CrossRef]
17. Van Noordwijk, M.; Hoang, M.H.; Neufeldt, H.; Oborn, I.; Yatich, T. *How Trees and People Can Co-Adapt to Climate Change: Reducing Vulnerability through Multifunctional Agroforestry Landscapes*; World Agroforestry Centre (ICRAF): Nairobi, Kenya, 2011.
18. Dewi, S.; Van Noordwijk, M.; Zulkarnain, M.T.; Dwiputra, A.; Hyman, G.; Prabhu, R.; Gitz, V.; Nasi, R. Tropical forest-transition landscapes: A portfolio for studying people, tree crops and agro-ecological change in context. *Int. J. Biodivers. Sci. Ecosyst. Serv. Manag.* **2017**, *13*, 312–329. [CrossRef]
19. Abdulai, I.; Vaast, P.; Hoffmann, M.P.; Asare, R.; Jassogne, L.; van Asten, P.; Rötter, R.P.; Graefe, S. Cocoa agroforestry is less resilient to sub-optimal and extreme climate than cocoa in full sun: Reply to Norgrove (2017). *Glob. Chang. Biol.* **2018**, *24*, e733–e740. [CrossRef] [PubMed]
20. Cerda, R.; Deheuvels, O.; Calvache, D. Contribution of cocoa agroforestry systems to family income and domestic consumption: Looking toward intensification. *Agrofor. Syst.* **2014**, *88*, 957–981. [CrossRef]

21. Somarriba, E.; Beer, J. Productivity of Theobroma cacao agroforestry systems with timber or legume service shade trees. *Agrofor. Syst.* **2011**, *81*, 109–121. [CrossRef]
22. Vaast, P.; Somarriba, E. Trade-offs between crop intensification and ecosystem services: The role of agroforestry in cocoa cultivation. *Agrofor. Syst.* **2014**, *88*, 947–956. [CrossRef]
23. World Atlas. Top 10 Cocoa Producing Countries—WorldAtlas [WWW Document]. Available online: https://www.worldatlas.com/articles/top-10-cocoa-producing-countries (accessed on 13 August 2019).
24. Tromba, A. What Stats Reveal about the Top 10 Cocoa Producing Countries [WWW Document]. ICCO. 2019. Available online: https://foodensity.com/cocoa-producing-countries/ (accessed on 13 August 2019).
25. Kroeger, A.; Koenig, S.; Thomson, A.; Streck, C. *Forest- and Climate-Smart Cocoa in Côte d'Ivoire and Ghana, Aligning Stakeholders to Support Smallholders in Deforestation-Free Cocoa*; World Bank: Washington, DC, USA, 2017.
26. PDSIP. *Outlook Kakao—Komoditas Pertanian Subsektor Perkebunan*; Kementerian Pertanian: Jakarta, Indonesia, 2016.
27. Ruf, F. The Sulawesi case: Deforestation, pre-cocoa and cocoa migrations. In *Beyond tropical Deforestation: From Tropical Deforestation to Forest Cover Dynamics and Forest Development*; Babin, D., Ed.; CIRAD: Montpellier, France, 2004; pp. 277–298.
28. Hartmann, M.; Niklaus, P.A.; Zimmermann, S.; Schmutz, S.; Kremer, J.; Abarenkov, K.; Lu, P. Resistance and resilience of the forest soil microbiome to logging-associated compaction. *ISME J.* **2013**, *8*, 226–244. [CrossRef]
29. Udawatta, R.P.; Rankoth, L.M.; Jose, S. Agroforestry and Biodiversity. *Sustainability* **2019**, *11*, 2879. [CrossRef]
30. Gijsman, A.J. *Deforestation and Land Use: Changes in Physical and Biological Soil Properties in Relation to Sustainability*; Agricultural University Wageningen, Department of Soil Science and Geology: Wageningen, The Netherlands, 1992.
31. Wang, Q.; Wang, S.; Yu, X. Decline of soil fertility during forest conversion of secondary forest to chinese fir plantations in subtropical China. *Land Degrad. Dev.* **2011**, *22*, 444–452. [CrossRef]
32. Mortimer, R.; Saj, S.; David, C. Supporting and regulating ecosystem services in cacao agroforestry systems. *Agrofor. Syst.* **2018**, *6*, 1639–1657. [CrossRef]
33. Rousseau, G.X.; Deheuvels, O.; Arias, I.R.; Somarriba, E. Indicating soil quality in cacao-based agroforestry systems and old-growth forests: The potential of soil macrofauna assemblage. *Ecol. Indic.* **2012**, *23*, 535–543. [CrossRef]
34. Fearnside, P.M. Global Warming and Tropical Land-Use Change: Greenhouse Gas Emission from Biomass Burning, Decomposition and Soils in Forest Conversion, Shifting Cultivation and Secondary Vegetation. *Clim. Chang.* **2000**, *46*, 115–158. [CrossRef]
35. Van Noordwijk, M.; Rahayu, S.; Hairiah, K.; Wulan, Y.C.; Farida, A.; Verbist, B. Carbon stock assessment for a forest-to-coffee conversion landscape in Sumber-Jaya (Lampung, Indonesia): From allometric equations to land use change analysis. *Sci. China Ser. C* **2002**, *45*, 75–86.
36. Schneider, B.M.; Andres, C.; Trujillo, G.; Alcon, F.; Amurrio, P.; Perez, E.; Weibel, F.; Milz, J. Cocoa and total system yields of organic and conventional agroforestry vs. monoculture systems in a long-term field trial in Bolivia. *Expl. Agric.* **2016**, *53*, 1–24. [CrossRef]
37. Guest, D.; Keane, P. Vascular-Streak Dieback: A New Encounter Disease of Cacao in Papua New Guinea and Southeast Asia Caused by the Obligate Basidiomycete Oncobasidium theobromae. *Phytopathology* **2007**, *97*, 1654–1657. [CrossRef]
38. Sari, R.R.; Saputra, D.D.; Hairiah, K.; Rozendaal, D.M.A.; Roshetko, J.; van Noordwijk, M. Gendered species preferences link tree diversity and carbon stocks in cacao agroforest in Southeast Sulawesi, Indonesia. *Land* **2020**, *9*, 108. [CrossRef]
39. Hairiah, K.; Dewi, S.; Agus, F.; Velarde, S.; Ekadinata, A.; Rahayu, S.; van Noordwijk, M. *Measuring Carbon Stocks: Across Land Use Systems: A Manual*; World Agroforestry Centre (ICRAF): Bogor, Indonesia, 2011.
40. Chave, J.; Andalo, C.; Brown, S.; Cairns, M.A.; Chambers, J.Q.; Eamus, D.; Fölster, H.; Fromard, F.; Higuchi, N.; Kira, T.; et al. Tree allometry and improved estimation of carbon stocks and balance in tropical forests. *Oecologia* **2005**, *145*, 87–99. [CrossRef]
41. Yuliasmara, F.; Wibawa, A.; Prawoto, A. Carbon stock in different ages and plantation systemof cocoa: Allometric approach. *Pelita Perkeb. Coffee Cocoa Res. J.* **2009**, *25*, 132. [CrossRef]

42. Arifin, J. *Estimasi Penyimpanan C Pada Berbagai Sistem Penggunaan Lahan di Kecamatan Ngantang*; Universitas Brawijaya: Malang, Indonesia, 2001.
43. Brown, S. *Estimating Biomass and Biomass Change of Tropical Forests: A Primer*; Food and Agriclture Organization: Rome, Italy, 1997.
44. Mokany, K.; Raison, R.J.; Prokushkin, A.S. Critical analysis of root: Shoot ratios in terrestrial biomes. *Glob. Chang. Biol.* **2006**, *12*, 84–96. [CrossRef]
45. Norgrove, L.; Hauser, S. Carbon stocks in shaded *Theobroma cacao* farms and adjacent secondary forests of similar age in Cameroon. *Trop. Ecol.* **2013**, *54*, 15–23.
46. Naelson, D.W.; Sommers, L.E. Carbon and Organic Matter. In *Methods of Soil Analysis, Part 3, Chemical Methods*; Page, A.L., Baker, D.E., Keeney, D.R., Eds.; American Society of Agronomy: Madison, WI, USA, 1996; pp. 1004–1005.
47. Blake, G.R.; Hartge, K.H. Bulk density. In *Methods of Soil Analysis: Part 1—Physical and Mineralogical Methods*; Klute, A., Ed.; Soil Science Society of America, American Society of Agronomy: Madison, WI, USA, 1986; pp. 363–375.
48. Gee, G.W.; Or, D. 2.4 Particle-size analysis. In *Methods of Soil Analysis: Part 4 Physical Methods*; Klute, A., Ed.; Soil Science Society of America, American Society of Agronomy: Madison, WI, USA, 2002; pp. 255–293.
49. Klute, A.; Dirksen, C. Hydraulic conductivity and diffusivity: Laboratory methods. In *Methods of Soil Analysis: Part 1—Physical and Mineralogical Methods*; Klute, A., Ed.; Soil Science Society of America, American Society of Agronomy: Madison, WI, USA, 1986; pp. 687–734.
50. Klute, A. Water retention: Laboratory methods. In *Methods of Soil Analysis: Part 1—Physical and Mineralogical Methods*; Klute, A., Ed.; Soil Science Society of America, American Society of Agronomy: Madison, WI, USA, 1986; pp. 635–662.
51. McIntyre, D.S. Sample preparation. In *Methods for Analysis of Irrigated Soils*; Loveday, J., Ed.; Commonwealth Agricultural Bureaux, Famham Royal: Bucks, UK, 1974; pp. 21–37.
52. Weil, R.; Brady, N. *The Nature and Properties of Soils*, 15th ed.; Prentice Hall: Upper Saddle River, NJ, USA, 2017.
53. Marshall, T.J.; Holmes, J.W. *Soil Physics*, 2nd ed.; Cambridge University Press: Cambridge, UK, 1988.
54. Cassel, D.K.; Nielsen, D.R. Field capacity and available water capacity. In *Methods of Soil Analysis: Part 1—Physical and Mineralogical Methods*; Klute, A., Ed.; Soil Science Society of America, American Society of Agronomy: Madison, WI, USA, 1986; pp. 901–926.
55. Ghaley, B.B.; Wösten, H.; Olesen, J.E.; Schelde, K.; Baby, S.; Karki, Y.K.; Børgesen, C.D.; Smith, P.; Yeluripati, J.; Ferrise, R.; et al. Simulation of Soil Organic Carbon Effects on Long-Term Winter Wheat (*Triticum aestivum*) Production Under Varying Fertilizer Inputs. *Front. Plant Sci.* **2018**, *9*, 1158. [CrossRef]
56. Van Noordwijk, M.; Gitz, V.; Minang, P.A.; Dewi, S.; Leimona, B.; Duguma, L.; Pingault, N.; Meybeck, A. People-Centric Nature-Based Land Restoration through Agroforestry: A Typology. *Land* **2020**, *9*, 251. [CrossRef]
57. Minasny, B.; Malone, B.P.; McBratney, A.B.; Angers, D.A.; Arrouays, D.; Chambers, A.; Chaplot, V.; Chen, Z.S.; Cheng, K.; Das, B.S.; et al. Soil carbon 4 per mille. *Geoderma* **2017**, *292*, 59–86. [CrossRef]
58. Juarez, R.I.N.; Hodnett, M.G.; Fu, R.; Goulden, M.L.; Randow, C. Control of Dry Season Evapotranspiration over the Amazonian Forest as Inferred from Observations at a Southern Amazon Forest Site. *J. Clim.* **2007**, *20*, 2827–2838. [CrossRef]
59. Olchev, A.; Ibrom, A.; Erasmi, S. Effects of land-use changes on evapotranspiration of tropical rain forest margin area in Central Sulawesi (Indonesia): Modelling study with a regional SVAT model. *Ecol. Model.* **2008**, *212*, 131–137. [CrossRef]
60. BPS Polman. *Kabupaten Polewali Mandar dalam Angka (Polewali Mandar Regency in Figures*; PEMDA Polman: Polewali, Indonesia, 2017.
61. Niether, W.; Schneidewind, U.; Armengot, L.; Adamtey, N.; Schneider, M.; Gerold, G. Spatial-temporal soil moisture dynamics under different cocoa production systems. *Catena* **2017**, *158*, 340–349. [CrossRef]
62. Jackson, N.A.; Wallace, J.S. Soil evaporation measurements in an agroforestry system in Kenya. *Agric. For. Meteorol.* **1999**, *94*, 203–215. [CrossRef]
63. Lin, B.B. Agroforestry management as an adaptive strategy against potential microclimate extremes in coffee agriculture. *Agric. For. Meteorol.* **2007**, *144*, 85–94. [CrossRef]

64. Abdulai, I.; Jassogne, L.; Graefe, S.; Asare, R.; Asten, P.; Van Vaast, P.; La, P. Characterization of cocoa production, income diversification and shade tree management along a climate gradient in Ghana. *PLoS ONE* **2018**, *13*, e0195777. [CrossRef] [PubMed]
65. Bayala, J.; Heng, L.K.; van Noordwijk, M.; Ouedraogo, S.J. Hydraulic redistribution study in two native tree species of agroforestry parklands of West African dry savanna. *Acta Oecol.* **2008**, *34*, 370–378. [CrossRef]
66. Van Noordwijk, M.; Lawson, G.; Hairiah, K.; Wilson, J. Root distribution of trees and crops: Competition and/or complementarity. In *Tree-Crop Interactions: Agroforestry in a Changing Climate*; Ong, C.K., Black, C.R., Wilson, J., Eds.; CAB International: Wallingford, UK, 2015; pp. 221–257. [CrossRef]
67. Hairiah, K.; Widianto, W.; Suprayogo, D.; van Noordwijk, M. Tree Roots Anchoring and Binding Soil: Reducing Landslide Risk in Indonesian Agroforestry. *Land* **2020**, *9*, 256. [CrossRef]
68. Wartenberg, A.C.; Blaser, W.J.; Roshetko, J.M.; van Noordwijk, M.; Six, J. Soil fertility and *Theobroma cacao* growth and productivity under commonly intercropped shade-tree species in Sulawesi, Indonesia. *Plant Soil* **2019**. [CrossRef]
69. Suprayogo, D.; van Noordwijk, M.; Hairiah, K.; Meilasari, N.; Rabbani, A.L.; Ishaq, R.M.; Widianto, W. Infiltration-Friendly Agroforestry Land Uses on Volcanic Slopes in the Rejoso Watershed, East Java, Indonesia. *Land* **2020**, *9*, 240. [CrossRef]
70. Koko, L.K.; Snoeck, D.; Lekadou, T.T.; Assiri, A.A. Cacao-fruit tree intercropping effects on cocoa yield, plant vigour and light interception in Côte d'Ivoire. *Agrofor. Syst.* **2013**, *87*, 1043–1052. [CrossRef]
71. Armengot, L.; Barbieri, P.; Andres, C.; Milz, J.; Schneider, M. Cacao agroforestry systems have higher return on labor compared to full-sun monocultures. *Agron. Sustain. Dev.* **2016**, *36*. [CrossRef]
72. Niether, W.; Schneidewind, U.; Fuchs, M.; Schneider, M.; Armengot, L. Below- and aboveground production in cocoa monocultures and agroforestry systems. *Sci. Total Environ.* **2019**, *657*, 558–567. [CrossRef] [PubMed]
73. Somarriba, E.; Cerda, R.; Orozco, L.; Cifuentes, M.; Dávila, H.; Espin, T.; Mavisoy, H.; Ávila, G.; Alvarado, E.; Poveda, V.; et al. Carbon stocks and cocoa yields in agroforestry systems of Central America. *Agric. Ecosyst. Environ.* **2013**, *173*, 46–57. [CrossRef]
74. Van Noordwijk, M. Agroforestry as nexus of sustainable development goals. *IOP Conf. Ser. Earth Environ. Sci.* **2020**, *449*, 012001. [CrossRef]
75. Saputra, D.D.; Sari, R.R.; Hairiah, K.; Roshetko, J.; Suprayogo, D.; van Noordwijk, M. Can cocoa agroforestry restore degraded soil structure following conversion from forest to agricultural use? *Agrofor. Syst.* **2020**, in press.

10

# Enhancing Vietnam's Nationally Determined Contribution with Mitigation Targets for Agroforestry: A Technical and Economic Estimate

**Rachmat Mulia [1,*], Duong Dinh Nguyen [2], Mai Phuong Nguyen [1], Peter Steward [3], Van Thanh Pham [1], Hoang Anh Le [4], Todd Rosenstock [3] and Elisabeth Simelton [1]**

1. World Agroforestry (ICRAF), Hanoi 100000, Vietnam; N.Maiphuong@cgiar.org (M.P.N.); P.ThanhVan@cgiar.org (V.T.P.); E.Simelton@cgiar.org (E.S.)
2. Institute of Geography, Vietnam Academy of Science and Technology, Hanoi 100000, Vietnam; ndduong@ig.vast.vn
3. World Agroforestry (ICRAF), P.O. Box 30677, Nairobi 00100, Kenya; P.Steward@cgiar.org (P.S.); T.Rosenstock@cgiar.org (T.R.)
4. Department of Science, Technology and Environment, Ministry of Agriculture and Rural Development, Hanoi 100000, Vietnam; anhlh.khcn@mard.gov.vn

* Correspondence: r.mulia@cgiar.org

**Abstract:** The Nationally Determined Contributions (NDCs) of several non-Annex I countries mention agroforestry but mostly without associated mitigation target. The absence of reliable data, including on existing agroforestry practices and their carbon storage, partially constrains the target setting. In this paper, we estimate the mitigation potential of agroforestry carbon sequestration in Vietnam using a nationwide agroforestry database and carbon data from the literature. Sequestered carbon was estimated for existing agroforestry systems and for areas into which these systems can be expanded. Existing agroforestry systems in Vietnam cover over 0.83 million hectares storing a 1346 ± 92 million ton $CO_2$ equivalent including above-, belowground, and soil carbon. These systems could be expanded to an area of 0.93–2.4 million hectares. Of this expansion area, about 10% is considered highly suitable for production, with a carbon sequestration potential of 2.3–44 million ton $CO_2$ equivalent over the period 2021–2030. If neglecting agroforestry's potential for modifying micro-climates, climate change can reduce the highly suitable area of agroforestry and associated carbon by 34–48% in 2050. Agroforestry can greatly contribute to Vietnam's 2021–2030 NDC, for example, to offset the greenhouse gas emissions of the agriculture sector.

**Keywords:** agriculture sector; carbon sequestration; cost efficiency; land suitability; potential expansion areas; representative concentration pathway

## 1. Introduction

In 2015, all signatory countries of the landmark Paris Agreement pledged to reduce their national greenhouse gas (GHG) emissions and enhance resilience to climate change. The Nationally Determined Contributions (NDCs) are the blueprints outlining mitigation and adaptation efforts, and the Paris Agreement requires each party to prepare, communicate, and pursue their NDCs with domestic or international support. NDCs describe post-2020 climate programs, and parties can prioritize sectors which substantially contribute to their national emissions. Most signatory countries submitted their first NDCs to the United Nations Framework Convention on Climate Change (UNFCCC) by 2016 [1], with the option to amend before a final submission in 2020.

Agriculture features prominently in the NDCs of non-Annex I countries, which are mostly developing countries [2]. About 40% of 148 non-Annex I countries include mitigation measures for the agriculture sector in their NDCs, with half of these mentioning integrated systems such as agroforestry [3]. However, the NDCs of many countries do not elaborate these mitigation measures into concrete actions and associated targets.

Agroforestry, an integrated agricultural system with crops and trees, can substantially reduce GHG emissions through carbon sequestration [4,5]. Agroforestry also increases farmer adaptation to climate change through, e.g., diversified products and sources of income, resource use efficiency, and improved micro-climates [6–9]. Yet the absence of reliable data, including on types and distribution of existing practices and their carbon storage, partially constrain estimation of mitigation and adaptation potential from agroforestry at national scale [10].

Agriculture is a significant source of emissions in Vietnam. Total GHG emissions from Vietnam's agricultural sector reached an 88.3 million ton $CO_2$ equivalent (mil $tCO_2e$) in 2010, accounting for 35.8% of total national emissions [11], and are projected to increase to 109 mil $tCO_2e$ by 2030 under baseline conditions. Vietnam's first NDC submitted to the UNFCCC in 2016 identified 15 mitigation measures for the agriculture sector but excluded agroforestry. The measures focused on improving the efficiency of inputs, plot management practices, such as transforming conventional water management in rice to alternate wetting drying for reducing methane production, and waste treatment, such as converting livestock waste into biogas, and are expected to reduce the projected 2030 baseline emissions by 6–42% [12]. Vietnam has recently included agroforestry in its revised NDC submitted to the UNFCCC in 2020 as part of the Land Use, Land Use Change, and Forestry (LULUCF) sector. However, the revised NDC only specifies the purpose of agroforestry measure, namely for "enhancing carbon stocks and conserving lands", without elaboration on activities and associated mitigation or adaptation targets.

A methodological framework for estimating mitigation and adaptation potential of agroforestry and associated targets at a national or sub-national scale for climate programs such as NDCs is necessary. In this paper, focusing on the mitigation potential, we describe such a framework that we used to provide a technical and economic estimate of mitigation potential from agroforestry in Vietnam by 2030 by assessing its capacity for offsetting GHG emissions through carbon sequestration. We took into account above-ground carbon (AGC), below-ground carbon (BGC), and soil carbon (SOC) sequestered in existing areas and potential expansion areas of agroforestry across the country. We considered the impact of future climate change scenarios on the expansion area and carbon sequestration potential from agroforestry. The economic estimate accounts for the investment cost required to fulfill the mitigation potential over 2021–2030.

## 2. Materials and Methods

### 2.1. Methodological Framework

Agroforestry mitigation potential was estimated at a national scale and can be disaggregated into a sub-national scale using eight ecological regions (Figure 1a), differentiated based on geographical characteristics, topographical features, ecosystem types, and climate [13]. These eight regions are North West, North East, Red River Delta, North Central Coast, South Central Coast, Central Highlands, South East, and Mekong River Delta. The detailed characteristics of each region, including soil condition and dominant land uses, are given in [13].

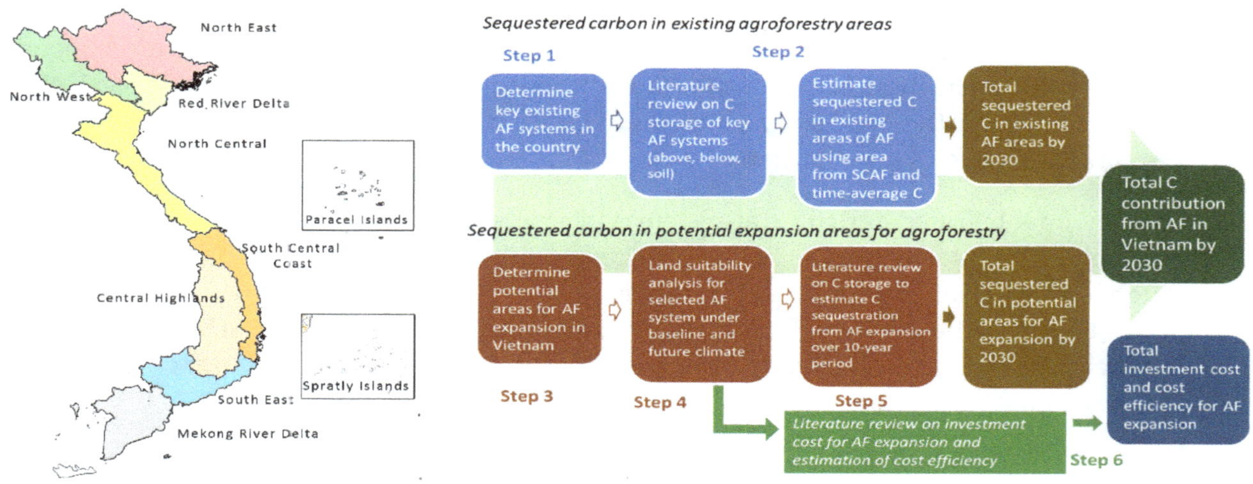


(a) b)

**Figure 1.** (**a**) Eight regions of Vietnam. (**b**) Methodological framework for estimating mitigation potential from agroforestry (AF) in Vietnam.

The assessment of agroforestry mitigation potential consists of six steps (Figure 1b). The first two steps pertain to existing agroforestry areas and the following four pertain to potential expansion areas for agroforestry. The sequestered carbon in the potential expansion areas was assessed over a ten-year period (2021–2030), and the impact of future climate scenarios on extent areas for agroforestry system expansion was investigated using representative concentration pathway (RCP) 4.5 and RCP 8.5 scenarios. The aggregate of sequestered carbon across existing and potential agroforestry areas constitutes the total carbon contribution from agroforestry by 2030.

We used the information on types and distribution of existing agroforestry practices in Vietnam from the Spatially Characterized Agroforestry (SCAF) database (http://scafs.worldagroforestry.org/), as a basis for estimating sequestration contributions from existing areas of agroforestry and for selecting agroforestry systems for expansion. For the carbon estimation, we used input carbon storage data reported in the literature or estimated carbon storage per land area using relevant allometric equations, stem diameter, and crop density data from the literature. The input AGC data are species-specific and mostly specific to Vietnam. On the other hand, the input BGC and SOC are system- but not species-specific. For example, the BGC of most of the assessed agroforestry systems was estimated using the BGC/AGC partitioning factor for agroforestry from [14], and the SOC using the SOC sequestration rate from [10].

### 2.1.1. Step 1: Determine Key Existing Agroforestry Systems in Vietnam

The SCAF database provides information on 48 agroforestry practices observed in 2013–2014 across 42 out of the 63 provinces in Vietnam. Most of the 48 practices can be classified into eight key systems based on their main perennial crop component, excluding the "other systems" (Table 1). The total area of the eight key systems is 820,000 hectares (ha) or about 91% of the total agroforestry system area. Examples of the eight key systems are illustrated in Figure 2. We cannot find a more recent database than SCAF for existing agroforestry systems in Vietnam.

**Table 1.** Existing agroforestry systems in Vietnam.

| Agroforestry System * | Total Area ($10^3$ ha) ** | Common System Components | Main Regions |
|---|---|---|---|
| Melaleuca (*Melaleuca cajuputi*)-based | 245.5 | Fresh-water inland forest with paddy rice, sugarcane, bananas, and fish | Mekong River Delta |
| Robusta coffee (*Coffea canephora*)-based | 245.3 | *Cassia siamea*, black pepper, fruit trees such as durian and avocado, and nuts such as macadamia | Central Highlands, South East |
| Rhizophora (*Rhizophora* spp.)-based | 149 | Mangrove system with shrimp farming | Mekong River Delta |
| Acacia-based | 129.5 | *Acacia mangium*, *Acacia auriculiformis*, or hybrid (*Acacia mangium* × *Acacia auriculiformis*), intercropped with cassava in the early years after tree planting until acacia canopy is closed | North East, Red River Delta, South Central Coast, Mekong River Delta |
| Rubber (*Hevea brasiliensis*)-based | 20.5 | Usually intercropped with cassava or maize in the early years after tree planting until rubber canopy is closed | North West, North Central Coast, Central Highlands |
| Arabica coffee (*Coffea arabica*)-based | 10.5 | *Leucaena leucocephala*, longan (*Dimocarpus longan*), mango (*Mangifera indica*), plum (*Prunus salicina*) as shade trees | North West |
| Cashew (*Anacardium occidentale*)-based | 10.4 | Intercropped with maize, black pepper, or Robusta coffee | Central Highlands, South East |
| Tea (*Camellia sinensis*)-based | 9.5 | *Acacia mangium* or hybrid, *Cassia siamea*, or *Illicum verum* as shade trees | North East, North Central Coast |
| Other systems | 79.8 | Various fruit- or timber tree-based systems with relatively small areas | Spread across regions |

* Ordered by total area; ** Total area in the country based on the Spatially Characterized Agroforestry (SCAF) database.

**Figure 2.** *Cont.*

**Figure 2.** Examples of key common agroforestry systems in Vietnam: (**a**) Arabica and leucaena in the North West region, (**b**) robusta, cassia, and black pepper in the Central Highlands, (**c**) tea and cassia in the North Central Coast, (**d**) acacia and cassava in the North East, (**e**) rubber, potato, and maize in the North Central Coast, (**f**) cashew, robusta, and black pepper in the South East, (**g**) melaleuca and rice in the Mekong River Delta, and (**h**) Rhizophora and shrimp farming in the Mekong River Delta. Source of photos, (**a**–**g**) (SCAF), (**h**) (https://nongnghiep.vn/tom---rung-voi-tang-truong-xanh-d233998.html).

### 2.1.2. Step 2: Estimate Sequestered Carbon in Existing Areas of Agroforestry

The existing areas of agroforestry in Vietnam refer to the 820,000 ha occupied by the eight key agroforestry systems. Due to a lack of information on crop ages, the total AGC sequestered in these areas was estimated using time-average AGCs obtained from the literature (Table A1). For the six key agroforestry systems beside the two in wetlands, we generally used the BGC/AGC biomass partitioning factor for agroforestry from [14] to estimate the BGC, and the response ratio from [10] to estimate the time-average SOC assuming all areas of agroforestry were converted from logged-over forests. For the two systems in wetlands, species-specific BGC and SOC are available from the literature (Table A1). The aggregate of AGC, BGC, and SOC from the eight key agroforestry systems constitutes the carbon contribution from existing areas of agroforestry. We assumed the total area and sequestered carbon in the existing agroforestry systems are constant until 2030.

### 2.1.3. Step 3: Determine Potential Areas for Agroforestry Expansion

Due to the lack of a recent national land cover map from a reliable institution in Vietnam, we used Vietnam's 2018 land cover map from [15] to determine potential areas for agroforestry expansion. Among other land cover types in the map, we selected croplands as potential areas for agroforestry expansion. The croplands are defined by [15] as "lands with herbaceous and shrubby crops followed by harvest and bare soil period". They exclude orchards, annual crops with trees, forest plantations, and wet and low-land paddy fields. We excluded wet paddy lands because they are the main source of staple food for Vietnam, forest plantations and other tree-based systems because they are high-biomass land uses, and barren forest lands because they can be used for forest restoration. Croplands are spread across the country and have a total area of about 3.6 million ha (Figure 3a). Due to the absence of a spatial boundary around existing agroforestry areas, and because croplands exclude all systems with trees, we assumed the existing areas of agroforestry and croplands (hereafter called the expansion domain) are thoroughly separated.

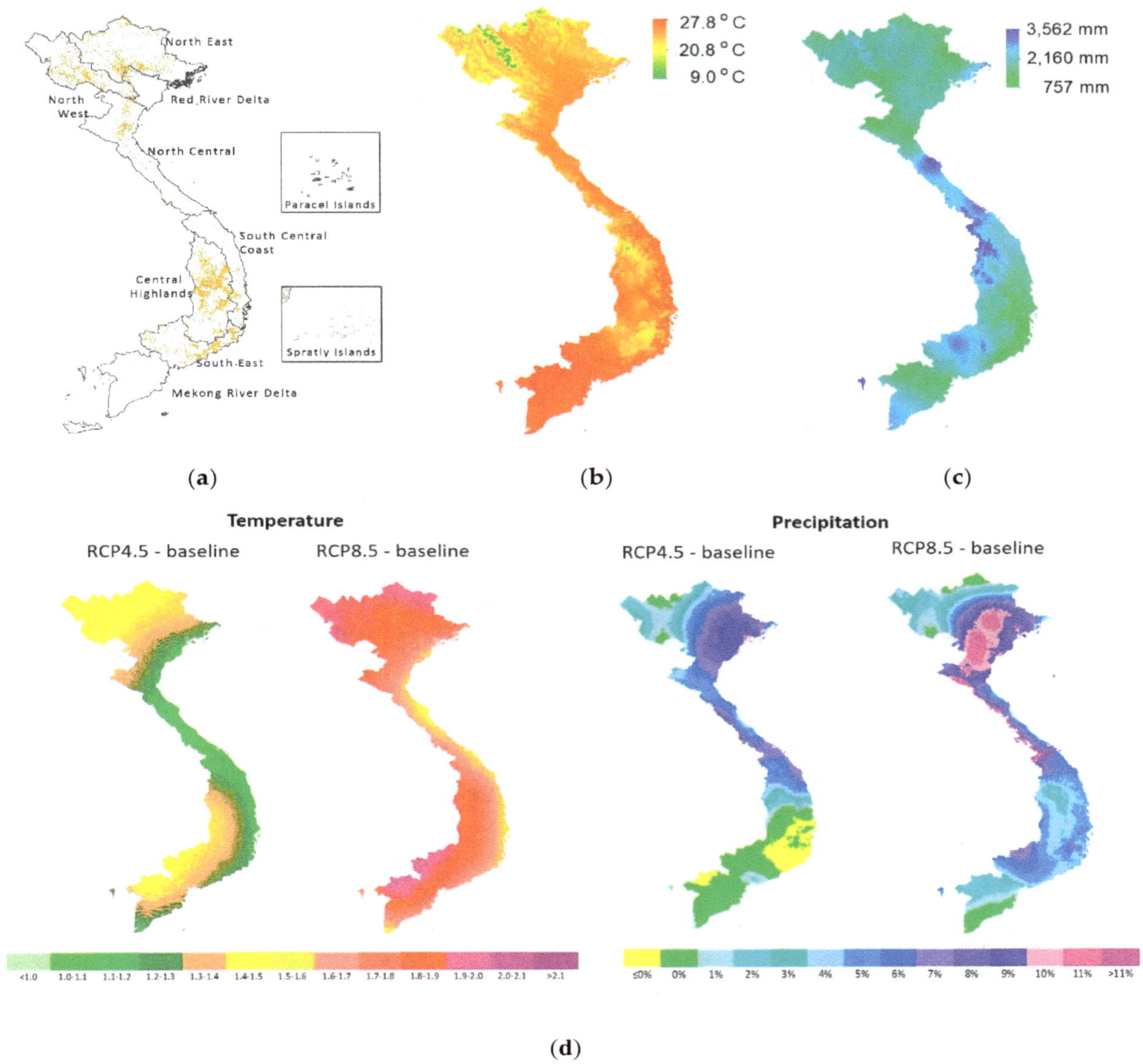


**Figure 3.** (**a**) Potential expansion domain of agroforestry in Vietnam; (**b**) baseline average annual temperature; (**c**) baseline average annual precipitation; (**d**) changes in temperature and precipitation under representative concentration pathway (RCP) 4.5 and RCP 8.5.

#### 2.1.4. Step 4: Select Agroforestry Systems for Expansion and Land Suitability Analysis

We selected five out of the eight key agroforestry systems for expansion, excluding rubber-, Rhizophora-, and Melaleuca-based systems. For rubber, the Vietnam's Master Plan on Agricultural Production Development to 2020 vision to 2030 focuses on strengthening the processing industry instead of area expansion. This orientation also applies to robusta coffee and tea; however, coffee- and tea-based agroforestry provide diverse products such as timber, nuts, or fruits, for which the country is still unable to meet national demand. We excluded Rhizophora- and Melaleuca-based systems because considerable further research is necessary. Spatial data for determining their potential expansion areas is scarce and we lack information such as inundation frequency, water salinity, and tide intensity, that is crucial for land suitability analysis of wetland systems e.g., see [16]. Moreover, we could not find facts about the suitable growing conditions of these two systems from a reliable institution in the country. The most popular acacia-based system in Vietnam is the short-rotation (3–5 years) type for pulp and paper with annual crops such as cassava in the first or second year after tree planting [17]. However, the system for expansion is long-rotation (8–12 years) for timber purposes to help minimize Vietnam's dependence on timber importation. For the cashew, the selected system for expansion is alley cropping with annual crops such as maize, not the perennial shade system with coffee. In the alley cropping system, cashew plants have a higher density to maximize production for national and export markets.

Based on the guideline of land evaluation from [18,19] and the available spatial data, the land suitability analysis for each agroforestry system for expansion considered the topographical, soil, and climatic conditions within the expansion domain. We used slope as an indicator for topographical conditions, soil depth and type for soil conditions, and average annual temperature and precipitation for climatic conditions. The suitability of areas within the expansion domain depends on the indicator values. The assignment of suitability levels involved two steps: first, the indicator values within the expansion domain were compared with thresholds of the growing conditions of the agroforestry systems. We utilized thresholds from a three-tiered ranking of growing conditions reported by reliable institutions in Vietnam (Table 2). Ranking s2 implies little to no limitation of enabling factors for sustaining crop productivity; s1 indicates moderate to severe limitation of such factors, requiring modest or substantial additional inputs or plot management practices for sustaining crop productivity; and s0 describes conditions in which the specified crop cannot grow even with any additional inputs or plot management practices. Subsequently, each area was classified as "highly suitable" if all its indicators met the s2 growing condition, "not suitable" if all indicators met the s0 growing condition, and "less suitable" if all indicators met the s1 growing condition, or if not all indicators met the s2 or s0 condition. For simplicity, we only assessed the suitability of the main perennial crop species in each system, e.g., tea for the tea-based system.

**Table 2.** Growing condition of the main crop species of selected agroforestry systems for expansion.

| | Acacia * | Cashew | Robusta | Arabica | Tea |
|---|---|---|---|---|---|
| Reference | [20,21] | NIAPP ** (data unpublished) | NIAPP (data unpublished) | NIAPP (data unpublished) | [19] |
| **s2** | | | | | |
| Soil type *** | Ac, Fl, RhFe, Gl | RhFe, Fe, Ac | RhFe, XaFe, Fe, Ac | RhFe, Fe | RhFe, Fe |
| Soil depth (m) | >1 | >1 | >1 | >1 | >1 |
| Slope (°) | <8 | <8 | <8 | <8 | <8 |
| Annual rainfall (mm) | 1500–2500 | 2100–2500 | 1600–2000 | 1000–2000 | >1800 |
| Annual temperature (°C) | 23–26 | 22–25 | 22–24 | 18–22 | >22–25 |
| **s1** | | | | | |
| Soil type | Lu, Fe | Ac | Ac, Fl | Ac | Ac, HuFe |
| Soil depth (m) | 0.5–1 | 0.5–1 | 0.5–1 | 0.5–1 | 0.5–1 |
| Slope (°) | 8 to 35 | 8–<25 | 8–20 | 8–20 | 8 to 20 |
| Annual rainfall (mm) | 800–<1500; >2500–3500 | 1300–<2100; >2500 | 1200–1600; >2000 | 800–1000; >2000 | 1000–1800 |
| Annual temperature (°C) | 20–<23; >26 | 18–<22; >25 | –8–22, >24 | 14–<18; >22–24 | 15–22; >25–35 |
| **s0** | | | | | |
| Soil type | Others | Others | Others | Others | Others |
| Soil depth (m) | <0.5 | <0.5 | <0.5 | <0.5 | <0.5 |
| Slope (°) | >35 | >=25 | >20 | >20 | >20 |
| Annual rainfall (mm) | <800; >3500 | <1300 | <1200 | <800 | <1000 |
| Annual temperature (°C) | <20 | <18 | <18 | <14; >24 | <15; >35 |

* Common thresholds for *Acacia mangium, Acacia auriculiformis, Acacia crassicarpa,* or acacia hybrid, ** Vietnam National Institute of Agricultural Planning and Projection (NIAPP), *** Ac: acrisols; Fe: ferrasols; RhFe: rhodic ferralsols; XaFe: xantic ferralsols; Fl: fluvisols; Gl: gleysols; HuFe: humic ferralsols.

We used the 1960–1990 climate data from WorldClim 1.4 [22] as the baseline climate (Table 3). The World Meteorological Organization recommended 1960–1990 as the baseline period [23]. To investigate the impact of future climate on suitable areas for agroforestry expansion, we selected the CNRM-CM5 climate model under RCP 4.5 and RCP 8.5 for the period 2041–2060, also from the WorldClim 1.4, which is available at 30 arc seconds resolution. WorldClim 2.0 provides climate data for the period 2021–2040 but at coarser resolution, and a stronger impact of climate change on the area for expansion and carbon sequestration potential from agroforestry can be expected using the 2041–2060 data. CNRM-CM5 is one of the best climate models for South East Asia [24]. Because we only considered the main perennial crop component of each agroforestry system for expansion in the land suitability analysis, the impact of future climate was also assessed for the main crop component only to represent the impact on each agroforestry system, neglecting the potential of agroforestry to modify micro-climates through an integration of multiple plant components that can reduce the intensity of climate change impact. Due to the absence of a spatial boundary to existing areas of agroforestry, we only investigated the impact of future climate on the potential expansion areas. All spatial inputs were standardized into one arc second, considered as a suitable resolution for country analysis.

**Table 3.** Input maps used for the land suitability analysis.

| Input Maps | Resolution | Coordinate System | Date | Source |
|---|---|---|---|---|
| Land cover | 1 arc second | Lat/Long World Geodetic System (WGS) 84 | 2018 | [15] |
| Digital Elevation Model (DEM) | 1 arc second | Lat/Long WGS 84 | 2019 | Advanced Land Observing Satellite (ALOS) Global Digital Surface Model (AW3D30) version 2.2 |
| Slope | 1 arc second | Lat/Long WGS 84 | 2019 | Generated from DEM |
| Soil type | 1: 1,000,000 | WGS 84 UTM Zone 48N | 2010 | NIAPP |
| Soil depth | 1: 1,000,000 | WGS 84 UTM Zone 48N | 2010 | NIAPP |
| Baseline average annual precipitation and temperature | 30 arc seconds | Lat/Long WGS 84 | 1960–1990 | WorldClim 1.4 |
| Future average annual precipitation and temperature | 30 arc seconds | Lat/Long WGS 84 | 2041–2060 | WorldClim 1.4 |

Based on the baseline climate data from WorldClim 1.4, the northern part of the country generally has lower average annual temperatures and precipitation than the central and southern parts (Figure 3b,c). A similar pattern was reported by, e.g., [13]. This pattern will potentially change in the future, especially under the RCP 8.5 scenario, due to substantial increases in temperature and precipitation in the northern part of the country (Figure 3d).

#### 2.1.5. Step 5: Estimate Sequestered Carbon in Potential Expansion Areas

We estimated the carbon sequestered through agroforestry expansion by subtracting the AGC and BGC stored in croplands from the total carbon accumulated over ten years in areas suitable for agroforestry expansion. We used the average AGC of croplands of 5 tC ha$^{-1}$ for "cropland containing annual crops" according to the Intergovernmental Panel on Climate Change (IPCC) [25] and a BGC/AGC partitioning factor of 0.05–0.2 for "permanent cropland" [14]. The SOC is 53–158 tC ha$^{-1}$ for "permanent cropland" [14]. We assumed that the conversion of croplands into agroforestry retained the SOC in the soil, but we excluded this SOC from the estimation of accumulated SOC in the ten-year agroforestry expansion. All input AGC, BGC, and SOC data for the five agroforestry systems for expansion are given in Table A1.

There were five scenarios of agroforestry expansion, one for each agroforestry system. In each scenario, the system expands into highly suitable areas only, or into 10% and, at most, 25% of the less suitable areas. We assumed that agroforestry systems that have expanded into less suitable areas accumulate AGC, BGC, and SOC at the same rates as in highly suitable areas.

The highly and less suitable areas for expansion were gradually converted into agroforestry across the ten years at a constant conversion rate. Therefore, in the first year, only 10% of suitable areas was converted into agroforestry, to reach the total suitable area in the tenth year. For simplification, we applied a constant carbon accumulation rate across the years for AGC, BGC, and SOC.

To assess the advantages of using agroforestry to offset the GHG emissions from the agriculture sector, we compared the accumulated carbon levels in the expansion domains from agroforestry and from sole crop plantations. We used three commodities in the comparison, namely arabica coffee, robusta coffee, and tea. Acacia and cashew crops were not considered because the sole plantation of the two commodities are forestry, not agricultural systems. The comparison was made for highly suitable areas and AGC only because the BGC and SOC input data are not species-specific. All estimated carbon values were converted into $CO_2$e with a factor of 3.67.

#### 2.1.6. Step 6: Estimate the Cost Efficiency of Agroforestry Expansion

We defined cost efficiency as the investment cost required to sequester one ton of $CO_2$ equivalent. We calculated cost efficiency using the total investment cost to establish and maintain agroforestry systems for ten years under the baseline climate conditions. The cost efficiency will not change under the future climate conditions because the change in investment cost is proportionate to that in the potential expansion area. We estimated all investment costs according to their monetary worth in 2019 (Table 4) by using the annual inflation rate of Vietnam. We compared cost efficiency among agroforestry expansion scenarios, and between agroforestry and sole plantations of arabica, robusta, and tea.

**Table 4.** Input investment costs for agroforestry and sole crop expansion.

| System | Investment Cost (USD $ha^{-1}$ $year^{-1}$) | Source |
|---|---|---|
| Acacia agroforestry | 173 ± 4.6 | [26] for the case of South Central Coast |
| Arabica agroforestry | 2587 | SCAF database for the case of North West |
| Cashew agroforestry | 213 | SCAF database for the case of South East |
| Robusta agroforestry | 2124 ± 574 | SCAF for the case of Central Highlands |
| Tea agroforestry | 2806 | SCAF database for the case of North East |
| Arabica sole plantation | 1835 | [27] for the case of North Central Coast |
| Robusta sole plantation | 941 ± 24 | [28] for the case of Central Highlands |
| Tea sole plantation | 2642 | [29] for the case of Central Highlands |

## 3. Results

### *3.1. Sequestered Carbon in Existing Areas of Agroforestry*

The total (TOC) of sequestered AGC, BGC, and SOC in the existing areas of agroforestry, from the eight key systems in Vietnam, reaches 1346 ± 92 mil $tCO_2$e. The two systems in wetlands, namely Rhizophora- and Melaleuca-based, contribute about 82% to the TOC (Figure 4) thanks to their high carbon storage per hectare and large areas. About 52% of the TOC is SOC, 27% is BGC, and 21% is AGC. The share of TOC from BGC is higher than from AGC due to a large contribution from the root biomass of the Rhizophora-based system. The AGC, BGC, and SOC sequestered by agroforestry systems are shown in Table A2.

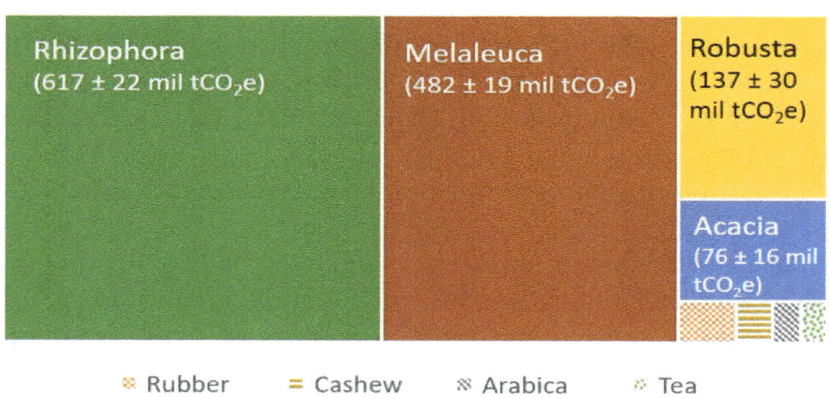


**Figure 4.** Main contributions to the total sequestered carbon in existing areas of agroforestry.

### *3.2. Suitable Areas for Agroforestry Expansion*

Highly suitable areas for the five agroforestry systems for expansion range from 24 to 419 thousand ha nationwide under the baseline climate (Table 5). Cashew has the most limited area, while acacia has the largest. However, in terms of aggregate of highly and less suitable areas, the most limited is arabica. Neglecting the capacity of agroforestry system to modify micro-climate that can reduce the impact of climate change especially warming climate, the highly suitable areas for agroforestry expansion were substantially affected by the potential change in climate (Table 5). For example, large areas across the country are projected to have an increase in average annual temperature by 1.8–2 °C by 2050 under RCP 8.5, and this reduces the highly suitable area of the arabica coffee-based system by 89% compared to baseline. For most of the agroforestry systems, the change in climate transformed the highly suitable areas into less suitable areas for expansion. The two coffee-based systems are the most severely affected, while acacia is relatively resistant to the climate change. On average, for all agroforestry systems, the highly suitable areas are potentially reduced by 34% and 48% in 2050 under RCP 4.5 and RCP 8.5, respectively.

**Table 5.** Suitable areas for agroforestry expansion under baseline and future climate conditions.

| | Acacia | Arabica | Robusta | Cashew | Tea |
|---|---|---|---|---|---|
| | | Highly suitable (thousand ha) | | | |
| Baseline | 419 | 54 | 189 | 24 | 181 |
| RCP 4.5 | 435 | 15 | 36 | 21 | 170 |
| RCP 8.5 | 389 | 6 | 32 | 17 | 127 |
| | | Combined (highly and less) suitable (thousand ha) | | | |
| Baseline | 2407 | 937 | 1985 | 1789 | 1787 |
| RCP 4.5 | 2447 | 362 | 1980 | 1761 | 1784 |
| RCP 8.5 | 2468 | 268 | 1997 | 1827 | 1797 |
| | | Impact of climate change on highly suitable area (%) | | | |
| RCP 4.5 | 4% | −72% | −81% | −14% | −6% |
| RCP 8.5 | −7% | −89% | −83% | −32% | −30% |

Among the eight regions, Central Highlands contains the largest area that is highly suitable for agroforestry expansion (Figure 5a) due to favorable soil and climate conditions for crop cultivation. The South Central Coast and Mekong River Delta have the smallest suitable areas due to higher temperatures, lower precipitation, or unsuitable soil types for agroforestry systems' expansion. These two regions, along with the Red River Delta, also have the smallest suitable area for agroforestry expansion when both highly and less suitable areas are combined (Figure 5b). Suitable areas for agroforestry systems organized by region under baseline and future climate conditions are given in Table A3.

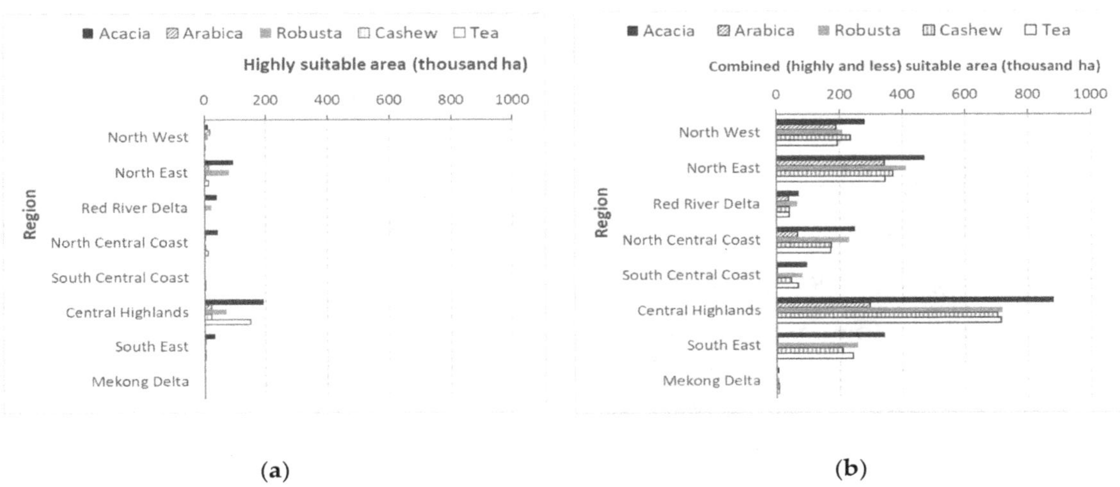


**Figure 5.** (**a**) Highly and (**b**) combined (highly and less suitable) areas by species and region.

### 3.3. Sequestered Carbon in the Agroforestry Expansion Areas

Among the five scenarios, expansion using arabica- and cashew-based agroforestry in highly suitable areas accumulated the smallest TOC over the ten-year expansion period under baseline climate conditions (Figure 6). Acacia-based AF accumulated the highest TOC that reaches 44 ± 4.5 mil $tCO_2e$ by 2030. The inclusion of 10% or 25% of the less suitable areas substantially increased the TOC on average by a factor of 3.3 and 6.8, respectively (Figure 6b,c), compared to the TOC from highly suitable areas only. The sequestered TOC values under baseline and future climate conditions with or without the inclusion of less suitable areas are provided in Table A4.

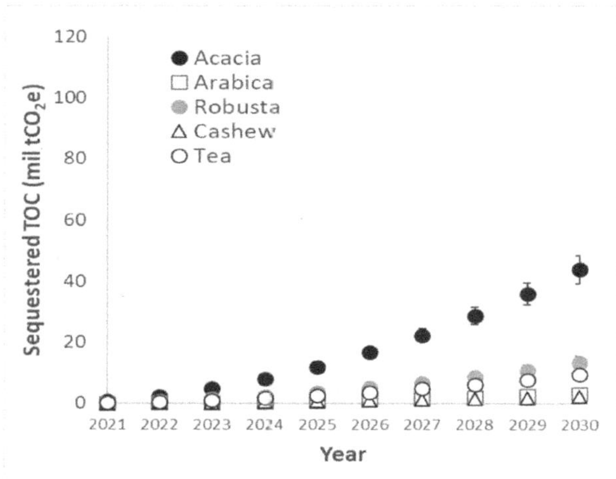


(a)

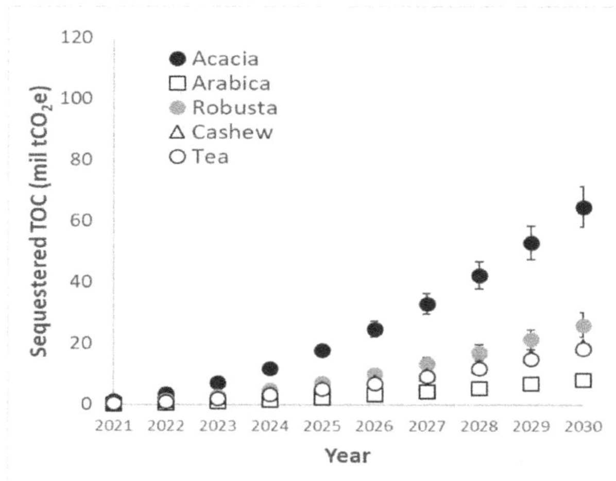


(b)

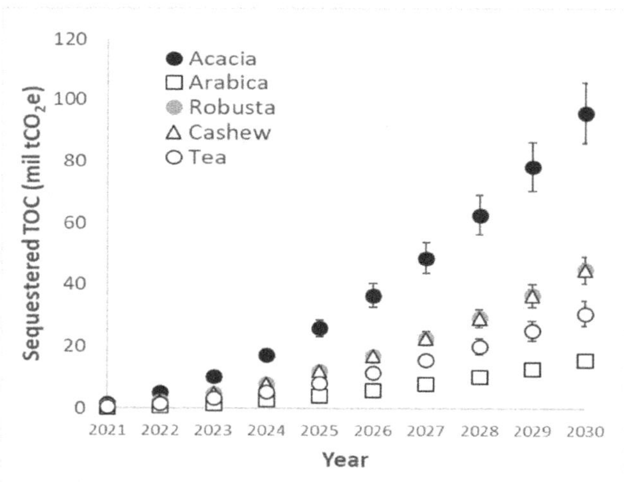


(c)

**Figure 6.** Sequestered total carbon (TOC) over ten-year agroforestry expansion in (**a**) highly suitable and (**b**) highly suitable and 10% or (**c**) 25% of less suitable areas under baseline climate conditions.

### 3.4. Cost Efficiency of Agroforestry Expansion

The investment cost for agroforestry expansion in the highly suitable areas under baseline climate conditions ranges from USD 28 to 2790 million for the ten-year period (Figure 7a). The cost doubled when the combined suitable areas were included. Among the five agroforestry systems, the expansion of tea-based systems into the total highly suitable area of 180,000 ha requires the largest investment cost. For the purpose of sequestering carbon, neglecting potential economic returns, acacia- and cashew-based systems are the most cost efficient among the five agroforestry systems for expansion. The cost efficiency of the two systems is USD 8–12 per $tCO_2e$. The cost efficiencies of sole crop plantations are USD 9121, 1041, and 934 per $tCO_2e$ for arabica, robusta, and tea, respectively. Agroforestry expansion is 1.3–17 times more cost-efficient for sequestering carbon than sole crop plantations.

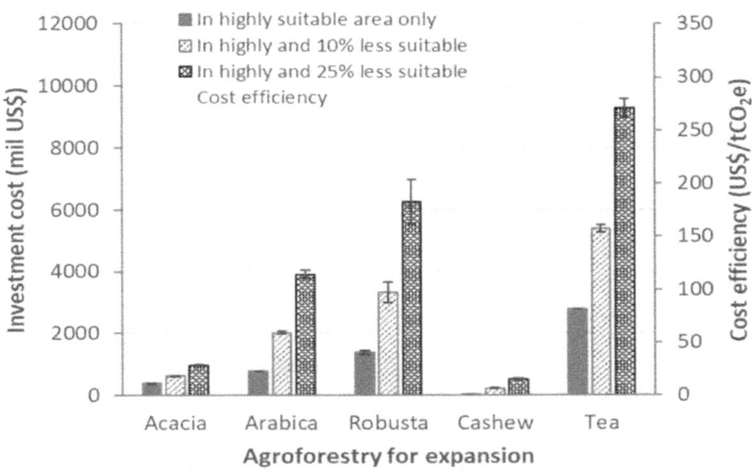


(a)

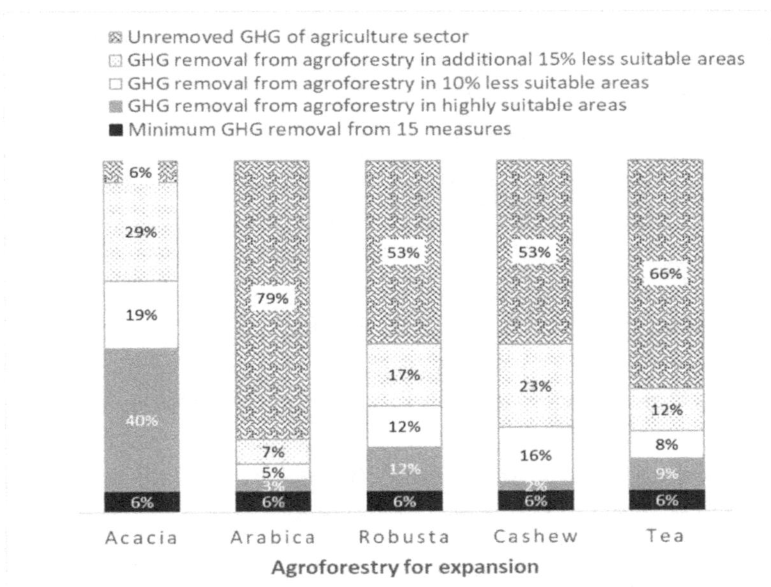


(b)

**Figure 7.** (**a**) Total investment cost and cost efficiency of the five agroforestry expansion scenarios, (**b**) mitigation contribution from agroforestry expansion for removing greenhouse gas (GHG) emission of Vietnam's agriculture sector.

*3.5. Mitigation Contribution to Agriculture Sector*

The sequestered TOC in the existing areas of agroforestry that reaches 1346 ± 92 mil $tCO_2e$ can thoroughly offset the projected GHG emissions of the agriculture sector by 2030. However, if only the carbon contribution from post-2020 programs is considered, the sequestered TOC in agroforestry expansion areas under baseline climate can remove 15–88% of the total GHG emissions of the agriculture sector compared to if 25% of the less suitable areas is included in the carbon assessment (Figure 7b). The acacia-based expansion provides the largest contribution, while the arabica-based provides the smallest. If the 15 mitigation measures in Vietnam's first NDC provide the minimum contribution of

6%, about 6–79% of projected emissions by 2030 would remain unremoved. A higher contribution from agroforestry can be expected if more than 25% of the less suitable areas were considered in the assessment.

## 4. Discussion

### *4.1. Agroforestry in Vietnam's 2020 NDCs*

In Vietnam's 2020 NDC[1], agroforestry is mentioned in Section 2.4.3 as part of measures for the LULUCF sector. Agroforestry and forest protection and restoration are most likely placed as part of the same sector due to their potential usefulness for carbon sequestration and land conservation. Agroforestry-related activities and targets will likely be elaborated in the NDC action plan that is still under development by the government at this time.

To offset sectoral emissions, it is, however, more effective to include agroforestry as part of the agriculture sector. By 2010, in Vietnam, forests and other land uses, excluding agriculture, such as grasslands had generated a net negative emission of 19.2 mil $tCO_2e$ with a projected emission of −42.5 and −45.3 mil $tCO_2e$ by 2020 and 2030, respectively [11]. Moreover, the country has committed to the UNFCCC's Koronivia Joint Work on Agriculture (KJWA) to promote and enhance investment in climate-smart agriculture such as agroforestry [30]. Enhancing terrestrial and soil carbon is among the priorities of the KJWA [31], along with improved nutrient and water management for food security and resilience to climate change, for which agroforestry can also generate relevant benefits.

As in its first NDC, Vietnam's 2020 NDC focused on improving the efficiency of inputs, plot management practices, and waste treatment as mitigation measures for the agriculture sector.

### *4.2. Agroforestry Systems for Expansion and Impact of Climate Change*

Several studies in the literatures [32–35] also projected a strong impact of climate change on suitable area and production of arabica and robusta coffee. Globally, climate change potentially reduces the suitable area for coffee by 50% across climate scenarios, and Vietnam is one among several coffee producing countries that will be severely affected [32,34]. The change in climate, especially warming temperature beyond the optimal threshold for growth, affects the coffee biological process that results in reduced photosynthesis, a slower or halted ripening process, or flower abortion [36]. In addition, the change in temperature can trigger pest and disease outbreak. The impact of climate change on suitable area and production of other species, such as tea, has also been reported in the literature [37–39].

Our study only assessed the impact of climate change on the main perennial crop species of each agroforestry system for expansion. However, the projected impact does not necessarily represent impact on agroforestry that uses that species as the main crop. The presence of other plant components in agroforestry can potentially modify the micro-climate, which reduces the intensity of climate change impact. For example, shading trees or ground cover crops in coffee agroforestry systems can keep soil moisture high and soil temperature low, and these benefits are absent in sole coffee plantation [40–42]. The role of coffee agroforestry systems for mitigating the impact of a warming climate has been demonstrated in the southeast Brazil [33]. The suitable area for sole coffee plantation is projected to decrease by 60% in 2050, driven by a 1.7 °C ± 0.3 increase in the average annual temperature. However, coffee agroforestry systems with a 50% shade cover can reduce the average annual temperatures within the systems, and 75% of the total area of coffee agroforestry in the region will still be suitable for production by 2050. Therefore, the strong impact of the warming climate on robusta and arabica coffee projected in our study clearly suggests the need for prioritizing coffee agroforestry rather than sole

---

1 https://www4.unfccc.int/sites/ndcstaging/PublishedDocuments/Viet%20Nam%20First/Viet%20Nam_NDC_2020_Eng.pdf submitted on September 2020.

coffee plantation for expansion in Vietnam and promoting a gradual conversion of existing sole coffee plantations into agroforestry. The projected impact of the warming climate on other species such as tea, as shown in our study, also suggests the need for modifying the micro-climate through agroforestry.

Vietnam's Master Plan on agricultural production development to 2020, vision to 2030, specifies the Central Highlands, South East, and North Central Coast as the main production regions for robusta coffee. Our land suitability analysis shows that the three regions have suitable, both highly and less suitable combined, areas for robusta coffee, each above 200,000 ha, and thus supports the specification. The other regions such as the North West and North East also have large suitable areas for robusta coffee. However, the Central Highlands has been the production house of robusta coffee in Vietnam for decades, with strong market access, role of private sectors, and supporting infrastructure, including the processing industry [43]. Therefore, the Plan is likely focusing the production of robusta coffee in regions around the Central Highlands. The Plan also specifies main production regions for tea, namely the North East, North West, and Central Highlands. Our land suitability analysis suggests that the South East and North Central Coast are also suitable for tea expansion. Therefore, aside from existing infrastructures and market access for tea that are currently concentrated in the North East, North West, and Central Highlands [44], expansion of tea agroforestry in the other two regions can become an option. For cashew, the Plan specifies the Central Highlands, South East, and South Central Coast. As for robusta coffee and tea, the specification likely considers existing infrastructures and access to market as main factors. Putting those factors aside, our analysis shows that other regions such as North West, North East, and North Central Coast have larger suitable areas for cashew than that of the South Central Coast. Another factor that drives the Plan to consider the South Central Coast is likely its territorial bordering with the Central Highlands. The Plan does not specify regions for arabica coffee and acacia. According to our analysis, three regions, namely the North West, North East, and Central Highlands, have suitable areas for arabica coffee, each above 180,000 ha, and according to [43], the main arabica cultivations in Vietnam can be found in the North West and Central Highlands. Therefore, the expansion of arabica-based agroforestry can target the North West and Central Highlands, with the North East as an option. Acacia can grow in a wide range of terrain, soil, and climate conditions and is relatively easy to manage [45]. In addition, it is a nitrogen-fixing species that can help restore soil fertility. Our analysis shows that acacia is suitable in all regions except the Mekong River Delta.

Rhizophora-based agroforestry should be considered as an agroforestry system for expansion, although our study could not assess its potential area for expansion and sensitivity to climate change due to lack of input data. Rhizophora-based agroforestry can potentially reconcile mangrove restoration and livelihood improvement through combining mangrove plantation and shrimp farming. The area of mangrove forests in Vietnam, especially in the Mekong River Delta, has decreased by 35,000 ha, or 32% of their initial area, during the past three decades (1988–2018) [46], mainly due to anthropogenic activities. About 7300 ha of degraded mangrove forests have been regenerated from 2013 to 2018, but stronger efforts for mangrove restoration in Vietnam are necessary to counter the escalated impact of sea water intrusion amidst livelihood pressure from surrounding populations. By 2050, about 13% of existing rice agricultural areas in the Mekong River Delta are likely to be converted to shrimp farms due to the high economic benefit and impact of sea water intrusion [47]. An effective land share between mangrove plantations and shrimp farms within Rhizophora-based agroforestry systems has been suggested by several studies, e.g., [48,49], for the purpose of optimizing economic return from the shrimp farming and ecological contribution from the mangrove. A recent study [49] claims that mangrove density has no ecological impacts on shrimp farming, and the recommended mangrove coverage for shrimp farming is about 60%. Rhizophora-based agroforestry systems are also a solution for restoring mangrove biodiversity and coastal food chains [49], apart from their substantial contribution to carbon sequestration for GHG removal.

### 4.3. Advantages of Using Agroforestry Rather than Sole Crop Plantation for NDCs

Our study shows that agroforestry is more cost efficient in sequestering carbon than sole crop plantation (see also [12]). Consequently, the cost required to remove GHG emissions is lower when using agroforestry than sole crop plantation. For example, expansion of robusta-based agroforestry in Vietnam can potentially remove 41% of total GHG emissions of the country's agriculture sector by 2030 (see Figure 7b), equivalent to about 45 mil $tCO_2e$. The expansion requires an investment cost of about USD 6.3 billion (see Figure 7a). Removing a similar amount of GHG emissions using sole robusta plantations will require an additional cost of about USD 41 billion. For the case of arabica coffee and tea, the required additional costs are about USD 145 billion and 20 billion, respectively.

Agroforestry can offer more climate change mitigation and adaptation co-benefits as compared to sole crop plantations. For example, robusta-based agroforestry systems in the Central Highlands region were reported to sequester, on average, 0.16 $tCO_2e$ per ton coffee produced thanks to sequestered carbon in shade trees and contribution from nitrogen-fixing trees to reduce chemical inputs. On the other hand, sole coffee plantations were net GHG emitters with, on average, 0.37 $tCO_2e$ per ton coffee produced [50]. In terms of economic return, crop diversification in agroforestry can increase and stabilize farmers' incomes [7]. For example, in the Central Highlands, the net income of a robusta-macadamia (*Macadamia integrifolia*) system reached USD 2500 $ha^{-1}$ $year^{-1}$ compared to USD 1793 $ha^{-1}$ $year^{-1}$ for sole robusta plantations [29]. For tea, the net annual income from a sole tea plantation in the Central Highlands was USD 720 $ha^{-1}$ $year^{-1}$ [29], while a tea-*Acacia mangium* system in the North East generated USD 1688 $ha^{-1}$ $year^{-1}$, as reported in the SCAF database. There is also evidence of agroforestry's role in stabilizing farmers' annual incomes from fruit tree-based systems in the North West [51] and acacia-based systems in the North Central Coast [26,52]. Higher and more stable incomes from agroforestry mean that farmers can reinvest in improved adaptation strategies.

### 4.4. Caveats in the Carbon and Cost Assessment for Agroforestry

In our study, due to limited information from the literature, we assumed that the sequestered carbon per ha of agroforestry in highly and less suitable areas, when used as inputs for the national-scale estimation of carbon potential from agroforestry expansion, are comparable. Likewise, the input investment costs per ha for agroforestry expansion were assumed to be comparable for all suitable areas. To overcome the scarcity of input data, future studies can consider the use of tools for simulating soil–climate–crop interaction in agroforestry—e.g., see [53]—to obtain projection of plot-level sequestered carbon and investment cost in different growing conditions, for example, in highly and less suitable areas. The projected carbon and cost per ha can be used to better estimate carbon sequestration potential and associated cost of agroforestry expansion. For example, if the projected carbon per ha of agroforestry in less suitable areas is reduced by 10% compared to that in highly suitable areas, agroforestry expansion under baseline climate can remove 13–83% instead of 15–88% of the total GHG emissions of the agriculture sector by 2030 if 25% of the less suitable areas were included in the calculation. Future studies on the return on investment from agroforestry that include sensitivity analyses of climate change impacts are also necessary. For such assessment, the studies can use the projected impact of future climate on suitable areas for agroforestry expansion as described in the current study and the projected plot-level investment cost, including level of crop production from the simulation tools for highly and less suitable areas of agroforestry.

### 4.5. Ways Forward to Foster Agroforestry in Vietnam's NDC

Vietnam's 2020 NDC outline national climate programs spanning from 2021 to 2030, with an update to the UNFCCC required by 2025. Among efforts needed to enhance the role of agroforestry in climate mitigation and adaptation informed by the revised NDC, including the inclusion of agroforestry as part of measures for the agriculture sector, are continuous improvement of the database on agroforestry practices, further investigation on their potential mitigation and adaptation benefits, and the development of reliable monitoring, reporting, and verification systems. Vietnam's 2020 NDCs also emphasize that proposed measures need to demonstrate synergy and co-benefits among climate change adaptation, mitigation, and sustainable development goals, including the promotion of gender equality. For the agriculture sector, Section 3.1.4 of the country's NDC underlines that "women's decision-making power within the family is often limited which constrains them to apply their experience and knowledge to selecting varieties and cultivation techniques". Additional scientific evidence is needed to demonstrate advantages of agroforestry over other agricultural practices in contributing to such synergies, including empowering women in decision making. Several studies have shown certain gendered preferences when selecting tree species for agroforestry, e.g., [54–56]. Similar studies for investigating such preference in Vietnamese farmers are necessary. Understanding the reasoning and factors that drive these preferences can have long-term impacts on interventions for climate change mitigation and adaptation, as well as enabling progress on gender equality.

The inclusion of agroforestry in mechanisms that can reward carbon sequestration can also foster agroforestry in Vietnam's NDC. For example, some studies claimed that agroforestry can be a direct or indirect target of Reducing Emissions from Deforestation and Forest Degradation (REDD+) [57], which, to date, remains as the UNFCCC's sole framework that suggests a carbon-reward mechanism. Globally, about half of the 73 developing countries that have REDD+ strategies cited agroforestry as a potential measure for reducing forest degradation and deforestation [58]. Vietnam's REDD+ action plan for 2030 does not, however, explicitly mention agroforestry. Agroforestry is likely relevant for measure 4.1.2 on "promoting sustainable and deforestation-free agriculture"[2], but no elaboration of that measure is provided in the action plan, which creates uncertainty on the relevancy. Recently, Vietnam's national policy on payments for forest ecosystem services is under amendment to include forest carbon in addition to water services. The policy applies a mandatory payment to beneficiaries of forest water service and, very much in the same way, the proposed scheme of payment for carbon services suggests large GHG emitters such as the cement industry and coal-generated power plants would have to pay forest communities and landowners to support forest protection and expansion [59]. Some types of agroforestry, such as acacia-based systems for timber purposes with temporary intercrops, probably classify under the category of forestry land uses that can receive such payments. A verification from relevant authorities is, however, necessary for timber-based systems with permanent intercrops. Furthermore, a feasible monitoring, reporting, and verification system to track progress towards agroforestry mitigation and adaptation targets should be developed to foster agroforestry in Vietnam's NDC. In the current study, the variety and areas of existing agroforestry practices in Vietnam were known from the SCAF database, relying on provincial partners as the principal data sources. Technical, financial, and institutional challenges for developing a monitoring, reporting, and verification system for agroforestry in Vietnam should be identified and properly addressed.

## 5. Conclusions

The suitable area of each agroforestry system for expansion at the national and sub-national scales and the estimated mitigation potential under baseline and future climate conditions described in this study can support Vietnam in specifying agroforestry activities, mitigation targets, and associated costs for developing the action plan of its 2020 NDC. Among the crop species assessed for = agroforestry expansion, the two varieties of coffee, namely robusta and arabica, will be the most severely affected by climate conditions in 2050. To reduce the impact of climate change, future agricultural expansion in Vietnam should consider these crop species in agroforestry, which has potential to modify micro-climates and has other co-benefits such as being more cost efficient in sequestering carbon compared to sole crop plantation. Thanks to these benefits, agroforestry will bring less investment risks than sole crop plantation.

---

2 http://vietnam-redd.org/Upload/CMS/Content/Library-GovernmentDocuments/419%20NRAP%202030%20En.pdf

The 2020 NDC for Vietnam includes agroforestry as part of measures for the LULUCF sector. However, for the purpose of offsetting sectoral GHG emissions, it is more effective to include agroforestry as part of agriculture sector. Excluding the sequestration contribution from the existing areas of agroforestry, the total AGC, BGC, and SOC sequestered in the potential areas for agroforestry expansion can remove 15–88% of the total GHG emissions of the agriculture sector. This is achieved when both highly suitable and 25% of less suitable areas for agroforestry expansion were included in the carbon assessment.

Efforts to foster the implementation of Vietnam's NDC through agroforestry would benefit from continuous data provision for estimating agroforestry's mitigation and adaptation benefits, capacity to achieve synergy and co-benefits between climate change adaptation, mitigation and sustainable development goals, and to promote gender equality. In addition, a reliable monitoring, reporting, and verification system is necessary to track progress towards agroforestry's mitigation and adaptation targets. If all these further efforts generate at least preliminary outputs accessible to relevant authorities within the next 3–4 years, we can expect agroforestry to have a broader role in climate mitigation and adaptation, informed by Vietnam's revised NDC by 2025.

**Author Contributions:** Conceptualization, R.M., M.P.N., P.S. and T.R.; data curation, R.M., D.D.N., M.P.N., P.S., V.T.P., H.A.L., T.R. and E.S.; formal analysis, D.D.N., M.P.N. and P.S.; funding acquisition, R.M. and T.R.; investigation, R.M., D.D.N., M.P.N., P.S. and V.T.P.; methodology, R.M., D.D.N., M.P.N., P.S., V.T.P. and E.S.; software, D.D.N.; validation, R.M., D.D.N., M.P.N., P.S., H.A.L., T.R. and E.S.; visualization, D.D.N. and V.T.P.; writing—original draft preparation, R.M.; writing—review and editing, R.M., D.D.N., M.P.N., P.S., V.T.P., H.A.L., T.R. and E.S. All authors have read and agreed to the published version of the manuscript.

**Acknowledgments:** This work was implemented as part of the CGIAR Research Program on Climate Change, Agriculture and Food Security (CCAFS), which is carried out with support from the CGIAR Trust Fund and through bilateral funding agreements. We specifically thank USAID for funding this work. For details, please visit https://ccafs.cgiar.org/donors. The views expressed in this document cannot be taken to reflect the official opinions of these organizations. We sincerely thank Delia Catacutan, Nguyen Quang Tan, Vu Tan Phuong, Chu Van Chuong, Bui My Binh, Nguyen Phu Hung, and Meine van Noordwijk for numerous and valuable inputs to the study. We also thank valuable comments from three anonymous reviewers.

## Appendix A

**Table A1.** Input data for estimation of sequestered carbon in agroforestry systems.

| Agroforestry System | Tree Density (trees $ha^{-1}$) * | AGC ** | BGC | SOC *** | Source |
|---|---|---|---|---|---|
| Tea-based | Agroforestry (AF): 17,000 (tea), 150–200 (shade trees)<br>Sole: 18,000–25,000 | TA: 13.3 tC $ha^{-1}$<br>SR AF: 1.7 ± 0.18 tC $ha^{-1}$ $year^{-1}$<br>SR sole: 1.4 ± 0.14 tC $ha^{-1}$ $year^{-1}$ | BGC/AGC: 0.10–0.34 | RR-TA: 1.0 ± 0.2<br>SR: 1.59 ± 0.4 tC $ha^{-1}$ $year^{-1}$ | AGC-TA and AGC-SR AF: [60] for the case in North East Vietnam. AGC-SR sole: calculated from [61]. BGC/AGC for AF [14]. RR-TA from 'forest to shade perennial' for tropical region [10]. SOC-SR from 'croplands to multistrata system' for tropical region [10] |
| Coffee robusta-based | AF: 500–1100 (coffee), 85–150 (shade trees)<br>Sole: 750–1500 | TA: 13.4 ± 0.3 tC $ha^{-1}$<br>SR AF: 2.63 ± 0.55 tC $ha^{-1}$ $year^{-1}$<br>SR sole: 0.75 ± 0.01 tC $ha^{-1}$ $year^{-1}$ | BGC/AGC: 0.10–0.34 | RR-TA: 1.0 ± 0.2<br>SR: 1.59 ± 0.4 tC $ha^{-1}$ $year^{-1}$ | AGC-TA [50]. AGC-SR AF: calculated from [62]. AGC-SR sole and calculated using allometric equation from [63]. All for the case in Central Highlands. BGC/AGC for AF [14]. RR-TA and SOC-SR similar as for tea. |
| Coffee arabica-based | AF: 3000–5000 (coffee), 100–200 (shade trees)<br>Sole: 4000–8000 | TA: 13.5 ± 5.5 tC $ha^{-1}$<br>SR AF: 1.9 ± 0.24 tC $ha^{-1}$ $year^{-1}$<br>SR sole: 0.5 ± 0.06 tC $ha^{-1}$ $year^{-1}$ | BGC/AGC: 0.10–0.34 | RR-TA: 1.0 ± 0.2<br>SR: 1.59 ± 0.4 tC $ha^{-1}$ $year^{-1}$ | AGC-TA and AGC-SR AF: [64] for the case in North West. AGC-SR sole: calculated using allometric equation from [65] for the case in North West. RR-TA and SOC-SR similar as for robusta. |
| Cashew-based | 100–200 | TA: 54.2 tC $ha^{-1}$<br>SR: 4.2 ± 0.35 tC $ha^{-1}$ $year^{-1}$ | TA: 9.2 tC $ha^{-1}$<br>SR: 0.7 ± 0.06 tC $ha^{-1}$ $year^{-1}$ | RR-TA: 0.95<br>SR: 0.84 ± 0.26 tC $ha^{-1}$ $year^{-1}$ | AGC-TA, AGC-SR, BGC-TA, BGC-SR: stem diameter from [66] and allometry equation from [67]. RR-TA from forest to silvoarable for tropical region [10]. SOC-SR from 'croplands to silvoarable' for tropical region [10]. |
| Rubber-based | 500 | TA: 25.3 ± 2.76 tC $ha^{-1}$ | TA: 4.5 ± 0.3 tC $ha^{-1}$ | RR-TA: 0.95 | AGC-TA: stem diameter [68], allometric equation: [69]. BGC-TA: using AGC-BGC relation from [69]. RR-TA is similar as for cashew [10] |
| Acacia spp.-based # | Short-rotation (3–5 years): 4500–8000<br>Long-rotation (8–12 years): 600–1200 | TA: 25.3 ± 6.2 tC $ha^{-1}$<br>SR: 4.0 ± 0.37 tC $ha^{-1}$ $year^{-1}$ | BGC/AGC: 0.10–0.33 | RR-TA: 0.95<br>SR: 0.84 ± 0.26 tC $ha^{-1}$ $year^{-1}$ | AGC-TA and AGC-SR: [26] for the case in South Central Coast. BGC/AGC for tree plantation [14]. RR-TA is similar as for rubber [10]. SOC-SR similar as for cashew. |
| Rhizophora-based | 8000–10,000 | TA: 156.4 tC $ha^{-1}$ | TA: 568.4 tC $ha^{-1}$ | TA: 386 tC $ha^{-1}$ | AGC-TA and BGC-TA calculated based on [70] for the case in Mekong River Delta. SOC-TA [71] |
| Melaleuca-based | 5500–6500 | TA: 178.4 ± 14.6 tC $ha^{-1}$ | TA: 44.6 ± 4.3 tC $ha^{-1}$ | TA: 312.2 ± 25.4 tC $ha^{-1}$ | AGC-TA, BGC-TA, SOC-TA: calculated based on [72] for the case in Mekong River Delta |

* Common tree density, not specifically those reported in the cited literature, ** TA: time-average, SR-AF: sequestration rate of AF system, SR-sole: sequestration rate of sole crop system. In the C estimation, for the AGC inputs without standard error, we applied a 10% standard error from the mean. AGC input without standard error *** RR-TA: response ratio to estimate SOC of the eight key existing AFs. The land cover before conversion into those AFs was assumed as logged over forest with SOC ranges from 68–205 tC $ha^{-1}$ [14]. # AGC-TA is for short-rotation (3–5 years) as the current most popular acacia systems in Vietnam. AGC-SR is for long-rotation (8–12 years), for expansion.

**Table A2.** Sequestered carbon in the existing areas of agroforestry.

| Agroforestry System * | Area ($10^3$ ha) | AGC (mil $tCO_2e$) Total | AGC SE | BGC (mil $tCO_2e$) Total | BGC SE | SOC (mil $tCO_2e$) Total | SOC SE | TOC ** (mil $tCO_2e$) Total | TOC SE |
|---|---|---|---|---|---|---|---|---|---|
| Melaleuca-based | 246 | 161 | 6.19 | 40 | 1.83 | 281 | 10.8 | 482 | 18.8 |
| Robusta-based | 245 | 12 | 0.12 | 2.4 | 0.57 | 123 | 29.1 | 137 | 29.5 |
| Rhizophora-based | 149 | 86 | 4.04 | 320 | 8.93 | 211 | 9.97 | 617 | 22.4 |
| Acacia-based | 130 | 12 | 1.40 | 2.4 | 0.57 | 62 | 14.6 | 76.1 | 15.6 |
| Rubber-based | 21 | 1.9 | 0.10 | 0.4 | 0.09 | 9.8 | 2.3 | 12.0 | 2.4 |
| Arabica-based | 11 | 0.5 | 0.06 | 0.1 | 0.02 | 5.3 | 1.25 | 5.9 | 1.3 |
| Cashew-based | 10 | 2.1 | 0.10 | 0.4 | 0.03 | 4.9 | 1.17 | 7.4 | 1.25 |
| Tea-based | 10 | 0.5 | 0.04 | 0.1 | 0.02 | 4.8 | 1.13 | 5.4 | 1.17 |
| All systems | 820 | 275 | 12.0 | 366 | 12.1 | 701 | 70.4 | 1343 | 92.4 |

* ordered by the total area in the country, ** total of AGC, BGC and SOC.

**Table A3.** Suitable areas for agroforestry expansion by species and region under baseline and future climate.

| ER * | Acacia Baseline | Acacia RCP 4.5 | Acacia RCP 8.5 | Arabica Baseline | Arabica RCP 4.5 | Arabica RCP 8.5 | Robusta Baseline | Robusta RCP 4.5 | Robusta RCP 8.5 | Cashew Baseline | Cashew RCP 4.5 | Cashew RCP 8.5 | Tea Baseline | Tea RCP 4.5 | Tea RCP 8.5 |
|---|---|---|---|---|---|---|---|---|---|---|---|---|---|---|---|
| | | | | | | | Highly suitable area (thousand ha) | | | | | | | | |
| NW | 12 | 22 | 23 | 19 | 6 | 3 | 11 | 8 | 6 | | | 0.01 | 4 | 8 | 6 |
| NE | 94 | 112 | 109 | 13 | 3 | 1 | 81 | 9 | 5 | 0.01 | 0.01 | 0.01 | 13 | 38 | 32 |
| RRD | 41 | 42 | 42 | | | | 23 | 0.01 | | | | | | 8 | 1 |
| NCC | 42 | 55 | 47 | 0.01 | | | 3 | | | 0.01 | 0.01 | 0.01 | 12 | 1 | 1 |
| SCC | 2 | 0.01 | 1 | | | | 0.01 | 0.01 | 0.01 | 0.01 | 0.01 | 0.01 | 0.01 | 0.01 | 0.01 |
| CH | 192 | 202 | 166 | 23 | 6 | 2 | 72 | 19 | 21 | 23 | 20 | 16 | 150 | 114 | 86 |
| SE | 35 | 1 | 1 | 0.01 | | | 0.01 | | 0.01 | 1 | 0.01 | | 2 | 0.01 | |
| MRD | 0.01 | | | | | | | | | | | | | | |
| **Total** | **419** | **435** | **389** | **54** | **15** | **6** | **189** | **36** | **32** | **24** | **21** | **17** | **181** | **170** | **127** |
| | | | | | | | Combined (highly and less) suitable area (thousand ha) | | | | | | | | |
| NW | 282 | 307 | 309 | 190 | 141 | 120 | 210 | 210 | 210 | 236 | 236 | 236 | 194 | 194 | 194 |
| NE | 472 | 491 | 493 | 342 | 102 | 80 | 411 | 413 | 414 | 369 | 373 | 373 | 345 | 345 | 345 |
| RRD | 72 | 72 | 72 | 39 | | | 68 | 68 | 68 | 41 | 41 | 41 | 40 | 40 | 40 |
| NCC | 250 | 250 | 250 | 66 | 5 | 2 | 232 | 232 | 232 | 175 | 176 | 176 | 173 | 173 | 173 |
| SCC | 99 | 99 | 99 | 1 | 0 | 0 | 82 | 77 | 84 | 48 | 42 | 52 | 69 | 68 | 72 |
| CH | 881 | 884 | 885 | 298 | 114 | 65 | 718 | 718 | 718 | 704 | 679 | 723 | 715 | 715 | 715 |
| SE | 344 | 337 | 353 | 1 | 0 | 0 | 257 | 254 | 263 | 209 | 207 | 218 | 243 | 241 | 251 |
| MRD | 8 | 8 | 8 | | | | 8 | 8 | 8 | 8 | 8 | 8 | 7 | 7 | 7 |
| **Total** | **419** | **435** | **389** | **54** | **15** | **6** | **189** | **36** | **32** | **24** | **21** | **17** | **181** | **170** | **127** |

* North West (NW), North East (NE), Red River Delta (RRD), North central Coast (NCC), South Central Coast (SCC), Central Highlands (CH), South East (SE) and Mekong River Delta (MRD).

**Table A4.** Sequestered TOC in highly suitable with/out partial less-suitable areas under baseline and future climate.

| | Sequestered TOC (mil $tCO_2e$) by 2030 * | | | | | | | | |
|---|---|---|---|---|---|---|---|---|---|
| | Highly Suitable | | | Highly and 10% Less Suitable | | | Highly and 25% Less Suitable | | |
| Scenario ** | Baseline | RCP4.5 | RCP8.5 | Baseline | RCP4.5 | RCP8.5 | Baseline | RCP4.5 | RCP8.5 |
| 1 | 44 ± 4.5 | 46 ± 4.5 | 40 ± 4.1 | 65 ± 7.1 | 67 ± 7.2 | 62 ± 6.6 | 96 ± 11 | 98 ± 11 | 95 ± 11 |
| 2 | 3.2 ± 0.4 | 0.9 ± 0.1 | 0.4 ± 0.1 | 8.3 ± 1.1 | 3.0 ± 0.4 | 1.8 ± 0.2 | 16 ± 2.1 | 6.0 ± 0.8 | 4.0 ± 0.5 |
| 3 | 13 ± 2.1 | 2.5 ± 0.4 | 2.3 ± 0.4 | 26 ± 1.9 | 16 ± 1.8 | 16 ± 1.9 | 45 ± 4.5 | 37 ± 4.4 | 37 ± 4.5 |
| 4 | 2.3 ± 0.2 | 1.9 ± 0.2 | 1.6 ± 0.2 | 19 ± 4.2 | 19 ± 2.5 | 19 ± 2.5 | 45 ± 7.4 | 44 ± 5.7 | 45 ± 5.7 |
| 5 | 9.6 ± 1.2 | 8.9 ± 1.1 | 6.7 ± 0.9 | 18 ± 2.4 | 18 ± 2.3 | 16 ± 2.1 | 31 ± 4.2 | 31 ± 4.1 | 29 ± 3.8 |

* Average and standard error ** 1: expansion of acacia-, 2: arabica-, 3: robusta-, 4: cashew-, and 5: tea-based AF over the country.

## References


1. UNFCCC. *Aggregate Effect of the Intended Nationally Determined Contributions: An Update*; UNFCCC: Marrakech, Morocco, 2016.
2. UNFCCC. Parties. Available online: https://unfccc.int/process/parties-non-party-stakeholders/parties-convention-and-observer-states?field_national_communications_target_id%5B514%5D=514 (accessed on 15 September 2020).
3. FAO. *The Agriculture Sectors in the Intended Nationally Determined Contribution: Analysis*; FAO: Italy, Rome, 2016.
4. Feliciano, D.; Ledo, A.; Hillier, J.; Nayak, D.R. Which agroforestry options give the greatest soil and above ground carbon benefits in different world regions? *Agric. Ecosyst. Environ.* **2018**, *254*, 117–129. [CrossRef]
5. Zomer, R.J.; Neufeldt, H.; Xu, J.; Ahrends, A.; Bossio, D.; Trabucco, A.; Van Noordwijk, M.; Wang, M. Global tree cover and biomass carbon on agricultural land: The contribution of agroforestry to global and national carbon budgets. *Sci. Rep.* **2016**, *6*, 29987. [CrossRef] [PubMed]
6. Lasco, R.D.; Delfino, R.J.P.; Espaldon, M.L.O. Agroforestry systems: Helping smallholders adapt to climate risks while mitigating climate change. *Wiley Interdiscip. Rev. Clim. Chang.* **2014**, *5*, 825–833. [CrossRef]
7. Catacutan, D.C.; van Noordwijk, M.; Nguyen, T.H.; Öborn, I.; Mercado, A.R. *Agroforestry: Contribution to Food Security and Climate-Change Adaptation and Mitigation in Southeast Asia*; World Agroforestry Centre (ICRAF) Southeast Asia Regional Program; ASEAN-Swiss Partnership on Social Forestry and Climate Change: Bogor, Indonesia, 2017.
8. van Noordwijk, M.; Bayala, J.; Hairiah, K.; Lusiana, B.; Muthuri, C.; Khasanah, M.; Mulia, R. Agroforestry solutions for buffering climate variability and adapting to change. In *Climate Change Impact and Adaptation in Agricultural Systems*; Fuhrer, J., Gregory, P.J., Eds.; CAB International: Wallingford, UK, 2014; pp. 216–232.
9. Simelton, E.; Dam, B.V.; Catacutan, D. Trees and agroforestry for coping with extreme weather events: Experiences from northern and central Vietnam. *Agrofor. Syst.* **2015**, *89*, 1065–1082. [CrossRef]
10. Cardinael, R.; Umulisa, V.; Toudert, A.; Olivier, A.; Bockel, L.; Bernoux, M. Revisiting IPCC Tier 1 coefficients for soil organic and biomass carbon storage in agroforestry systems. *Environ. Res. Lett.* **2018**, *13*, 124020. [CrossRef]
11. MONRE. *The Initial Biennial Updated Report of Vietnam to the United Nations Framework Convention on Climate Change*; MONRE: Hanoi, Vietnam, 2014.
12. Escobar-Carbonari, D.; Grosjean, G.; Läderach, P.; Nghia, T.D.; Sander, B.O.; McKinley, J.; Sebastian, L.; Tapasco, J. Reviewing Vietnam's Nationally Determined Contribution: A new perspective using the marginal cost of abatement. *Front. Sustain. Food Syst.* **2019**, *3*, 14. [CrossRef]
13. Vu, T.P.; Nguyen, T.M.L.; Nguyen, N.L.; Do, D.S.; Nguyen, X.Q.; Tran, V.L.; Ngo, D.Q.; Tran, V.C.; Nguyen, D.K.; Lai, V.C.; et al. *Final Report on Forest Ecological Stratification in Vietnam*; Research Centre for Forest Ecology and Environment: Hanoi, Vietnam, 2011.
14. Ziegler, A.D.; Phelps, J.; Yuen, J.Q.; Webb, E.L.; Lawrence, D.; Fox, J.M.; Bruun, T.B.; Leisz, S.J.; Ryan, C.M.; Dressler, W.; et al. Carbon outcomes of major land-cover transitions in SE Asia: Great uncertainties and REDD+ policy implications. *Glob. Chang. Biol.* **2012**, *18*, 3087–3099. [CrossRef]
15. Saah, D.; Tenneson, K.; Poortinga, A.; Nguyen, Q.; Chishtie, F.; Aung, K.S.; Markert, K.N.; Clinton, N.; Anderson, E.R.; Cutter, P.; et al. Primitives as building blocks for constructing land cover maps. *Int. J. Appl. Earth Obs. Geoinf.* **2020**, *85*, 101979. [CrossRef]
16. Shih, S.S. Spatial habitat suitability models of mangroves with *Kandelia obovata*. *Forests* **2020**, *11*, 477. [CrossRef]
17. Pistorius, T.; Ho, D.T.H.; Tennigkeit, T.; Merger, E.; Wittmann, N.; Conway, D. *Business Models for the Restoration of Short-Rotation Acacia Plantations in Vietnam*; UNIQUE Forestry and Land Use GmbH: Hanoi, Vietnam, 2016.
18. FAO. *A Framework for Land Evaluation*; Food and Agriculture Organization of the United Nations: Rome, Italy, 1976.
19. MARD. *National Guideline Number TCVN 8409:2010 on Land Evaluation Procedure for Land Use Planning at District Level*; MARD: Hanoi, Vietnam, 2010.
20. Nguyen, V.L. *Use of GIS Modelling in Assessment of Foresty Land's Potential in Thua Thien Hue Province of Central Vietnam*; Georg-August-University: Gottingen, Germany, 2008.
21. VAFS. *Determine Appropriate Regional Planting of A. Mangium × A. Auriculiformis for Timber Supply in North Central*; VAFS: Hanoi, Vietnam, 2010.

22. Hijmans, R.J.; Cameron, S.E.; Parra, J.L.; Jones, P.G.; Jarvis, A. Very high resolution interpolated climate surfaces for global land areas. *Int. J. Climatol.* **2005**, *25*, 1965–1978. [CrossRef]
23. IPCC-TGCIA. *General Guidelines on the Use of Scenario Data for Climate Impact and Adaptation Assessment*; Finnish Environment Institute: Helsinki, Finland, 2007.
24. Kamworapan, S.; Surussavadee, C. Performance of CMIP5 global climate models for climate simulation in Southeast Asia. In Proceedings of the IEEE Region 10 Annual International Conference, Proceedings/TENCON, Penang, Malaysia, 5–8 November 2017; pp. 718–722.
25. Saah, D.; Battles, J.; Gunn, J.; Buchholz, T.; Schmidt, D.; Roller, G.; Romsos, S. *Technical Improvements to the Greenhouse Gas (GHG) Inventory for California Forests and Other Lands*; Spatial Informatics Group: Pleasanton, CA, USA, 2015.
26. Mulia, R.; Khasanah, N.; Catacutan, D.C. Alternative forest plantation systems for Southcentral Coast of Vietnam: Projections of growth and production using the WaNuLCAS model. In *Towards Low-Emission Landscapes in Vietnam*; Mulia, R., Simelton, E., Eds.; World Agroforestry (ICRAF) Vietnam, World Agroforestry (ICRAF) Southeast Asia Regional Program: Hanoi, Vietnam, 2018; pp. 45–60.
27. Kuit, M.; Jansen, D.M.; Nguyen, V.T. *Manual for Arabica Cultivation*; Tan Lam Agricultural Product Joint Stock Company: Quang Tri, Vietnam, 2004.
28. Catacutan, D.C.; Huy, B.; Pham, T.V.; Mulia, R. *Gender, Land Use and Land Use Change in Vietnam: Coffee as Commodity Driver of Land Use Change in the Central Highlands of Vietnam*; World Agroforestry (ICRAF): Hanoi, Vietnam, 2015.
29. Mulia, R.; Do, H.T.; Pham, T.V.; Nguyen, Q.K.; Nguyen, T.K.D. *Green Growth Action Plan for Lam Dong Province for the Period of 2021–2030, Vision to 2050*; World Agroforestry (ICRAF): Hanoi, Vietnam, 2019.
30. UNFCCC. *Submission from the Socialist Republic of Vietnam in Response to Decision 4/CP.23*; UNFCCC: New York, NY, USA, 2018.
31. Chiriacò, M.V.; Perugini, L.; Bombelli, A.; Bernoux, M.; Gordes, A.; Kaugure, L. *Koronivia Joint Work on Agriculture: Analysis of Submissions on Topic 2(a)–Modalities for Implementation of the Outcomes of the Five In-Session Workshops*; Food Agriculture Organization of United Nations: Rome, Italy, 2018.
32. Bunn, C.; Läderach, P.; Ovalle Rivera, O.; Kirschke, D. A bitter cup: Climate change profile of global production of Arabica and Robusta coffee. *Clim. Chang.* **2015**, *129*, 89–101. [CrossRef]
33. Gomes, L.C.; Bianchi, F.J.J.A.; Cardoso, I.M.; Fernandes, R.B.A.; Filho, E.I.F.; Schulte, R.P.O. Agroforestry systems can mitigate the impacts of climate change on coffee production: A spatially explicit assessment in Brazil. *Agric. Ecosyst. Environ.* **2020**, *294*, 106858. [CrossRef]
34. Ovalle-Rivera, O.; Läderach, P.; Bunn, C.; Obersteiner, M.; Schroth, G. Projected Shifts in *Coffea arabica* Suitability among Major Global Producing Regions Due to Climate Change. *PLoS ONE* **2015**, *10*, e0124155. [CrossRef] [PubMed]
35. Pham, Y.; Reardon-Smith, K.; Mushtaq, S.; Cockfield, G. The impact of climate change and variability on coffee production: A systematic review. *Clim. Chang.* **2019**, *156*, 609–630. [CrossRef]
36. UTZ. *Climate Change and Vietnamese Coffee Production-Manual on Climate Change Mitigation and Adaptation in the Coffee Sector for Local Trainers and Coffee Farmers*; UTZ: Amsterdam, The Netherlands, 2016.
37. Duncan, J.M.A.; Saikia, S.D.; Gupta, N.; Biggs, E.M. Observing climate impacts on tea yield in Assam, India. *Appl. Geogr.* **2016**, *77*, 64–71. [CrossRef]
38. Han, W.; Li, X.; Yan, P.; Zhang, L.; Ahammed, G. Tea cultivation under changing climatic conditions. In *Global Tea Science: Current Status and Future Needs*; Sharma, V.S., Gunasekare, M.T.K., Eds.; Burleigh Dodds Science Publishing: Cambridge, UK, 2018; pp. 1–18.
39. Marx, W.; Haunschild, R.; Bornmann, L. Global Warming and Tea Production—The Bibliometric View on a Newly Emerging Research Topic. *Climate* **2017**, *5*, 46. [CrossRef]
40. Lin, B.B. The role of agroforestry in reducing water loss through soil evaporation and crop transpiration in coffee agroecosystems. *Agric. For. Meteorol.* **2010**, *150*, 510–518. [CrossRef]
41. Moreira, S.L.S.; Pires, C.V.; Marcatti, G.E.; Santos, R.H.S.; Imbuzeiro, H.M.A.; Fernandes, R.B.A. Intercropping of coffee with the palm tree, macauba, can mitigate climate change effects. *Agric. For. Meteorol.* **2018**, *256–257*, 379–390. [CrossRef]
42. de Souza, H.N.; de Goede, R.G.M.; Brussaard, L.; Cardoso, I.M.; Duarte, E.M.G.; Fernandes, R.B.A.; Gomes, L.C.; Pulleman, M.M. Protective shade, tree diversity and soil properties in coffee agroforestry systems in the Atlantic Rainforest biome. *Agric. Ecosyst. Environ.* **2012**, *146*, 179–196. [CrossRef]

43. ICC. *Country Coffee Profile: Vietnam*; ICC: Nairobi, Kenya, 2019.
44. Hopkins, G. Vietnamese Tea: Origin, Geography, Cultivars, Current Affairs. Available online: https://specialtyteaalliance.org/world-of-tea/vietnamese-tea-origin/#:~{}:text=Themaincentresforoolongbyvolumeofteaproduced (accessed on 12 October 2020).
45. Nambiar, E.S.; Harwood, C.E.; Kien, N.D. Acacia plantations in Vietnam: Research and knowledge application to secure a sustainable future. *South. For.* **2015**, *77*, 1–10. [CrossRef]
46. Mokievsky, V.O.; Son, T.; Dobrynin, D.V. The Dynamics of Mangroves in the Mekong Delta (Vietnam): From Degradation to Restoration. *Dokl. Earth Sci.* **2020**, *494*, 745–747. [CrossRef]
47. Zhu, T.; Trinh, M. Climate Change Impacts on Agriculture in Vietnam. Available online: https://www.slideshare.net/tingjuzhu/climate-change-impacts-on-agric (accessed on 5 September 2020).
48. Binh, C.T.; Phillips, M.J.; Demaine, H. Integrated shrimp-mangrove farming systems in the Mekong delta of Vietnam. *Aquac. Res.* **1997**, *28*, 599–610. [CrossRef]
49. Truong, T.D.; Do, L.H. Mangrove forests and aquaculture in the Mekong river delta. *Land Use Policy* **2018**, *73*, 20–28. [CrossRef]
50. Kuit, M.; Guinée, L.; Jansen, D.C.S. *The Carbon Footprint of Vietnam Robusta Coffee Source or Sink?* IDH: Utrecht, The Netherlands, 2019.
51. Do, V.H.; La, N.; Mulia, R.; Bergkvist, G.; Dahlin, A.S.; Nguyen, V.T.; Pham, H.T.; Öborn, I. Fruit Tree-Based Agroforestry Systems for Smallholder Farmers in Northwest Vietnam—A Quantitative and Qualitative Assessment. *Land* **2020**, *9*, 451. [CrossRef]
52. Catacutan, D.C.; Do, T.H.; Simelton, E.; Hanh, V.T.; Hoang, T.L.; Patton, I.; Hairiah, K.; Le, T.T.; van Noordwijk, M.; Nguyen, M.P.; et al. *Assessment of the Livelihoods and Ecological Conditions of Bufferzone Communes in Song Thanh National Reserve, Quang Nam Province, and Phong Dien Natural Reserve, Thua Thien Hue Province*; World Agroforestry (ICRAF): Hanoi, Vietnam, 2017.
53. Luedeling, E.; Smethurst, P.J.; Baudron, F.; Bayala, J.; Huth, N.I.; van Noordwijk, M.; Ong, C.K.; Mulia, R.; Lusiana, B.; Muthuri, C.; et al. Field-scale modeling of tree-crop interactions: Challenges and development needs. *Agric. Syst.* **2016**, *142*, 51–69. [CrossRef]
54. Mulyoutami, E.; Roshetko, J.M.; Martini, E.; Awalina, D. Janudianto Gender roles and knowledge in plant species selection and domestication: A case study in South and Southeast Sulawesi. *Int. For. Rev.* **2015**, *17*, 99–111. [CrossRef]
55. Sari, R.; Saputra, D.; Hairiah, K.; Rozendaal, D.; Roshetko, J.; van Noordwijk, M. Gendered species preferences link tree diversity and carbon stocks in cacao agroforest in Southeast Sulawesi, Indonesia. *Land* **2020**, *9*, 108. [CrossRef]
56. Ureta, J.U.; Evangelista, K.P.A.; Habito, C.M.; Lasco, R. Exploring gender preferences in farming system and tree species selection: Perspectives of smallholder farmers in southern Philippines. *J. Environ. Sci. Manag.* **2016**, *19*, 56–73.
57. Minang, P.A.; Duguma, L.A.; Bernard, F.; Mertz, O.; van Noordwijk, M. Prospects for agroforestry in REDD+ landscapes in Africa. *Curr. Opin. Environ. Sustain.* **2014**, *6*, 78–82. [CrossRef]
58. Rosenstock, T.S.; Wilkes, A.; Jallo, C.; Namoi, N.; Bulusu, M.; Suber, M.; Mboi, D.; Mulia, R.; Simelton, E.; Richards, M.; et al. Making trees count: Measurement and reporting of agroforestry in UNFCCC national communications of non-Annex I countries. *Agric. Ecosyst. Environ.* **2019**, *284*, 106569. [CrossRef]
59. Clouse, C. *Vietnam's New Conservation Plan Prioritizes Trees and People. Emissions? Not So Much*; Mongabay: Brooklyn, NY, USA, 2020.
60. Mulia, R.; Nguyen, M.P.; Do, H.T. Forest and crop-land intensification in the four agro-ecological regions of Vietnam: Impact assessment with the FALLOW model. In *Towards Low-Emission Landscapes in Vietnam*; Mulia, R., Simelton, E., Eds.; World Agroforestry (ICRAF) Vietnam, World Agroforestry (ICRAF) Southeast Asia Regional Program: Hanoi, Vietnam, 2018; pp. 89–108.
61. Kalita, R.; Kalita, R.M.; Das, A.K.; Nath, A.J. Carbon stock and sequestration potential in biomass of tea agroforestry system in Barak Valley, Assam, North East India. *Int. J. Ecol. Environ. Sci.* **2017**, *42*, 107–114.
62. Phan, V.T.; Huynh, V.Q.; Dang, T. Determination of raw biomass, yield and income from intercropping models of coffee (*Coffea canephora* Pierre) compare with perennial trees in tropical climate. *AJCS* **2020**, *14*, 1835–2707. [CrossRef]
63. Guillemot, J.; le Maire, G.; Munishamappa, M.; Charbonnier, F.; Vaast, P. Native coffee agroforestry in the Western Ghats of India maintains higher carbon storage and tree diversity compared to exotic agroforestry. *Agric. Ecosyst. Environ.* **2018**, *265*, 461–469. [CrossRef]

64. Nguyen, M.P.; Mulia, R.; Nguyen, Q.T. *Fruit Tree Agroforestry in Northwest Vietnam: Sample Cases of Current Practices and Benefits*; World Agroforestry (ICRAF): Hanoi, Vietnam, 2020.
65. Segura, M.; Kanninen, M.; Suárez, D. Allometric models for estimating aboveground biomass of shade trees and coffee bushes grown together. *Agrofor. Syst.* **2006**, *68*, 143–150. [CrossRef]
66. Avtar, R.; Suzuki, R.; Sawada, H. Natural forest biomass estimation based on plantation information using PALSAR data. *PLoS ONE* **2014**, *9*, e86121. [CrossRef]
67. Mlagalila, H.E. *Assessment of Volume, Biomass, and Carbon Stock of Cashew Nuts Trees in Liwale District, Tanzania*; Sokoine University of Agriculture: Morogoro, Tanzania, 2016.
68. Vietnam Rubber Association. *Vietnam's Rubber Industry on International Integration*; Vietnam Rubber Association: Ho Chi Minh City, Vietnam, 2015.
69. Brahma, B.; Sileshi, G.W.; Nath, A.J.; Das, A.K. Development and evaluation of robust tree biomass equations for rubber tree (*Hevea brasiliensis*) plantations in India. *For. Ecosyst.* **2017**, *4*, 14. [CrossRef]
70. Dang, T. *Biomass of Rhizophora Apiculata Forest*; Minh Hai Centre for Mangrove Forest Technique Application: Ho Chi Minh City, Vietnam, 2011.
71. Bridgham, S.D.; Megonigal, J.P.; Keller, J.K.; Bliss, N.B.; Trettin, C. The carbon balance of North American wetlands. *Wetlands* **2006**, *26*, 889–916. [CrossRef]
72. Phan, T.; Subasinghe, S.M.C.U. An assessment of the carbon stocks of Melaleuca forests in the lower U Minh National Park in Ca Mau province of Southern Vietnam. *Am. J. Eng. Res.* **2018**, *7*, 305–315.

# Fruit Tree-Based Agroforestry Systems for Smallholder Farmers

**Van Hung Do [1,2,*], Nguyen La [1], Rachmat Mulia [1], Göran Bergkvist [2], A. Sigrun Dahlin [3], Van Thach Nguyen [1], Huu Thuong Pham [1] and Ingrid Öborn [2,4]**

1 World Agroforestry (ICRAF) Vietnam, 249A Thuy Khue Street, Thuy Khue Ward, Tay Ho District, Hanoi, Vietnam; l.nguyen@cgiar.org (N.L.); r.mulia@cgiar.org (R.M.); n.thach@cgiar.org (V.T.N.); p.thuong@cgiar.org (H.T.P.)

2 Department of Crop Production Ecology, Swedish University of Agricultural Sciences (SLU), P.O. Box 7043, 750 07 Uppsala, Sweden; goran.bergkvist@slu.se (G.B.); ingrid.oborn@slu.se (I.Ö.)

3 Department of Soil and Environment, Swedish University of Agricultural Sciences (SLU), P.O. Box 7014, 750 07 Uppsala, Sweden; sigrun.dahlin@slu.se

4 World Agroforestry Centre (ICRAF) Headquarters, UN Avenue, P.O. Box 30677, Nairobi 00100, Kenya

* Correspondence: d.hung@cgiar.org

**Abstract:** Rapid expansion of unsustainable farming practices in upland areas of Southeast Asia threatens food security and the environment. This study assessed alternative agroforestry systems for sustainable land management and livelihood improvement in northwest Vietnam. The performance of fruit tree-based agroforestry was compared with that of sole cropping, and farmers' perspectives on agroforestry were documented. After seven years, longan (*Dimocarpus longan* Lour.)-maize-forage grass and son tra (*Docynia indica* (Wall.) Decne)-forage grass systems had generated 2.4- and 3.5-fold higher average annual income than sole maize and sole son tra, respectively. Sole longan gave no net profit, due to high investment costs. After some years, competition developed between the crop, grass, and tree components, e.g., for nitrogen, and the farmers interviewed reported a need to adapt management practices to optimise spacing and pruning. They also reported that agroforestry enhanced ecosystem services by controlling surface runoff and erosion, increasing soil fertility and improving resilience to extreme weather. Thus, agroforestry practices with fruit trees can be more profitable than sole-crop cultivation within a few years. Integration of seasonal and fast-growing perennial plants (e.g., grass) is essential to ensure quick returns. Wider adoption needs initial incentives or loans, knowledge exchange, and market links.

**Keywords:** fruit tree-based agroforestry; economic benefits; ecosystem services; farmer perspectives; resource competition; systems improvement; uptake and expansion

---

## 1. Introduction

The United Nations sustainable development goals and Agenda 2030 include poverty eradication, ending hunger, and environmental restoration, among other objectives [1]. Related targets are to implement resilient agricultural practices that increase productivity and production, and to maintain ecosystems that strengthen the capacity for adaptation to climate change and risks and improve land health [2]. Agroforestry, a planned combination of trees and crops with or without livestock on the same land, is increasingly being recognised as a sustainable system to reconcile agricultural production and environmental protection [3,4]. When combined with contour planting on sloping uplands, agroforestry is an effective land-use system to reduce soil erosion and maintain soil fertility [5,6].

In addition, as an integrated and more permanent farming system, agroforestry can generate diverse economic, ecological, and social benefits [3,7] beyond those provided by sole-crop farming systems.

Mountainous areas in the lower Mekong region are experiencing severe forest and land degradation, driven by expansion of unsustainable farming practices [8]. For example, in northwest Vietnam, sole-maize cultivation is widespread over hills and fragile sloping land [9,10]. The northwest region is home to ethnic minorities with a poverty rate of about 14% in 2016, or 8% higher than the average poverty rate at the national level, according to the 2017 statistic book of Vietnam. Around 60% of land in the region has a slope of ≥ 30% [11]. Soil degradation in the region is acute, resulting in low crop productivity [12–16].

Driven by high economic benefits, smallholder fruit-tree cultivation has recently expanded in several provinces in northwest Vietnam [17]. For example, the total area of fruit-tree plantations in Dien Bien, Yen Bai, and Son La provinces reached 58,464 ha in 2018, a 51.4% increase compared with 2015. The main fruit commodities are longan (*Dimocarpus longan* Lour.), mango (*Mangifera indica* L.) and plum (*Prunus domestica* L.). There is also some production of son tra (*Docynia indica* (Wall.) Decne), also called H'Mong apple, which is native to the region and one of 50 special fruits of Vietnam [18]. Son tra is a multipurpose tree, restoring natural forest cover and producing fruit [19].

Despite recent developments, farmers in the northwest region generally lack technical knowledge of agroforestry [9,20], including fruit tree-based agroforestry, in terms of adequate species composition, optimal plant arrangement and spacing, and management practices to optimise delivery of products and ecosystem services over time. Good management could better utilise potential economic, social, and environmental benefits of diversified tree-based farming systems. Farmers in the region usually develop "temporary" agroforestry by combining fruit trees and annual crops such as maize or cassava, and vegetables, in the early years of planting before tree canopy closure, most often in the first to third year after tree planting [21]. Reliable scientific-based information on permanent combinations of fruit trees and annual crops is necessary to promote agroforestry systems that can offer long-term and diverse income sources through product diversification to farmers in the region.

This study assessed the performance of two fruit-tree agroforestry systems in order to obtain knowledge on sustainable farming systems for the region. Quantitative and qualitative approaches were used to assess the agroforestry systems: longan–maize–forage grass and son tra–forage grass. Specific objectives were (i) to evaluate the productivity and profitability of agroforestry systems compared with sole-tree and annual crop systems over the seven years after establishment and (ii) to survey farmers on the performance of fruit tree-based systems to identify possibilities for improvement and wider-scale development.

## 2. Materials and Methods

### 2.1. Site Description

On-farm experiments with two agroforestry systems, longan (*Dimocarpus longan* Lour.)–maize (*Zea mays* L.)–forage grass and son tra (*Docynia indica* (Wall.) Decne.) –forage grass, were carried out on three farms each, using farms as replicates. The farms were situated in Van Chan district (21.56° N, 104.56° E; 374 m a.s.l.) in Yen Bai province and Tuan Giao district (21.56° N, 103.50° E; 1267 m a.s.l.) in Dien Bien province, northwest Vietnam (Figure 1). The climate at both sites is sub-humid tropical, with a rainy season from April to October and a dry season from November to March. Mean annual temperature is 18.6 °C and 21 °C; and annual rainfall is 1200–1600 mm and 1700–2000 mm in Tuan Giao and Van Chan, respectively. Mean slope of the experimental plots was 27% at both sites.

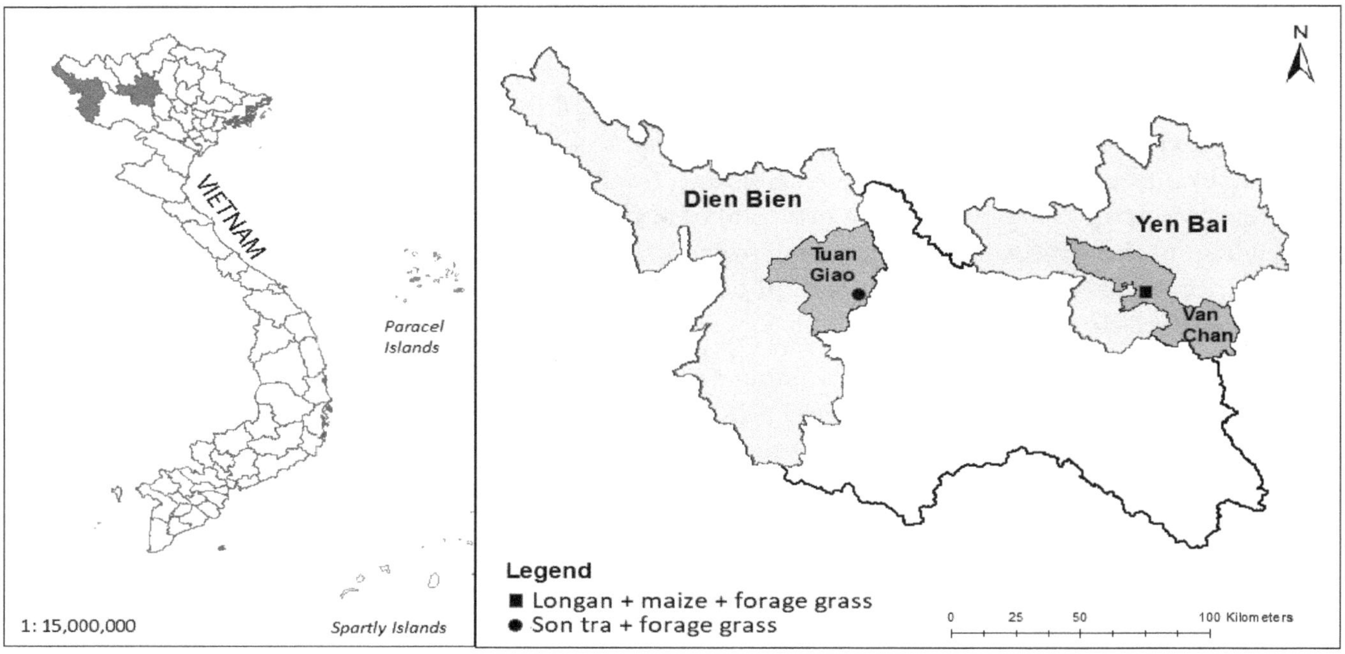


**Figure 1.** Location of the agroforestry experiments with longan-maize-forage grass in Van Chan District, Yen Bai province, and son tra-forage grass in Tuan Giao District, Dien Bien province, north-west Vietnam. Replicate trials were established on three farms in each district.

The soil profile at each site was characterised at the start of the experiments. The soils at Van Chan were silty clay loams, with, on average, pH 4.7, 1.7% soil organic matter (SOM), 0.12% total nitrogen (N), 0.02% total phosphorus (P), and 0.50% total potassium (K). The soil at Tuan Giao was silty clay, with, on average, pH 4.6, 3.8% SOM, and total N, P, and K of 0.24%, 0.02%, and 0.85%, respectively. SOM and total N, and P, and K were determined by the Walkley–Black method [22], Kjeldahl method [23], and digestion with mixed strong acids [24,25], respectively. Available soil P (Bray II) [26] was 5 mg $kg^{-1}$ at Van Chan and 9.2 mg $kg^{-1}$ at Tuan Giao.

### 2.2. Field Experiment Design

Both experiments were designed as randomised complete blocks with three replicates on three different farms. At Van Chan, the experiment lasted seven years (2012–2018). The agroforestry system consisted of longan, maize and guinea grass (*Panicum maximum* Jacq.) (LMG) and was compared with sole-crop maize (SM) and sole-crop longan (SL) (Figure 2a). The sole-crop longan was planted with 5 m row spacing and 5 m spacing between trees within rows (400 trees $ha^{-1}$). In the LMG system, longan was planted at 5 m spacing in double rows along contour lines, with 15 m between two double rows (240 trees $ha^{-1}$). Guinea grass was planted in double rows 0.5 m from the trees, and the distance between two rows was 0.5 m. The seed rate, row spacing, and distance between plants for sole-crop maize was 15 kg $ha^{-1}$, 0.65 m, and 0.3 m, respectively. The seed rate was 10–20% lower in the LMG system, since maize was not sown in the grass strips or within 0.5 m from the canopy of longan, so maize plants were sown with the same row spacing and plant distance in both systems. The longan variety used in the experiment was late maturing. The maize variety used in all cropping systems was the hybrid PAC 999.

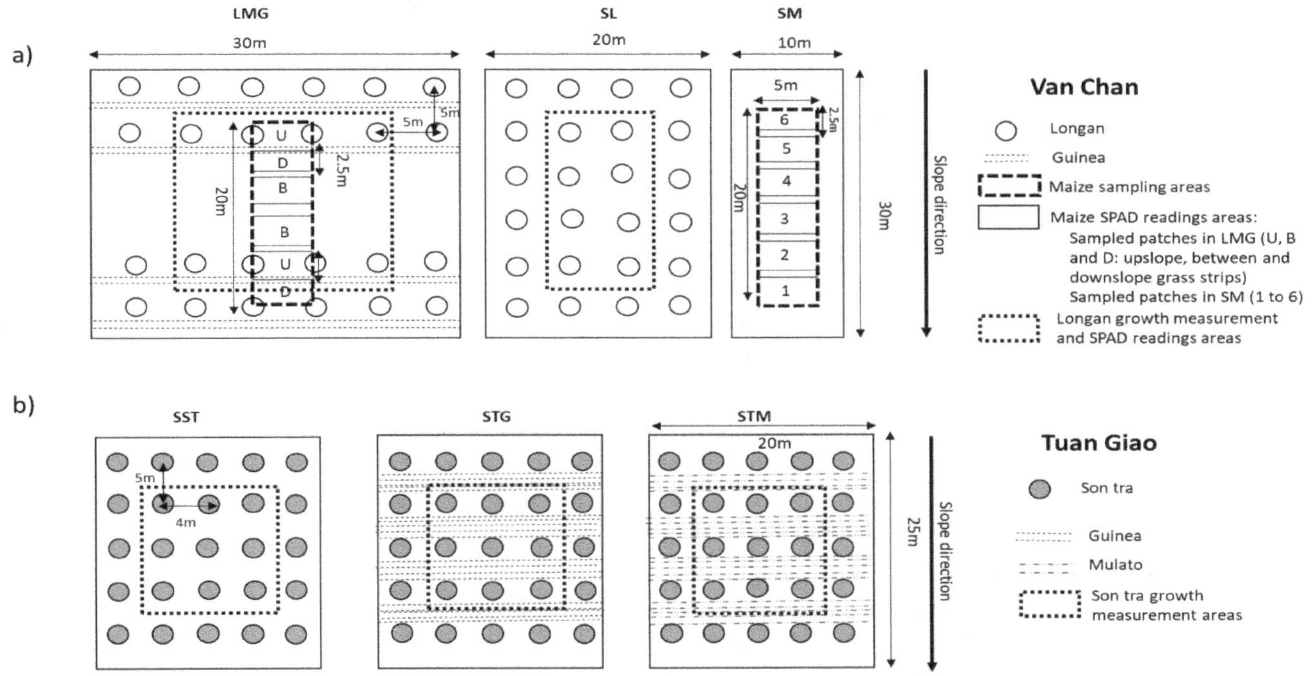


**Figure 2.** Design of field experiments: (**a**) Van Chan: sole-crop maize (SM), sole-crop longan (SL), and longan–maize–forage grass (LMG), U: upslope grass strips, D: downslope grass strips, B: between grass strips. The plot area was 300 $m^2$ for sole-crop maize, 600 $m^2$ for sole-crop longan, and 900 $m^2$ for the LMG agroforestry system; (**b**) Tuan Giao: sole-crop son tra (SST), son tra–guinea grass (STG), and son tra–mulato grass (STM). Plot area was 500 $m^2$.

The experiment at Tuan Giao lasted six years (2013–2018) and comprised three treatments: sole-crop son tra (SST), son tra–guinea grass (STG), and son tra–mulato grass (*Brachiaria* sp.) (STM). In all treatments, son tra was planted with 5 m row spacing and with 4 m spacing between trees within rows (500 trees $ha^{-1}$). Seven rows of guinea grass or mulato grass were planted between two rows of son tra in the STG and STM system, respectively (Figure 2b). The distance between the grass rows was 0.5 m and the strips were 1 m from the son tra rows. Grafted son tra seedlings were used, while guinea grass and mulato grass cuttings were obtained from a nursery.

Mineral NPK fertiliser was applied annually to maize in SM and LMG (NPK 5–10–3) as a basal application, with a topdressing with urea (46% N) and potassium chloride (48.6% K) at maize stage 6–7 fully expanded leaves (50%) and before silking (50%). In the SL and LMG treatments, 15 kg of composted animal manure and 1 kg of mineral fertiliser (NPK 5–10–3) were applied per longan tree in year 1. In years 2–7, 1 kg mineral fertiliser (NPK 5–10–3) was applied per tree, while in years 5–7, 20 kg of animal manure were applied per tree. In SST, STG, and STM, son tra received 15 kg composted animal manure and 1 kg mineral fertiliser (NPK 5–10–3) per tree in year 1 and an annual topdressing of 0.9 kg mineral fertiliser (NPK 5–10–3) per tree in years 2–6. In both experiments, the purpose of planting grass strips was to utilise nutrients in runoff, and therefore no nutrients were applied to the forage grasses. For more information about the experiments in Van Chan and Tuan Giao, see Table S1 in Supplementary Materials (SM).

### 2.3. *Data Collection in the Field Trials*

#### 2.3.1. Tree Growth and Tree/Maize/Grass Yield Determination

Eight longan and nine son tra trees in each plot were measured every three months for the whole experimental period to determine base diameter (consistently measured at a height of 10 cm from soil surface because the trees were still small in the early years of experiments), canopy diameter, and plant height. Fresh weight biomass production of forage grasses was measured monthly by harvesting a

5 m forage grass strip per plot and weighing the biomass. Maize grain production was measured by harvesting a 5 × 20 m sub-area within each plot, air-drying the cobs outdoors before shelling and weighing. Fruit yield per plot was determined by collecting and weighing the fruit of all trees at harvest.

#### 2.3.2. Competition for Resources in the Longan–Maize–Forage Grass System

An in-depth study of the variation in plant N concentration, growth, and productivity was carried out in year 7 of the experiment at Van Chan. Maize stover (stems, leaves, cobs, and covers) and grain were harvested at physiological maturity and weighed to determine their fresh weight. Fresh sub-samples of these materials were weighed and dried to constant weight. The ratio between fresh and dry weight was calculated and used to calculate the total harvested dry weight of each material. Within the LMG plots, measurements and sampling were performed in duplicate patches at three positions on the plots; 2.5 m upslope of the grass strips, between grass strips (4 m distance), and 2.5 m downslope of the grass strips (marked U, B, and D, respectively, in Figure 2a). The sampled area of each patch was 2.5 × 5 m. Similar sampling of patches was carried out in SM.

Plant N status was monitored in LMG and SM using a soil plant analysis development (SPAD) 502 Plus chlorophyll meter to determine the amount of chlorophyll present in plant leaves [27], as a proxy for N concentration [28]. The SPAD readings and maize plant height measurements were carried out at four vegetative stages of the maize crop (3–4, 6–7, and 10–11 fully expanded leaves, and silking). In each sampled patch, five maize plants along a diagonal were used for measurements on each occasion. The third, sixth, ninth, and index leaves were used as standard leaves for the stages 3–4, 6–7, and 10–11 fully expanded leaves and silking, respectively. The SPAD readings were taken at two-thirds of the distance from the leaf tip towards the stem [29]. In grass, the SPAD readings were carried out on 10 new fully expanded leaves [30] and height measurements were made on 10 grass plants every month in a 5 m section of each grass strip before cutting during the maize season. For longan, the SPAD readings were taken on eight longan trees within LMG and SL (Figure 2a) at the beginning and end of the maize season. One fully expanded mature leaf on the east, west, south, and north side of each tree was selected. The third leaflet position from the terminal leaf of each fully expanded mature leaf was used as the standard leaf for SPAD readings [31].

#### 2.3.3. Land Equivalent Ratio

A land equivalent ratio (LER) was used to compare yields in the different treatments, with LER greater than 1.0 indicating that the mixed system (intercrop) was more advantageous than the sole crop. LER was calculated as [32]:

$$\text{LER} = \text{Intercrop1/Sole crop1} + \text{Intercrop2/Sole crop2} + \ldots\ \ldots \qquad (1)$$

The fresh yield of sole-crop guinea grass and sole-crop mulato grass was calculated from their average reported dry biomass yield, i.e., 30 ton ha$^{-1}$ year$^{-1}$ [33] and 18.5 ha$^{-1}$ year$^{-1}$ [34], respectively, assuming a dry matter content of 23% [35] and 21% [36], respectively. The LER of the LMG, STG, and STM systems was calculated annually.

#### 2.3.4. Profitability

Cost-benefit analysis was performed for each agroforestry and sole-crop treatment, taking into account details of investment costs, maintenance costs, and revenue from products sold across monitoring years. Net profit was calculated by subtracting all input costs from gross income. Annual inputs included fertiliser, pesticide, labour, planting materials, etc. Total annual income was calculated based on yield and the price obtained for the different products at harvest. Data on the cost of inputs and market prices for products were obtained from the provincial extension departments covering the study sites (see Table S2 in Supplementary Materials). Net profits of each system were calculated as:

$$N = T - I \quad (2)$$

where $N$ is net profit, $T$ is total income, and $I$ is total cost of all inputs, all in USD ha$^{-1}$ year$^{-1}$.

### 2.4. Selection of Participants for Farmer Group Discussions

Farmers' perceptions and aspirations for the agroforestry systems involving longan–maize–forage grass (in Yen Bai) and son tra–forage grasses (in Dien Bien) were documented in group discussions carried out in January 2020. For each agroforestry system, two villages were selected: one village that hosted an experiment (experiment-hosting village) and a nearby village (non-hosting village) (Table S3 in Supplementary Materials). In each village, farmers who were familiar with or had observed the agroforestry system in the field experiment were selected and divided into three groups based on resources and gender (poor female, poor male, non-poor mixed female and male). Farmers hosting the experiments were interviewed individually, using the same open-ended questions as in the group discussions. In total, there were six different farmer groups at each study site, three in the experiment-hosting village and three in the non-hosting village, plus the three farmers hosting the experiments at each site (experiment-hosting famers). The Vietnamese government's poverty scale [37] was used to capture responses from farmers experiencing different levels of poverty. The questions (see Table S4 in Supplementary Materials) were posed by an interview team, including three researchers from World Agroforestry (ICRAF) in Vietnam who served as facilitators. All interviews were recorded and the responses were transcribed and translated into English by the researchers after each group discussion. The responses from farmers belonging to the different groups were then analysed to identify the consensus or most common responses to each question within each group. Thus, responses from individual farmers are not presented. The main ideas expressed in responses were identified and grouped into themes/categories reflecting farmers' perceptions of the two agroforestry systems tested in terms of tree, maize, and grass performance related to competition for resources, economic and ecological benefits, markets, constraints to adoption, and potential of agroforestry as a future option for the region.

### 2.5. Statistical Analysis

The software R (version 3.6.1) was used for all statistical analyses. Repeated measures ANOVA with the mixed model was used to assess the effects of the different treatments on maize, grass, and tree performance; yield; and profitability over the years. Log-transformation was used to normalise the data where necessary. When a significant difference was indicated in F-tests, lsmeans was used to identify significant ($p < 0.05$) differences between means. Repeated measures ANOVA was also applied to compare SPAD values and growth of maize in LMG and SM plots in year 7 of the experiment at Van Chan. ANOVA was used to compare the yield of maize at different positions relative to the grass strips within LMG in the last year, and then Tukey's HSD test was used to identify positions that were significantly different from other positions.

## 3. Results

### 3.1. Tree Growth

There was a significant effect by cropping systems on growth of longan trees. Base diameter, canopy diameter, and height in the sole-crop (SL) system were significantly greater ($p < 0.05$) than in the LGM system (Figure 3a). By the end of year 7, the base diameter of longan in SL and LMG had increased by 9 and 7 cm, respectively, since planting, and the height of longan trees was about 148 cm in SL and 121 cm in LGM, i.e., a height increase of 36 and 32 cm year$^{-1}$ in SL and LGM, respectively.

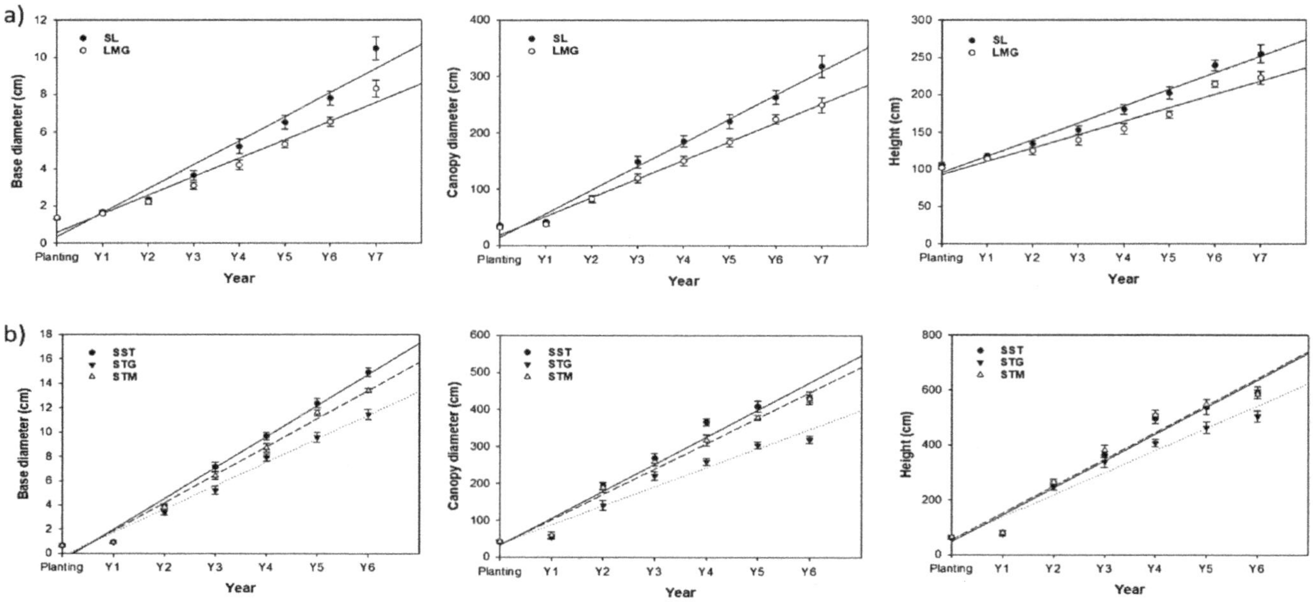


**Figure 3.** Regression lines describing tree growth (mean and standard error): (**a**) Growth of longan in the sole-tree system (SL) and longan–maize–forage grass (LMG) system; (**b**) growth of son tra in the sole-tree system (SST), son tra–guinea grass (STG) system, and son tra–mulato grass (STM) system.

The base diameter of son tra trees was significantly greater ($p < 0.05$) in the sole-tree system than in the systems with forage grass (STM and STG) (Figure 3b). Both tree height and canopy diameter were affected by the cropping system, with an interaction between cropping system and year ($p < 0.05$). Three years after planting, the canopy diameter and height of son tra trees were similar in the agroforestry and sole-tree systems. However, from year 4 to 6, canopy diameter and tree height were significantly higher ($p < 0.05$) in the sole-tree and STM systems than in the STG system (Figure 3b).

### 3.2. *Yield and Land Equivalent Ratio*

During the first three years, the products in LMG were primarily maize cobs and forage-grass biomass (Table 1). The grass started yielding from year 2. The products became more diversified from year 4, when longan started to bear fruit, and yield increased during subsequent years. There was no significant effect from the cropping system, or interaction between treatments and year, on maize yield. However, the yield of longan was significantly higher in the sole-tree system than in LMG, and there was a significant interaction between treatment and year ($p < 0.05$). From year 2 to 7, LER of the LMG system ranged from 1.1 to 1.9 (Figure 4a).

In the STG and STM agroforestry systems, the guinea grass and mulato grass were harvested from year 2 (2014), with high yield (Table 1). The agroforestry practices had more products from year 3, when son tra started to bear fruit. However, there was a significant effect from the cropping system on the productivity of son tra ($p < 0.05$), with fruit yield being significantly lower in agroforestry than in the sole-crop system. LER of the agroforestry practices from year 2 to 6 ranged from 0.5 to 1.1 for STG and 0.6 to 1.8 for STM (Figure 4b).

**Table 1.** Yield of maize (dry grain), longan, and son tra (fresh fruit) in sole-crop/tree systems (SM, SL, SST) and of these crops and forage grasses (fresh matter) in agroforestry systems (LMG, STG, STM) in the seven years of the field experiments.

| | | Yield (Ton ha$^{-1}$) | | | | | | | |
|---|---|---|---|---|---|---|---|---|---|
| **Crop/Trees** | **Cropping System** | **2012** | **2013** | **2014** | **2015** | **2016** | **2017** | **2018** | **Mean** |
| Maize | SM | 5.9 (±0.1) | 4.7 (±0.4) | 4.1 (±0.3) | 4.2 (±0.3) | 4.3 (±0.1) | 4.2 (±0.2) | 4.6 (±0.5) | 4.6 (±0.2) |
| | LMG | 5.5 (±0.3) | 5.3 (±0.1) | 3.9 (±0.2) | 4.1 (±0.2) | 4.1 (±0.1) | 4.0 (±0.2) | 4.2 (±0.4) | 4.5 (±0.3) |
| **By cropping system** | | | | | *p*-value = 0.33 | | | | |
| **Cropping system x year** | | | | | *p*-value = 0.35 | | | | |
| Longan | SL | | | | 0.35 (±0.2) | 0.32 (±0.3) | 0.47 (±0.1) | 3.04 (±1.1)a | 1.04 (±0.7)a |
| | LMG | | | | 0.06 (±0.03) | 0.18 (±0.2) | 0.38 (±0.2) | 0.90 (±0.3)b | 0.30 (±0.1)b |
| **By cropping system** | | | | | *p*-value = 0.02 | | | | |
| **Cropping system x year** | | | | | *p*-value = 0.03 | | | | |
| Guinea grass | LMG | | 4.4 (±4.4) | 19.5 (±2.3) | 15.9 (±1.2) | 18.2 (±1.1) | 18 (±1.5) | 14.6 (±4.1) | 15 (±2.3) |
| **Son tra** | SST | na | | | 0.6 (±0.4) | 5.6 (±4.1) | 2.1 (±1.6) | 8.7 (±7.6) | 4.2 (±1.8)a |
| | STG | na | | | 0.2 (±0.1) | 1.8 (±0.8) | 0.2 (±0.2) | 1.8 (±1.3) | 1.0 (±0.5)b |
| | STM | na | | | 0.2 (±0.1) | 0.9 (±0.4) | 0.5 (±0.4) | 4.7 (±3.4) | 1.6 (±1)ab |
| **By cropping system** | | | | | *p*-value = 0.03 | | | | |
| **Cropping system x year** | | | | | *p*-value = 0.75 | | | | |
| Guinea grass | STG | na | | 67 (±26.3) | 61 (±11) | 55 (±6.3) | 56 (±2.9) | 67 (±5.9) | 61 (±2.6) |
| Mulato grass | STM | na | | 58 (±29) | 65 (±4) | 74 (±7.6) | 64 (±4.7) | 66 (±4) | 65 (±2.5) |
| **By cropping system** | | | | | *p*-value = 0.62 | | | | |
| **Cropping system x year** | | | | | *p*-value = 0.85 | | | | |

SM: sole-crop maize, SL: sole-crop longan, LMG: longan–maize–forage grass, SST: sole-crop son tra, STG: son tra–guinea grass, STM: son tra–mulato grass; na: not applicable since the experiment was established in 2013. Values are mean ± standard error; different letters indicate significant differences ($p < 0.05$).

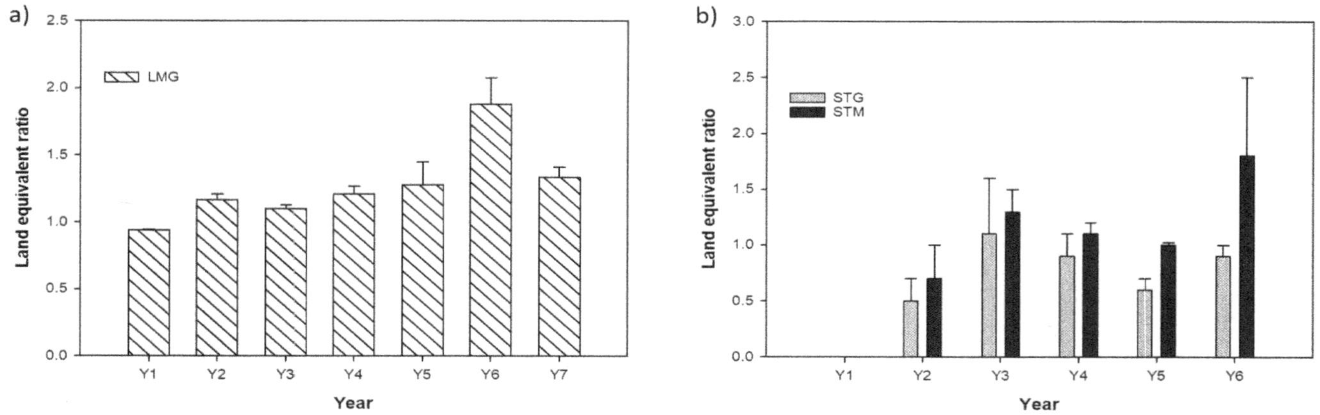


**Figure 4.** Land equivalent ratio (LER) of the agroforestry practices in each year of the experiment, expressed as mean and standard error (bars): (**a**) Longan–maize–forage grass (LGM); (**b**) son tra–guinea grass (STG) and son tra–mulato grass (STM).

### *3.3. Leaf Nitrogen Content and Competition in LMG*

The SPAD value was significantly higher in sole-crop maize than in the LMG system ($p < 0.05$) from maize development stages 6–7 to silking, while maize plant height was significantly higher from 10–11 fully expanded leaves to silking (Table 2). However, the biomass of maize, including grain and stover, was not significantly different between the sole-crop and agroforestry systems.

**Table 2.** Dry yield, height, and SPAD readings of maize in the longan–maize–forage grass system (LMG) and the sole-crop system (SM) in year 7 of the experiment.

| | **Maize Growth Stage** | | | | | **At Maturity** | | |
|---|---|---|---|---|---|---|---|---|
| | **Cropping System** | **3–4 Leaves** | **6–7 Leaves** | **10–11 Leaves** | **Silking** | | **Cropping System** | **Dry Yield (Ton Ha$^{-1}$)** |
| SPAD | SM | 38.0 | 44.6a | 52.1a | 57.7a | Grain | SM | 4.6 |
| | LMG | 38.5 | 41.3b | 47.8b | 54.3b | | LMG | 4.2 |
| *p*-value | | | <0.001 | | | *p*-value | | 0.25 |
| Height (cm) | SM | 28.4 | 65.2 | 112.1a | 230.4a | Stover | SM | 5.6 |
| | LMG | 26.9 | 61.4 | 96.3b | 218b | | LMG | 4.9 |
| *p*-value | | | <0.001 | | | *p*-value | | 0.09 |

Different letters indicate significant differences ($p < 0.05$).

The height of maize upslope, downslope, and between grass strips in LMG during year 7 was not significantly different from the height of maize in SM at the stages of 3–4 and 6–7 fully expanded leaves (Figure 5a). However, in later development stages, maize growth was significantly higher ($p < 0.05$) between two grass strips than immediately upslope or downslope of the grass. In stages 6–7 and 10–11 fully expanded leaves and silking, the SPAD readings of maize between grass strips were also significantly ($p < 0.05$) higher than those upslope and downslope of grass strips. The average SPAD readings for longan trees were not significantly different between LMG and SL (Figure 5a). Meanwhile, the average SPAD readings of guinea grass recorded 43.4 within LMG. This indicates that competition for N took place at positions where trees, crops, and grass were close to each other within the LMG system.

In LMG, the yield of maize grain between grass strips was 24% higher ($p < 0.05$) than in SM and about 62% higher than in upslope and downslope maize in LMG (Figure 5b). Yield of stover was also significantly higher (53–59%) between grass strips than for maize upslope and downslope of grass strips. Overall, the results clearly showed competition between grass, longan, and maize upslope and downslope of the grass strips within the LMG system in year 7.

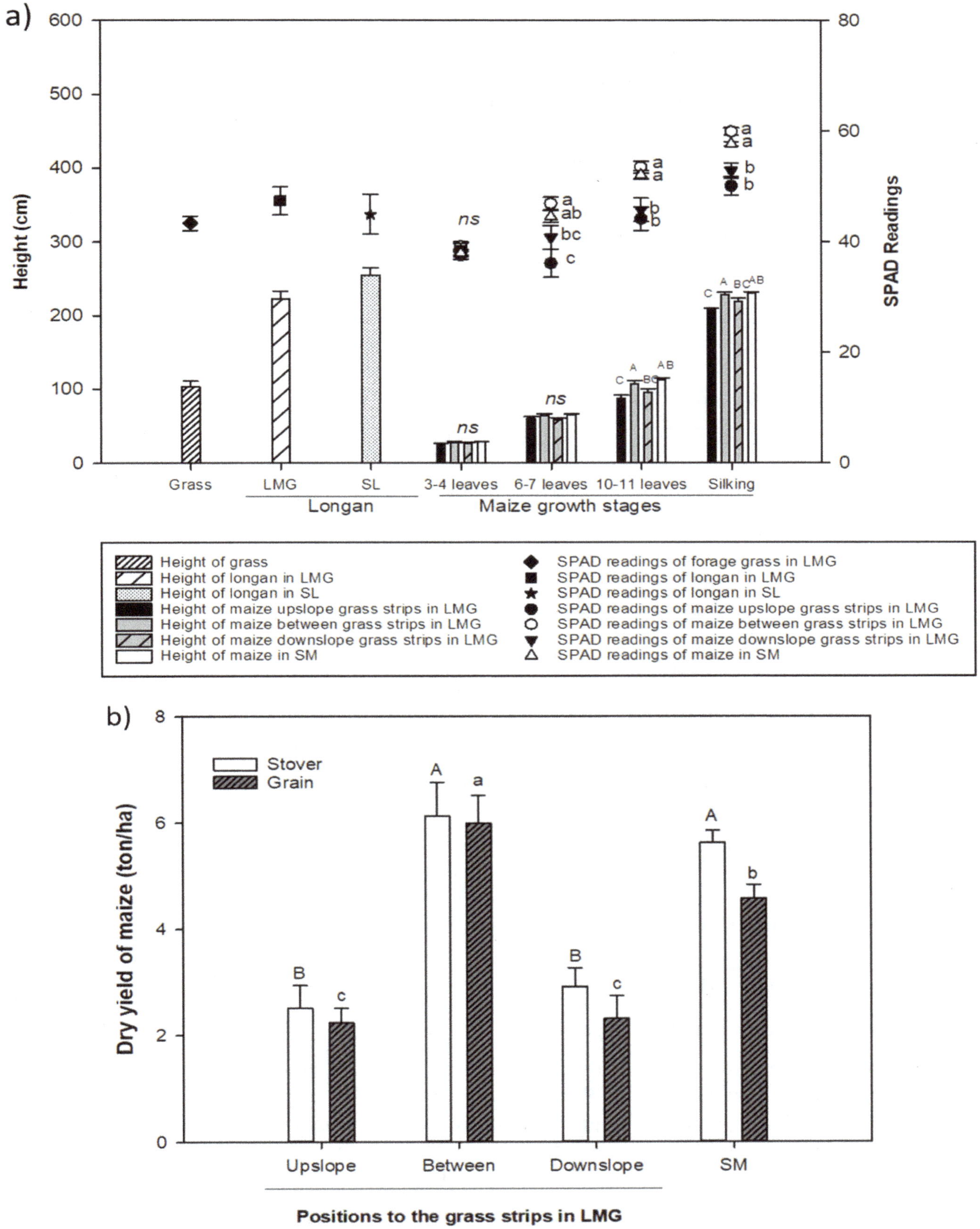


**Figure 5.** (**a**) Height of the tree and crop components and SPAD (soil plant analysis development) readings in the longan–maize–forage grass (LMG), sole-longan (SL) and sole-maize (SM); (**b**) dry yield of maize growing in different positions (upslope, between, downslope) relative to the grass strips within LMG in year 7. Values are means and standard errors. Bars with different upper case (stover) and lower-case (grain) letters indicate significant differences ($p < 0.05$).

### 3.4. *Profitability*

Sole maize had a mean annual investment cost of 670 USD ha$^{-1}$, while that of the sole-longan and the LMG system was 3.7-fold and 3.2-fold higher, respectively. The average maintenance cost of SL and LMG was 300 and 863 USD ha$^{-1}$ year$^{-1}$, respectively (Figure 6a). The net profit was related to the cropping system, with an interaction between cropping system and year ($p < 0.05$). The mean net profit of LMG (1018 USD ha$^{-1}$) was 2.4-fold higher than for SM, while the SL system only achieved a positive profit from year 6 (Table 3). The trend of decreasing net profit of SM across year was partially due to the decreasing selling price of maize over time (presented in the Supplementary Materials Table S2) and lower maize yield in the subsequent compared to the initial years of experiment. From year 2, the net profit from LGM was equal to that from SM, while from year 4 the net profit from LGM was significantly ($p < 0.05$) higher than for SM and SL. In addition, the cumulative profit from LMG was positive from year 2 and higher than that from SM from year 4 (Figure 6a). In contrast, the cumulative profit from SL was still negative in year 7.

**Table 3.** Net profit from the agroforestry systems and the corresponding sole crop/tree.

| | Net Profit (USD ha$^{-1}$) | | | | | | | |
|---|---|---|---|---|---|---|---|---|
| **Cropping System** | **2012** | **2013** | **2014** | **2015** | **2016** | **2017** | **2018** | **Mean (±SE)** |
| SM | 1118a | 611a | 388a | 246b | 233b | 196b | 190b | 425.9 (±118.9)b |
| SL | −2463c | −355b | −229b | 40b | −41b | 112b | 947ab | −284.4 (±336.8)c |
| LMG | −391b | 839a | 1550a | 1179a | 1380a | 1404a | 1168a | 1018.2 (±231.6)a |
| By cropping system | | | | | $p$-value < 0.001 | | | |
| Cropping system x year | | | | | $p$-value < 0.001 | | | |
| SST | na | −1422 | −290 | 42 | 2120 | 538 | 3238 | 704.5 (±632.4)b |
| STG | na | −1772 | 3297 | 2853 | 3069 | 2381 | 3570 | 2232.9 (±746.6)a |
| STM | na | −1772 | 2661 | 3018 | 4067 | 3147 | 4773 | 2648.7 (±857.1)a |
| By cropping system | | | | | $p$-value = 0.005 | | | |
| Cropping system x year | | | | | $p$-value = 0.72 | | | |

SM: sole-crop maize, SL: sole-crop longan, LMG: longan–maize–forage grass, SST: sole-crop son tra, STG: son tra–guinea grass, STM: son tra–mulato grass; na: not applicable since the experiment was established in 2013. Values are means; different letters indicate significant differences ($p < 0.05$).

In the year of establishment, the total input costs were approximately 1772 USD ha$^{-1}$ for both STG and STM, but lower (1422 USD ha$^{-1}$) for SST. In the following years, STG and STM required higher investment than the sole-tree system, mainly deriving from labour costs for forage-grass harvesting (Figure 6b). There was a significant effect ($p < 0.05$) of cropping system on net profit, with the mean net profit in STG (2233 USD ha$^{-1}$) and STM (2649 USD ha$^{-1}$) being around 3.2- and 3.7-fold higher, respectively, than in SST (Table 3). The SST system gave a positive net profit from year 3, but the cumulative profit from STG and STM was positive and higher than from SST from year 2 (Figure 6b).

### 3.5. *Farmers' Perceptions and Aspirations for Fruit Tree-Based Agroforestry*

#### 3.5.1. Tree and Crop Performance in Agroforestry

Most farmers were fully aware of possible effects of competition for resources (light, water, nutrients) on the performance of tree and crop components within the agroforestry systems (Figure 7). All interviewees in Van Chan noted that growth and productivity of maize in the longan–maize–forage grass system were lower than in sole-maize cultivation. They attributed this to close distance between trees, crops, and grass leading to competition in the agroforestry system. However, non-hosting village groups claimed that longan trees performed better in agroforestry than in sole cultivation since they believed that longan trees utilised the nutrients applied to the maize. However, the experiment-hosting village, the experiment-hosting farmers in Van Chan, and all interviewees in Tuan Giao reported that growth and productivity of trees in both agroforestry systems were lower than when trees were grown separately.

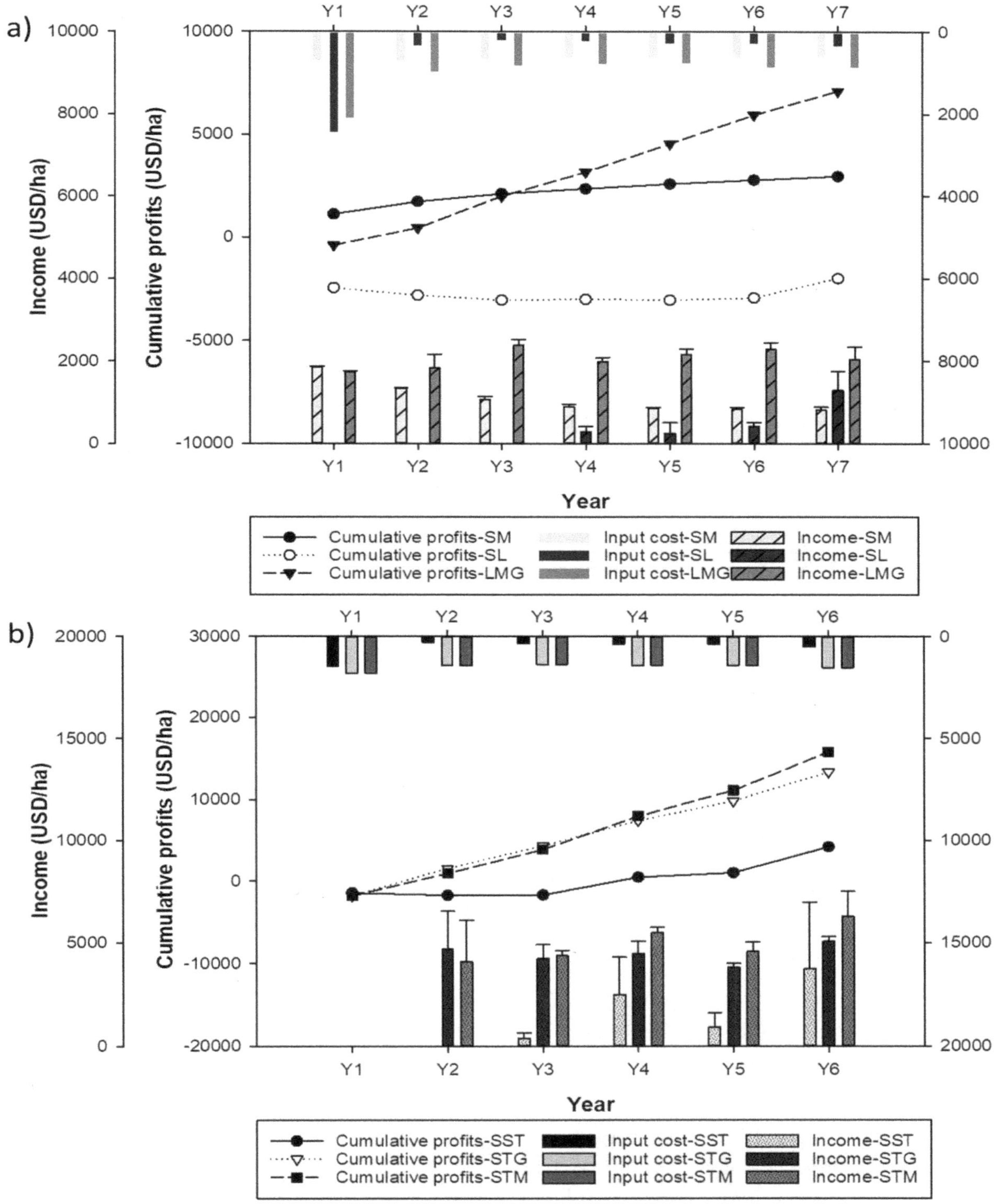


**Figure 6.** Input costs, income, and cumulative profit from: (**a**) the longan–maize–forage grass (LGM) compared with sole maize (SM) and sole longan (SL); (**b**) the son tra–guinea grass (STG) and son tra–mulato grass (STM) compared with sole son tra (SST).

The interviewees also suggested that the agroforestry systems could be optimised through better management of trees and crops (Figure 7). The groups proposed different solutions to improve the efficiency, such as adding more fertilisers to plants suffering from nutrient deficiency in areas where trees, crops, and grass affected each other's nutrient availability, reducing tree density and pruning to reduce shading. In addition, modifying the planting distance between trees and grass was suggested by groups from both sites. The farmers interviewed also suggested less-competitive crops for the agroforestry systems, e.g., legume species with biological N-fixation such as soybean and groundnut in LMG in Van Chan (3 of 7 groups), and upland rice or cucumber in STG and STM in Tuan Giao (2 of 7 groups).

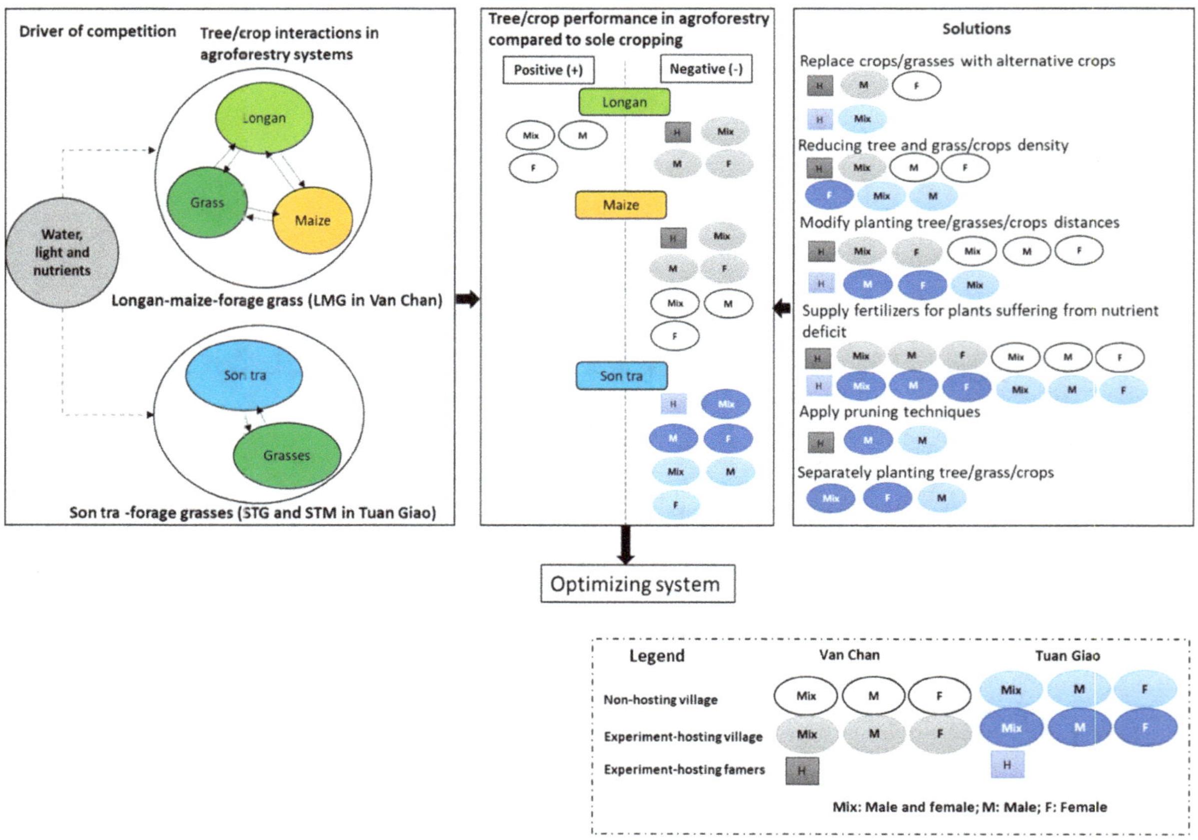


**Figure 7.** Farmers' perception of the performance of trees and crops in the agroforestry experiments in Van Chan and Tuan Giao compared with that of sole crops/trees. Open-ended questions were used in group interviews with non-hosting and experiment-hosting villages; experiment-hosting farmers were interviewed individually.

### 3.5.2. Benefits of the Agroforestry Systems

All famer groups in Van Chan and Tuan Giao shared the opinion that the experimental agroforestry systems produced earlier and more diverse products and gave higher economic benefit than the sole crop/tree (Figure 8a). Most interviewees reported that after 3–4 years, when the trees began to bear fruit, the income from agroforestry was much higher than from sole-crop cultivation. They also reported ecological benefits of the agroforestry systems in terms of reduced erosion, weed control, enhanced soil moisture and fertility, and greater resilience to extreme weather conditions (drought, snow, and frost) compared with sole-crop cultivation (Figure 8a). However, no group mentioned any benefits regarding pests and diseases, while only one group (the host farmers in Van Chan) mentioned terrace formation as an advantage (Figure 8a,b). Female and mixed groups in Tuan Giao claimed that the soil was less fertile and soil moisture lower in agroforestry than sole-tree cultivation, because the very dense forage grass used much water and nutrients within the agroforestry system (Figure 8b). Only the groups in Van Chan and the host farmer group in Tuan Giao expressed appreciation of the reduced labour requirement for harvesting forage from the grass strips in the agroforestry system (Figure 8b). These groups mentioned the possibility of using the forage to feed livestock, produce green manure, and provide earlier income when sold on the local market.

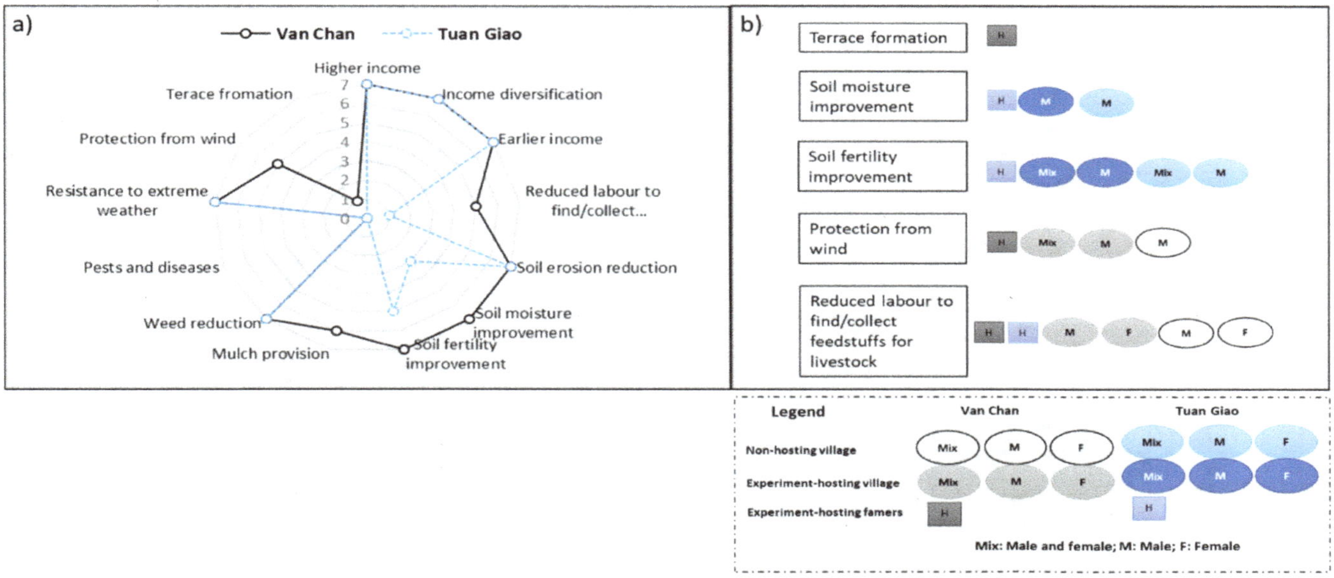


**Figure 8.** (**a**) Farmers' perceptions about benefits of agroforestry systems and number of farmer groups mentioning each of the identified benefits; (**b**) perceived benefits and the farmer groups in Van Chan and Tuan Giao that mentioned each. Open-ended questions were used in group interviews with non-hosting and experiment-hosting villages; experiment-hosting farmers were interviewed individually.

### 3.5.3. Constraints to Uptake of Agroforestry

Most of the farmer groups recognised and listed constraints to the uptake of agroforestry and proposed possible solutions to improve uptake in the local region (Figure 9). At both Tuan Giao and Van Chan, all groups indicated that the investment costs were higher than for sole-crop cultivation, making it difficult for poor households to adopt agroforestry. Management of pests and diseases in agroforestry was also more complicated, with more tree and crop components. An unstable market and low prices for products were other constraints to the uptake of agroforestry in the region.

All groups in Van Chan indicated that harsh weather events such as drought and lack of awareness among farmers of the benefits of agroforestry (4 of 7 groups) were the main drawbacks to the uptake of agroforestry. In Tuan Giao, all farmer groups considered that it would be difficult to combine traditional free grazing of livestock on crop residues with agroforestry. The forage grass was not considered valuable, since in this area with only free-grazing livestock farmers are not accustomed to collecting fodder. Extreme weather such as snow and frost and lack of techniques for implementing agroforestry were reported as other constraints to the adoption of agroforestry.

The farmers interviewed proposed solutions to address these issues (Figure 9). At Van Chan and Tuan Giao, all farmer groups mentioned training in agroforestry techniques, support in obtaining seedlings and fertilisers, and financial support or access to low-interest loans/credits as important incentives for implementing agroforestry. Development of market links for agroforestry products and a stable market were also considered key factors for agroforestry adoption by all farmer groups, but the suggested schemes differed. In Van Chan, the interviewees envisaged creating a stable market by building a farmers' cooperative to improve product quality to meet market demand and a processing factory to produce secondary products from longan fruit. The interviewees wanted maize replaced with other, higher-value annual crops. In Tuan Giao, the interviewees wanted a market link to a processing factory that would buy and add value to son tra and create a stable market.

All farmer groups in Van Chan and Tuan Giao saw a need for plant protection interventions to control pests and weeds as a way to reduce the labour costs of implementing agroforestry. According to farmers in Tuan Giao, shifting from free grazing to captive grazing and promoting livestock production to utilise the forage grass would increase the feasibility of agroforestry in the region. Although drought is a major concern in Van Chan, only the experiment-hosting farmers and the female farmer groups mentioned construction of water storage facilities as a solution. They saw a need for an electric pump and water tanks on the top of hills to supply water for tree/crops during drought periods.

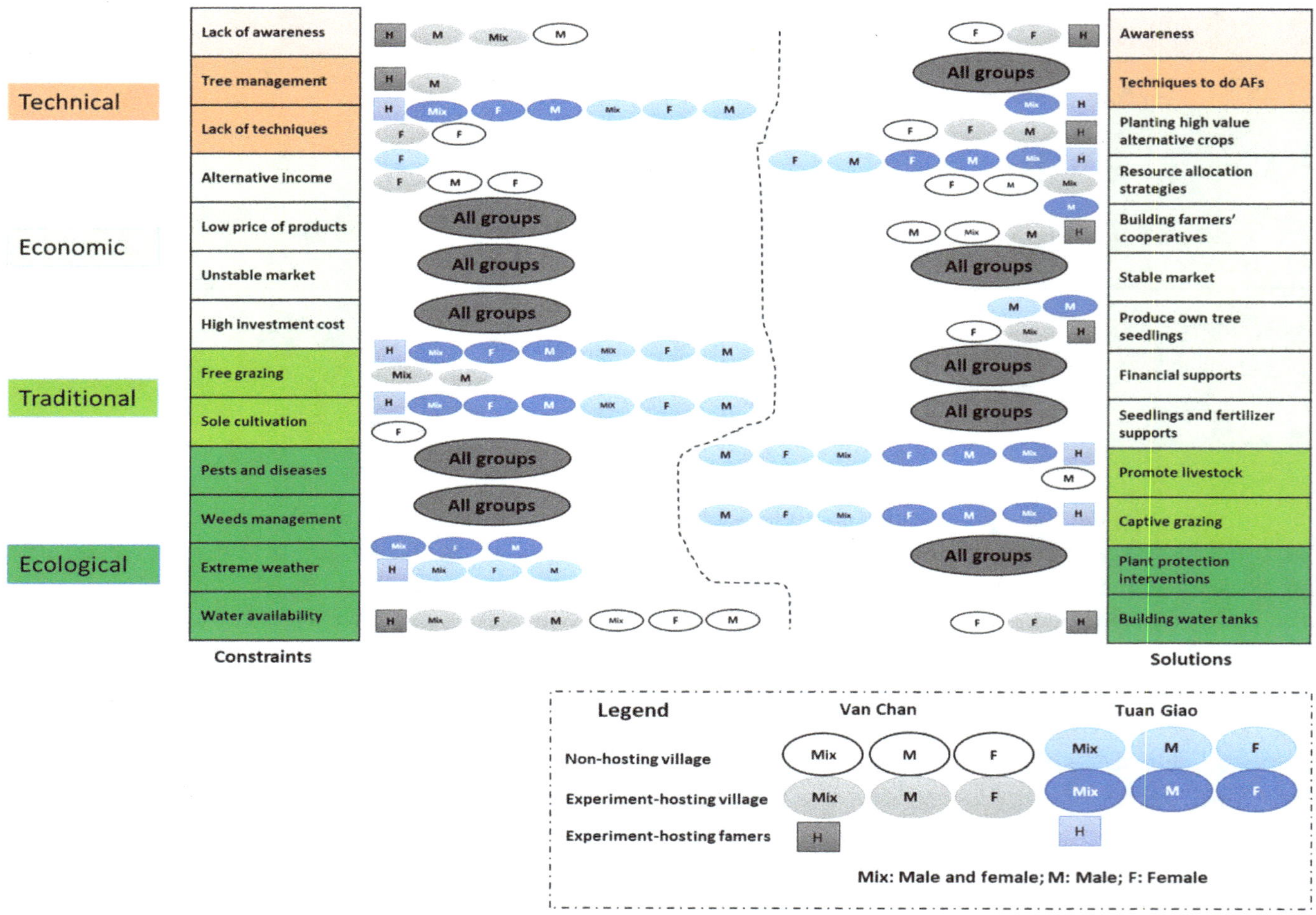


**Figure 9.** Farmers' perceptions of constraints (left) and solutions (right) to the uptake of agroforestry (AF) in Van Chan and Tuan Giao, and (centre) the farmer groups that mentioned the respective constraint/solution. Open-ended questions were used in group interviews with non-hosting and experiment-hosting villages; experiment-hosting farmers were interviewed individually.

All farmer groups interviewed mentioned a need to reduce the investment costs of agroforestry (Figure 9), e.g., by producing their own fruit-tree seedlings (3 of 7 groups in Van Chan and male groups in Tuan Giao). Some groups suggested offsetting the investment costs by planting higher-value crops to replace maize in Van Chan (4 of 7 groups) and forage grasses in Tuan Giao (2 of 7 groups). In addition, all farmers in Tuan Giao and 3 of 7 groups in Van Chan (Figure 9) indicated that resource allocation strategies could help reduce the maintenance cost of implementing agroforestry. They believed that during the first three years of the experiments, when the trees had not yet produced fruit, the farmers prioritised the annual crops and grasses to generate annual income. Later, when the trees were maturing and bearing fruit, farmers prioritised the trees.

### 3.5.4. Factors Enabling Expansion

The farmers at both Van Chan and Tuan Giao indicated that large-scale annual crop production on sloping land is an unstable system (land degradation, low yield). However, the ownership of land by local farmers is suited to implementing agroforestry. In addition, agroforestry has potential in both areas because it can bring economic and ecological benefits for local farmers. The local climate conditions are suitable for longan trees in Van Chan and for son tra trees in Tuan Giao, so both species can produce high yield. Recently, many farmers in Van Chan have shifted from sole-maize production to fruit trees and intercropping of fruit trees with annual crops, while farmers in Tuan Giao expressed interest in grafted son tra seedlings because they start to produce fruit rapidly. Local farmers saw potential for intercropping high-value trees (e.g., longan, mango, plum) and high-value crops (e.g., medicinal plants, soybean, green bean) in Van Chan, or amomum (*Amomun xanthioides* Wall.) in Tuan Giao (Figure 10).

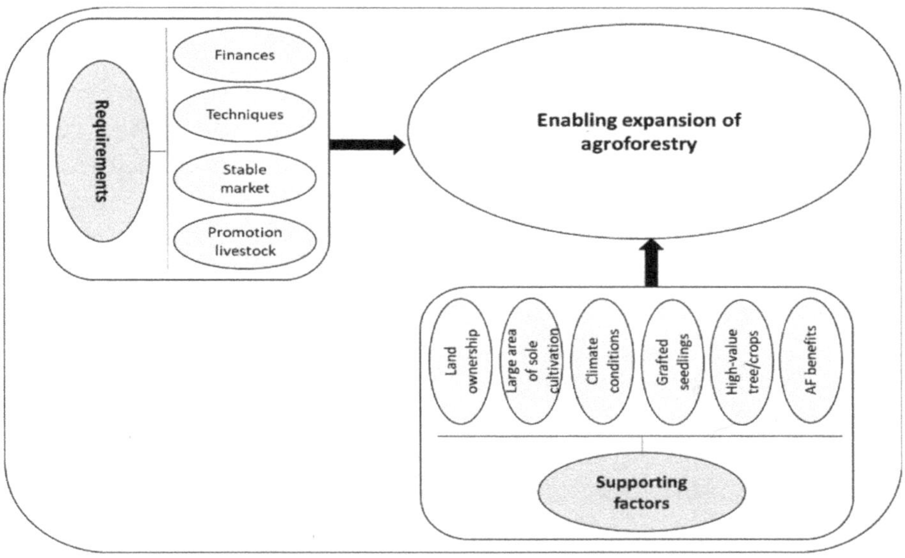


**Figure 10.** Farmers' perspectives about factors enabling expansion of agroforestry in Van Chan and Tuan Giao.

However, based on the interview responses, techniques to implement agroforestry, a stable market for products, and financial support for farmers in the establishment year(s), in combination with expansion of livestock production, would be required to expand agroforestry in northern upland areas of Vietnam (Figure 10).

## 4. Discussion

### 4.1. *Effects of Competition for Resources on Tree and Crop Performance in Agroforestry and Ways to Improve the Systems*

Total income was higher in the agroforestry systems than in the sole-cropping systems studied, but individual crop components generally grew more slowly in agroforestry systems than in sole-crop/tree systems, most likely due to competition for light, water, and nutrients [38]. The tree species in maize agroforestry systems may contribute differently to tree–crop interactions, e.g., leguminous tree species have been shown to compete less with maize for N than non-leguminous species [39–41]. The presence of tree roots, especially in the maize-cropping zone, also affects the competition with maize, and is determined by e.g., inherent rooting patterns, management, and soil conditions [41,42]. Conversely, maize restricts root development of trees in the cropping zone of agroforestry systems. A study on maize-based agroforestry systems in the sub-humid highlands of western Kenya indicated that the length of fine roots of intercropped trees (*Grevillea robusta* and *Senna spectabilis*) decreased in the maize root zone because of competition and damage to tree roots during weed hoeing [43]. In addition, maize uses the C4 photosynthetic pathway and is sensitive to shading [44] and may therefore be more negatively affected by tree shading in agroforestry systems than C3 species. Such competition was evident in the LMG system in our study, with slower growth and lower yield of longan and maize in areas where trees and crops were close to each other. This was particularly evident in year 7, when SPAD measurements showed competition for N between trees, crops, and grass growing close to each other (Table 2 and Figure 5).

In our experiments, the grass component of the agroforestry systems was competitive and negatively affected N uptake and growth of trees and maize. A previous study of maize intercropped with guinea grass in northwest Vietnam [45] found that aboveground biomass of maize at positions downslope and upslope of grass strips was around 60% and 40% lower, respectively, than that of maize 3 m from grass strips and sole maize, as we found for the LMG system (year 7). The farmer groups interviewed confirmed that maize downslope and upslope of grass strips showed lower growth and yield compared with maize farther from grass strips and sole maize, and that longan also had lower growth and yield as an intercrop than as a sole crop.

In our experiment, the yield from sole-longan planting was 2–4-ton ha$^{-1}$ at the seventh year after tree planting. However, higher yield can be expected with e.g., improved irrigation. For example, in Hung Yen province of the Red River Delta region of Vietnam, the longan yield could reach 20 ton per ha$^{-1}$ in the eighth year after tree planting [46]. Thanks to better market access including for export, partially due to the proximity to Hanoi as the country's capital and urban centre, the farmers in the province could derive high income from selling longan, and they partially allocate the income to improving irrigation systems [46]. The farmers in the province have been cultivating longan for decades.

However, the degree of competition may differ between grass species. A study in Costa Rica showed that when guinea grass and mulato grass were planted 0.9 m from *Eucalyptus deglupta* they produced similar grass biomass, but root length density (RLD) at 0–0.4 m depth was up to three-fold higher under guinea grass than under mulato grass [47]. At 0–0.4 m depth but 0.45 m from *E. deglupta* trees, RLD of guinea grass was up to four-fold higher than that of mulato grass. Thus *E. deglupta* growth was significantly reduced by the presence of guinea grass, and to a lesser extent by mulato grass, compared with sole-crop *E. deglupta* [47]. The STG and STM systems in our study confirmed the competition from guinea grass and mulato grass strips with the trees. In these systems, the forage grasses were planted 1 m from son tra rows, resulting in lower growth and yield of son tra trees with guinea grass than with mulato grass or sole-tree cultivation, while the two grasses produced similar grass biomasses.

It is possible to reduce competition between trees and crops by pruning the trees [41], as proposed by farmer groups in our study. Another option may be to intercrop C3 crops instead of C4 crops, as previous studies have indicated that yields of C3 crops are less reduced in agroforestry systems [48,49]. In our study, farmer groups suggested improving the efficiency of the agroforestry systems by planting legume species such as soybean and groundnut instead of maize in LMG, and by planting upland rice or cucumber to replace forage grasses in STG and STM. Greater planting distance between trees, crops, and grass strips would reduce competition. Supplying more fertiliser to plants suffering from nutrient deficiency in competition zones was also suggested in the group interviews.

### *4.2. Productivity Benefits and Ecosystem Services of Agroforestry Systems*

Evaluation of the agroforestry systems tested in this study indicated that they provided earlier products than sole-tree systems and more diverse products than sole-maize systems. They also gave higher total productivity for farmers than the sole-crop systems from the second year onwards. During the first three years, total productivity was mainly from forage grasses and maize, with the LMG, STG, and STM systems giving forage-grass biomass for farmers from the second year. The products became more diverse from year 4, when the trees started to bear fruit, with yield increasing in subsequent years.

We found that the LMG agroforestry system was more productive than sole maize and longan from year 2 onwards, as indicated by LER ranging from 1.1 to 1.9 (Figure 4a). In a previous study on agroforestry systems based on apple (*Malus domestica*), e.g., apple/maize, apple/peanut, and apple/millet, LER was found to be 1.2–1.3 after the apple trees started bearing fruit from year 6 [50]. In our study, LER of the STM system was >1.0 from year 3, when the son tra started bearing fruit. However, in the STG system LER was <1.0, which can probably be explained by competition, as previously shown [47]. Other studies on forage grasses have reported that guinea grass [33] produces more biomass than mulato grass [34] in sole-grass cultivation. It may therefore affect the LER of the STG agroforestry system. Management of tree and crop components of a fruit tree-based agroforestry system thus needs to change from the year of establishment to when trees are maturing and high-producing, so that farmers can overcome competition effects and optimise the efficiency of land use [50]. In this study, the farmer groups interviewed suggested that a resource allocation strategy could improve the productivity of different components of the agroforestry systems. In the first three years, when the trees had not yet produced fruit, their main priority was the annual crop and grasses, whereas they paid more attention to the trees when they started bearing fruit. The farmers needed the short-term income from annual crops to support the long-term benefits from the fruit trees.

Growing forage grasses can be an incentive to improve smallholder livestock production by improved the daily weight gain of cattle and reducing labour in finding feedstuffs [51]. In this study, farmer groups confirmed that growing forage grasses reduces the labour requirement for finding/collecting feedstuffs for livestock in areas where captive grazing is common practice. This may be particularly beneficial for rural women in the study region, as 60% of the workload in farming is carried out by women [11]. In areas where free grazing is common practice like in Tuan Giao district, farmers will be less motivated/perceive less benefit from growing forage grass. This can be a "temporary" constraint for agroforestry adoption in the areas because along with population growth and higher demand for agricultural lands, the area of free-gazing lands will become more limited in the future. Therefore, we strongly considered fodder grass as one of main components of the tested agroforestry systems. Moreover, agroforestry systems with grass have been identified as the most suitable practice for northwest Vietnam to reconcile livelihood and erosion control [9].

Sole-maize cultivation on steep slopes in the northwest region of Vietnam produced annual soil loss that reached up to 174 ton ha$^{-1}$ [15]. However, growing forage grass along the contour lines can play a significant role in reducing soil loss, especially on the steep slopes of the study region [15]. All experiments in our study were conducted in lands with about 27% slope, and measurement of soil erosion was not part of our study. However, a study in the northwest region that measured and compared soil erosion rate in agroforestry and sole-crop plantations clearly showed that soil erosion was substantially reduced in agroforestry [52]. The study found that the erosion rate in longan–mango–maize–forage grass agroforestry was 43% lower than that measured in sole-maize cultivations, and the rate in son tra–coffee–forage grass was 34% lower than that measured in sole-coffee plantations. All agroforestry systems and sole-coffee plantations observed in the study were three years old. A higher reduction in the soil erosion rate can be expected in more mature agroforestry such as in our experiments that have larger tree-canopy cover.

Ecological benefits or ecosystem services noted by farmers in this study were the effect of grass strips in reducing soil erosion and maintaining soil moisture and fertility, but also in forming terraces on the steep slopes [52]. In steeply sloping areas, the terraces formed could significantly increase agricultural productivity and enhance water-use efficiency when combined with other agricultural techniques [53].

### *4.3. Economic Benefits of Agroforestry Systems and Possibilities for Improvement*

The agroforestry systems evaluated here showed higher profitability than the sole-crop systems from year 2 onwards. However, the initial investment cost for agroforestry was high: 2122 USD ha$^{-1}$ for LMG and 1772 USD ha$^{-1}$ for STG and STM. Farmers in the region lack the financial resources to shift to new practices [10]. New practices thus need to be shown to be safe and ensure food security before smallholders risk changing from their current system. The main incentive for farmers to adopt agroforestry is increasing yield and stable prices for their products. When comparing production and profitability, a cycle of some years must be considered, because it takes longer to establish perennial trees than annual crops and the financial input is higher in agroforestry systems. Therefore, initial investment funding (possibly organised by farmers themselves), subsidies, or loans will be necessary to compensate for the high investment and maintenance costs in the first few years of agroforestry [16].

The farmer groups interviewed proposed some ways to make implementation of agroforestry more profitable. First, the establishment of agroforestry will require financial support or access to low-interest loans/credits. In addition, implementing agroforestry with fodder-grass strips would become more beneficial for local people if changing from free to captive grazing were promoted. To achieve both in the study region, local farmers can seek support from the Vietnamese government through e.g., the National Target Programme (NTP) on New Rural Development [54] or the NTP-Sustainable Poverty Reduction and 135 Programme [55]. In addition, they can seek loans (low interest rate) from formal actors such as the Vietnam Bank for Agriculture and Rural Development, the Vietnam Bank for Social Policy, and People's Credit Funds [56].

Second, the farmers interviewed suggested producing their own low-cost tree seedlings to reduce the investment cost. These could be grown in community nurseries, where all members share costs and provide inputs [57]. The project Agroforestry for Livelihoods of Smallholder Farmers in Northwest Vietnam (2012–2016), together with relevant stakeholders, has provided training for farmers on the establishment and management of smallholder and group nurseries, producing tree seedlings by seedling propagation, grafting, and marcotting techniques. The project has published various technical extension materials on producing different tree-species seedlings suitable for local conditions. These technical sources could be useful for local farmers producing their own tree seedlings [58].

Third, the interviewees believed that they could achieve stable production by forming growers' cooperatives and could improve product quality to meet market demand. The cooperatives could provide production services, including inputs for farm households, fertilisers, feed ingredients, plant protection chemicals, and vaccines for livestock. They could also mediate between entrepreneurs and farmers, representing and protecting the rights of farmer members in contracting to supply raw materials to processing enterprises and export agricultural products [11]. In rural development work, agricultural service cooperatives can make a significant contribution [11]. Recently, the Vietnamese government introduced a programme to develop 15,000 cooperatives and effective agricultural cooperative unions in rural areas, with the government providing institutions, mechanisms, and policies to support the programme [59]. This offers an opportunity for farmers in the region to develop cooperatives to ensure stable production of agricultural products.

## 5. Conclusions

Agroforestry systems based on fruit trees, grass, and crops had higher productivity, higher profitability, and earlier returns on investment than sole-crop fruit systems, but also higher initial investment costs. The agroforestry systems produced a diversity of products and provided ecosystem services such as erosion control and soil fertility improvement. However, challenges such as higher investment cost and an unstable market for agroforestry products make it uncertain whether agroforestry can be easily promoted in the area.

During development of the agroforestry systems, there were negative effects on growth and productivity of the different components, most likely due to competition. There was evidence of competition for nitrogen between tree, grass, and crop components at positions upslope and downslope of the grass strips. These competition effects need to be considered when designing agroforestry systems and formulating management regimes.

Future fruit tree-based agroforestry systems should apply adaptive management while the agroforestry system is maturing and consider measures such as widening the planting distance between trees, crops, and grass; supplying fertiliser to plant components suffering from nutrient deficiency; and pruning trees in competition zones. Introducing high-value crops or biological N-fixing species to reduce competition and support the growth of trees can also be considered in order to optimise the systems.

To enable uptake and expansion of agroforestry in northwest Vietnam, financial support to meet the higher investment costs for agroforestry and for better value chains with market stability are prerequisites for farmers. Local farmers can produce their own tree seedlings to reduce the investment cost for agroforestry in the region.

**Author Contributions:** Conceptualization, V.H.D., I.Ö., and N.L.; methodology, V.H.D., I.Ö., and N.L.; formal analysis, V.H.D.; investigation, V.H.D., H.T.P., and V.T.N; writing—original draft preparation, V.H.D.; writing—review and editing, I.Ö., G.B., A.S.D., R.M., N.L., H.T.P., and V.T.N. All authors have read and agreed to the published version of the manuscript.

**Acknowledgments:** We highly appreciate the support of researchers and technical staff from the AFLI project at World Agroforestry (ICRAF) and national partners in Vietnam. Our thanks to researchers at the Northern Mountainous Agriculture and Forestry Science Institute and Forest Science Centre of Northwest Vietnam for their assistance. We are grateful to Chu Van Tien, Tran Quang Khanh, and Vu Dinh Chi, who hosted experiments in Van Chan, to Giang Dung Vu, Mua A Do, and Vang A Sua, who hosted experiments in Tuan Giao, and to the local authorities and farmers in the study sites for their support during field research.

## References

1. United Nations General Assembly. Transforming Our World: The 2030 Agenda for Sustainable Development, A/RES/70/1, New York. 2015. Available online: https://www.un.org/ga/search/view_doc.asp?symbol=A/RES/70/1=E (accessed on 10 June 2020).
2. United Nations. Sustainable Development Goals: 17 Goals to Transform Our World. 2015. Available online: https://www.un.org/sustainabledevelopment/sustainable-development-goals/ (accessed on 27 October 2020).
3. Catacutan, D.C.; Van Noordwijk, M.; Nguyen, T.H.; Öborn, I.; Mercado, A.R. *Agroforestry: Contribution to Food Security and Climate-Change Adaptation and Mitigation in Southeast Asia. White Paper*; World Agroforestry Centre (ICRAF): Bogor, Indonesia, 2017.
4. Van Noordwijk, M.; Bayala, J.; Hairiah, K.; Lusiana, B.; Muthuri, C.; Khasanah, N.; Mulia, R. Agroforestry Solutions for Buffering Climate Variability and Adapting to Change. In *Climate Change Impact and Adaptation in Agricultural Systems*; Fuhrer, J., Gregory, P.J., Eds.; CABI Publishing: Wallingford, UK, 2014; pp. 216–232.
5. Roshetko, J.M.; Mercado, A.R.; Martini, E.; Prameswari, D. *Agroforestry in the Uplands of Southeast Asia. Policy Brief no. 77. Agroforestry Options for ASEAN Series no. 5*; World Agroforestry (ICRAF) Southeast Asia Regional Program: Bogor, Indonesia, 2017; Available online: http://www.awg-sf.org/www2/wp-content/uploads/2017/12/ANNEX-10.-PB5-Upland-AF-in-SEA.pdf/ (accessed on 27 October 2020).
6. Tacio, H.D. Sloping Agricultural Land Technology (SALT): A Sustainable Agroforestry Scheme for the Uplands. *Agrofor. Syst.* **1993**, *22*, 145–152. [CrossRef]
7. Nair, P.R.; Nair, V.D.; Kumar, B.M.; Showalter, J.M. Carbon Sequestration in Agroforestry Systems. *Adv. Agron.* **2010**, *108*, 237–307. [CrossRef]
8. Schreinemachers, P.; Holger, L.; Fröhlich, H.L.; Clemens, G.; Stahr, K. From Challenges to Sustainable Solutions for Upland Agriculture in Southeast Asia. In *Sustainable Land Use and Rural Development in Southeast Asia: Innovations and Policies for Mountainous Areas*; Fröhlich, H.L., Schreinemachers, P., Stahr, K., Clemens, G., Eds.; Springer: New York, NY, USA, 2013; pp. 3–23.
9. Hoang, L.T.; Roshetko, J.M.; Huu, T.P.; Pagella, T.; Mai, P.N. Agroforestry—The Most Resilient Farming System for the Hilly Northwest of Vietnam. *Int. J. Agric. Syst.* **2017**, *5*, 1–23. [CrossRef]
10. Zimmer, H.; Le Thi, H.; Lo, D.; Baynes, J.; Nichols, J.D. Why Do Farmers Still Grow Corn on Steep Slopes in Northwest Vietnam? *Agrofor. Syst.* **2017**, *92*, 1721–1735. [CrossRef]
11. Staal, S.; Tran, D.T.; Nguyen, D.P.; Vu, D.H.; Nguyen, S.; Nguyen, T.T.L.; Nguyen, T.S.; Le, N.T.; Ngo, T.H.; Truong, Q.C.; et al. *A Situational Analysis of Agricultural Production and Marketing, and Natural Resources Management Systems in Northwest Vietnam*; International Livestock Research Institute for CGIAR Research Program: Kenya, Nairobi, 2014; Available online: http://humidtropics.cgiar.org/wp-content/uploads/downloads/2014/09/Situational-Analysis-NW-Vietnam-Report.pdf/ (accessed on 10 September 2020).
12. Wezel, A.; Luibrand, A.; Thanh, L.Q. Temporal Changes of Resource Use, Soil Fertility and Economic Situation in Upland Northwest Vietnam. *Land Degrad. Dev.* **2002**, *13*, 33–44. [CrossRef]
13. Clemens, G.; Fiedler, S.; Cong, N.D.; Van Dung, N.; Schuler, U.; Stahr, K. Soil Fertility Affected by Land Use History, Relief Position, and Parent Material Under a Tropical Climate in NW-Vietnam. *Catena* **2010**, *81*, 87–96. [CrossRef]
14. Schmitter, P.; Dercon, G.; Hilger, T.; Le Ha, T.T.; Thanh, N.H.; Lam, N.; Vien, T.D.; Cadisch, G. Sediment Induced Soil Spatial Variation in Paddy Fields of Northwest Vietnam. *Geoderma* **2010**, *155*, 298–307. [CrossRef]
15. Tuan, V.D.; Hilger, T.; Macdonald, L.; Clemens, G.; Shiraishi, E.; Vien, T.D.; Stahr, K.; Cadisch, G. Mitigation Potential of Soil Conservation in Maize Cropping on Steep Slopes. *Field Crop. Res.* **2014**, *156*, 91–102. [CrossRef]

16. Do, H.; Luedeling, E.; Whitney, C. Decision Analysis of Agroforestry Options Reveals Adoption Risks for Resource-Poor Farmers. *Agron. Sustain. Dev.* **2020**, *40*, 1–12. [CrossRef]
17. General Statistics Office of Vietnam. *Statistical Yearbook of Dien Bien, Yen Bai and Son La province*; General Statistics Office of Vietnam: Hanoi, Vietnam, 2018.
18. Hoang, T.L.; Mamo, A.E. *Son tra, the H'mong Apple*; World Agroforestry (ICRAF): Ha Noi, Vietnam, 2015; Available online: http://blog.worldagroforestry.org/index.php/2015/07/01/son-tra-the-hmong-apple/ (accessed on 27 October 2020).
19. Ministry of Agriculture and Rural Development. *Decision No. 4961/QD-BNN-TCLN. Promulgating a List of Main Tree Species for Planting Production Forests and a List of Main Tree Species for Afforestation According to Forest Ecological Regions*; Ministry of Agriculture and Rural Development: Hanoi, Vietnam, 2014; Available online: https://thuvienphapluat.vn/van-ban/tai-nguyen-moi-truong/Quyet-dinh-4961-QD-BNN-TCLN-2014-Danh-muc-cay-chu-luc-trong-rung-san-xuat-vung-sinh-thai-lam-nghiep-259936.aspx/ (accessed on 30 September 2020).
20. Do, T.; Mulia, R. Constraints to Smallholder Tree Planting in the Northern Mountainous Regions of Vietam: A Need to Extend Technical Knowledge and Skills. *Int. For. Rev.* **2018**, *20*, 43–57. [CrossRef]
21. Nguyen, M.P.; Mulia, R.; Nguyen, Q.T. *Fruit Tree Agroforestry in Northwest Vietnam: Sample Cases of Current Practices and Benefits (A Report in Vietnamese Language)*; World Agroforestry (ICRAF): Ha Noi, Vietnam, 2020.
22. TCVN 8941:2011. Soil Quality—Determination of Total Organic Carbon–Walkley Black Method. 2011. Available online: https://vanbanphapluat.co/tcvn-8941-2011-chat-luong-dat-cac-bon-huu-co-tong-so-phuong-phap-walkley-black#van-ban-goc/ (accessed on 11 September 2020).
23. TCVN 6498:1999. Soil Quality—Determination of Total Nitrogen–Modified Kjeldahl Method. 1999. Available online: https://vanbanphapluat.co/tcvn-6498-1999-chat-luong-dat-xac-dinh-nito-tong-phuong-phap-kendan#van-ban-goc/ (accessed on 11 September 2020).
24. TCVN 8940:2011. Soil Quality—Determination of Total Phosphorus–Colorimetry Method. 2011. Available online: https://vanbanphapluat.co/tcvn-8940-2011-chat-luong-dat-xac-dinh-phospho-tong-so-phuong-phap-so-mau#van-ban-goc/ (accessed on 11 September 2020).
25. TCVN 8660:2011. Soil Quality—Method for Determination of Total Potassium. 2011. Available online: https://vanbanphapluat.co/tcvn-8660-2011-chat-luong-dat-phuong-phap-xac-dinh-kali-tong-so#van-ban-goc/ (accessed on 11 September 2020).
26. TCVN 8942:2011. Soil Quality—Determination of Available Phosphorus—Bray and Kurtz (Bray II) Method. 2011. Available online: https://vanbanphapluat.co/tcvn-8942-2011-chat-luong-dat-xac-dinh-phospho-de-tieu-phuong-phap-bray-kurtz#van-ban-goc/ (accessed on 7 July 2020).
27. Minolta. *SPAD-502 Owner's Manual*; Industrial Meter Division, Minolta Corporation: Ramsey, NJ, USA, 1989.
28. Wolfe, D.W.; Henderson, D.W.; Hsiao, T.C.; Alvino, A. Interactive Water and Nitrogen Effects on Senescence of Maize. II. Photosynthetic Decline and Longevity of Individual Leaves. *Agron. J.* **1998**, *80*, 865–870. [CrossRef]
29. Argenta, G.; Da Silva, P.R.F.; Sangoi, L. Leaf Relative Chlorophyll Content as an Indicator Parameter to Predict Nitrogen Fertilization in Maize. *Ciência Rural* **2004**, *34*, 1379–1387. [CrossRef]
30. Viana, M.C.M.; Da Silva, I.P.; Freire, F.M.; Ferreira, M.M.; Da Costa, É.L.; Mascarenhas, M.H.T.; Teixeira, M.F.F. Production and Nutrition of Irrigated Tanzania Guinea Grass in Response to Nitrogen Fertilization. *Rev. Bras. Zootec.* **2014**, *43*, 238–243. [CrossRef]
31. Sritontip, C.; Khaosumain, Y.; Changjeraja, S. Different Nitrogen Concentrations Affecting Chlorophyll and Dry Matter Distribution in Sand-Cultured Longan Trees. *Acta Hortic.* **2014**, 227–233. [CrossRef]
32. Mead, R.; Willey, R.W. The Concept of a 'Land Equivalent Ratio' and Advantages in Yields from Intercropping. *Exp. Agric.* **1980**, *16*, 217–228. [CrossRef]
33. Cook, B.G.; Pengelly, B.C.; Brown, S.D.; Donnelly, J.L.; Eagles, D.A.; Franco, M.A.; Hanson, J.; Mullen, B.F.; Partridge, I.J.; Peters, M.; et al. *Tropical Forages: An Interactive Selection Tool*; [CD-ROM]; CSIRO, DPI&F (Qld), CIAT, ILRI: Brisbane, Australia, 2005.
34. Argel, P.J.; Miles, J.W.; Guiot García, J.D.; Cuadrado Capella, H.; Lascano, C.E. *Cultivar Mulato II (Brachiaria hybrid CIAT 36087): A High-Quality Forage Grass, Resistant to Spittlebugs and Adapted to Well-Drained, Acid Tropical Soils*; International Center for Tropical Agriculture (CIAT): Cali, Colombia, 2008; pp. 1–3. Available online: http://ciat-library.ciat.cgiar.org/Articulos_CIAT/mulato_ii_ingles.pdf/ (accessed on 25 March 2020).

35. Aganga, A.A.; Tshwenyane, S. Potentials of Guinea Grass (*Panicum maximum*) as Forage Crop in Livestock Production. *Pak. J. Nutr.* **2004**, *3*, 1–4.
36. Bacorro, T.J.; Reyes, P.M.; Loresco, M.M.; Angeles, A.A. Herbage Dry Matter Yield, Nutrient Composition and In Vitro Gas Production of Mulato II Mombasa Grasses at 30-and 45-Day Cutting Intervals. *Philipp. J. Vet. Anim. Sci.* **2018**, *44*, 86–89.
37. Vietnamese Government. *Decision No. 59/2015/QD-TTg. Promulgating Multiple Dimensional Poverty Levels Applicable during 2016-2020*; Vietnamese Government: Hanoi, Vietnam, 2015. Available online: https://thuvienphapluat.vn/van-ban/Van-hoa-Xa-hoi/Decision-No-59-2015-QD-TTg-promulgating-multiple-dimensional-poverty-levels-applicable-301414.aspx/ (accessed on 9 September 2020).
38. Malezieux, E.; Crozat, Y.; Dupraz, C.; Laurans, M.; Makowski, D.; Lafontaine, H.O.; Rapidel, B.; De Tourdonnet, S.; Valantin-Morison, M. Mixing Plant Species in Cropping Systems: Concepts, Tools and Models. A Review. *Agron. Sustain. Dev.* **2009**, *29*, 43–62. [CrossRef]
39. Okorio, J.; Byenkya, S.; Wajja, N.; Peden, D. Comparative Performance of Seventeen Upperstorey Tree Species Associated With Crops in the Highlands of Uganda. *Agrofor. Syst.* **1994**, *26*, 185–203. [CrossRef]
40. Bertomeu, M. Growth and Yield of Maize and Timber Trees in Smallholder Agroforestry Systems in Claveria, Northern Mindanao, Philippines. *Agrofor. Syst.* **2011**, *84*, 73–87. [CrossRef]
41. Nyaga, J.; Muthuri, C.W.; Barrios, E.; Öborn, I.; Sinclair, F.L. Enhancing Maize Productivity in Agroforestry Systems Through Managing Competition: Lessons From Smallholders' Farms, Rift Valley, Kenya. *Agrofor. Syst.* **2017**, *93*, 715–730. [CrossRef]
42. Van Noordwijk, M.; Lawson, G.; Hairiah, K.; Wilson, J. Root Distribution of Trees and Crops: Competition and/or Complementarity. In *Tree-Crop Interactions: Agroforestry in a Changing Climate*, 2nd ed.; Ong, C.K., Black, C.R., Wilson., J., Eds.; CAB International: Wellington, UK, 2015; pp. 221–257.
43. Livesley, S.J.; Gregory, P.; Buresh, R. Competition in Tree Row Agroforestry Systems. 1. Distribution and Dynamics of Fine Root Length and Biomass. *Plant Soil* **2000**, *227*, 149–161. [CrossRef]
44. Chirko, C.P.; Gold, M.A.; Nguyen, P.; Jiang, J. Influence of Direction and Distance From Trees on Wheat Yield and Photosynthetic Photon Flux Density (Qp) in a Paulownia and Wheat Intercropping System. *For. Ecol. Manag.* **1996**, *83*, 171–180. [CrossRef]
45. Tuan, V.D.; Hilger, T.; Vien, T.D.; Cadisch, G. Nitrogen Recovery and Downslope Translocation in Maize Hillside Cropping as Affected by Soil Conservation. *Nutr. Cycl. Agroecosystems* **2014**, *101*, 17–36. [CrossRef]
46. Hung, V.; Hung, N. Effect of Pruning and Flower and Fruit Thinning on Growth, Yield and Quality of 'PHM 99.1.1' Late Longan in Hung Yen Province. In *VI International Symposium on Lychee, Longan and Other Sapindaceae Fruits*; Van Hau, T., Mitra, S.K., Drang Khanh, T., Eds.; Acta Hortic: Brussels, Belgium, 2019; pp. 143–148.
47. Schaller, M.; Schroth, G.; Beer, J.; Jiménez, F.; Jiménez, F. Root Interactions Between Young Eucalyptus Deglupta Trees and Competitive Grass Species in Contour Strips. *For. Ecol. Manag.* **2003**, *179*, 429–440. [CrossRef]
48. Rao, M.R.; Nair, P.K.R.; Ong, C.K. Biophysical Interactions in Tropical Agroforestry Systems. *Agrofor. Syst.* **1997**, *38*, 3–50. [CrossRef]
49. Thevathasan, N.; Gordon, A. Ecology of Tree Intercropping Systems in the North Temperate Region: Experiences from Southern Ontario, Canada. *Agrofor. Syst.* **2004**, *61*, 257–268. [CrossRef]
50. Xu, H.; Bi, H.; Gao, L.; Yun, L. Alley Cropping Increases Land Use Efficiency and Economic Profitability Across the Combination Cultivation Period. *Agronomy* **2019**, *9*, 34. [CrossRef]
51. Bush, R.D.; Young, J.R.; Suon, S.; Ngim, M.S.; Windsor, P.A. Forage Growing as an Incentive to Improve Smallholder Beef Production in Cambodia. *Anim. Prod. Sci.* **2014**, *54*, 1620–1624. [CrossRef]
52. Birgitta, S.; Hanna, T. Impact of Agroforestry on Soil Loss Mitigation in the Sloping Land of Northwest Vietnam. Master's Thesis, Swedish University of Agricultural Sciences, Uppsala, Sweeden, May 2020.
53. Chai, Q.; Gan, Y.; Turner, N.C.; Zhang, R.-Z.; Yang, C.; Niu, Y.; Siddique, K.H. Water-Saving Innovations in Chinese Agriculture. *Adv. Agron.* **2014**, *126*, 149–201. [CrossRef]
54. Vietnamese Government. *Decision No. 800/2010/QD-TTg. Approving the National Target. Program. on Building a New Countryside During 2010-2020*; Vietnamese Government: Hanoi, Vietnam, 2010. Available online: https://thukyluat.vn/vb/decision-no-800-qd-ttg-approving-the-national-target-program-on-building-1ad9a.html?hl=en#VanBanTA/ (accessed on 9 September 2020).

55. Vietnamese Government. *Decision No. 551/2013/QD-TTg. Approved the 135 Program. on Supporting Investment in Infrastructure, Production Development for Extremely Difficult Communes, Border Communes, Safety Zone Communes*; Vietnamese Government: Hanoi, Vietnam, 2013. Available online: http://vanban.chinhphu.vn/portal/page/portal/chinhphu/hethongvanban?class_id=2&_page=1&mode=detail&document_id=166543/ (accessed on 9 September 2020).
56. JICA. *Data Collection Survey on the Needs for Agriculture Productivity Improvement in Northern Six Provinces*; Japan International Cooperation Agency, Vietnam Ministry of Agriculture and Rural Development: Hanoi, Vietnam, 2016. Available online: https://openjicareport.jica.go.jp/pdf/12261343.pdf/ (accessed on 9 September 2020).
57. FAO. *Sustainable Forest Management (SFM) Toolbox*; The Food and Agriculture Organisation of the United Nations: Rome, Italy, 2018; Available online: http://www.fao.org/sustainable-forest-management/toolbox/modules/agroforestry/basic-knowledge/en/ (accessed on 9 September 2020).
58. La, N.; Catacutan, D.C.; Nguyen, M.P.; Do, V.H. *Agroforestry for Livelihoods of Smallholder Farmers in Northwest. Vietnam—Final Report, Canberra ACT 2601*; Australian Government, Australian Center for International Agricultural Research: Canberra, Australia, 2019; pp. 10–34. Available online: https://aciar.gov.au/publication/technical-publications/agroforestry-livelihoods-smallholder-farmers-northwest-Vietnam-final-report/ (accessed on 10 September 2020).
59. Vietnamese Government. *Decision No. 461/2018/QD-TTg. Approving the Scheme on the Development of 15,000 Cooperatives and Unions of Agricultural Cooperatives With Effective Operation up to 2020*; Vietnamese Government: Hanoi, Vietnam, 2018. Available online: https://thuvienphapluat.vn/van-ban/Doanh-nghiep/Quyet-dinh-461-QD-TTg-2018-De-an-phat-trien-15000-hop-tac-xa-lien-hiep-hop-tac-xa-nong-nghiep-380877.aspx/ (accessed on 9 September 2020).

# Permissions

All chapters in this book were first published in MDPI; hereby published with permission under the Creative Commons Attribution License or equivalent. Every chapter published in this book has been scrutinized by our experts. Their significance has been extensively debated. The topics covered herein carry significant findings which will fuel the growth of the discipline. They may even be implemented as practical applications or may be referred to as a beginning point for another development.

The contributors of this book come from diverse backgrounds, making this book a truly international effort. This book will bring forth new frontiers with its revolutionizing research information and detailed analysis of the nascent developments around the world.

We would like to thank all the contributing authors for lending their expertise to make the book truly unique. They have played a crucial role in the development of this book. Without their invaluable contributions this book wouldn't have been possible. They have made vital efforts to compile up to date information on the varied aspects of this subject to make this book a valuable addition to the collection of many professionals and students.

This book was conceptualized with the vision of imparting up-to-date information and advanced data in this field. To ensure the same, a matchless editorial board was set up. Every individual on the board went through rigorous rounds of assessment to prove their worth. After which they invested a large part of their time researching and compiling the most relevant data for our readers.

The editorial board has been involved in producing this book since its inception. They have spent rigorous hours researching and exploring the diverse topics which have resulted in the successful publishing of this book. They have passed on their knowledge of decades through this book. To expedite this challenging task, the publisher supported the team at every step. A small team of assistant editors was also appointed to further simplify the editing procedure and attain best results for the readers.

Apart from the editorial board, the designing team has also invested a significant amount of their time in understanding the subject and creating the most relevant covers. They scrutinized every image to scout for the most suitable representation of the subject and create an appropriate cover for the book.

The publishing team has been an ardent support to the editorial, designing and production team. Their endless efforts to recruit the best for this project, has resulted in the accomplishment of this book. They are a veteran in the field of academics and their pool of knowledge is as vast as their experience in printing. Their expertise and guidance has proved useful at every step. Their uncompromising quality standards have made this book an exceptional effort. Their encouragement from time to time has been an inspiration for everyone.

The publisher and the editorial board hope that this book will prove to be a valuable piece of knowledge for researchers, students, practitioners and scholars across the globe.

# List of Contributors

**Maya Sollen-Norrlin and Naomi Laura Jane Rintoul**
School of Psychology and Life Sciences, Canterbury Christ Church University, Canterbury CT1 1TE, UK

**Bhim Bahadur Ghaley**
Department of Plant and Environmental Sciences, Faculty of Science, University of Copenhagen, 2630 Taastrup, Denmark

**Daniela Rocha**
Laboratory of Agroecology and Biodiversity (LAB), Group of Agroecology and Environment (GAMA), University of Santiago de Chile, Santiago 7501015, Chile

**Santiago Peredo Parada and Claudia Barrera**
Laboratory of Agroecology and Biodiversity (LAB), Group of Agroecology and Environment (GAMA), University of Santiago de Chile, Santiago 7501015, Chile
Environment and Society Program, Pablo de Olavide University, 41013 Sevilla, Spain

**Sara Burbi**
Centre for Agroecology, Water and Resilience, Ryton Gardens Campus, Coventry University, Coventry CV8 3LG, UK

**Elok Mulyoutami and Betha Lusiana**
World Agroforestry Centre (ICRAF), Bogor 16001, Indonesia

**Sonya Dewi and Beria Leimona**
World Agroforestry (ICRAF), Bogor 16155, Indonesia

**Vincent Gitz, Nathanaël Pingault and Alexandre Meybeck**
Centre for International Forestry Research (CIFOR), Bogor 16115, Indonesia

**Peter A. Minang and Lalisa Duguma**
World Agroforestry (ICRAF), Nairobi 00100, Kenya

**Kurniatun Hairiah**
Department of Soil Science, Faculty of Agriculture, Brawijaya University, Malang 65415, Indonesia

**Rika Ratna Sari and Danny Dwi Saputra**
Soil Science Department, Faculty of Agriculture, Universitas Brawijaya, Malang 65145, Indonesia
Plant Production Systems Group, Wageningen University and Research, 6708 PB Wageningen, The Netherlands

**Danaë M. A. Rozendaal**
Plant Production Systems Group, Wageningen University and Research, 6708 PB Wageningen, The Netherlands
Centre for Crop Systems Analysis, Wageningen University and Research, 6708 PB Wageningen, The Netherlands

**Meine van Noordwijk**
Plant Production Systems Group, Wageningen University and Research, 6708 PB Wageningen, The Netherlands
World Agroforestry Centre, ICRAF Southeast Asia, Bogor 16001, Indonesia
Faculty of Agriculture, Brawijaya University, Jl Veteran, Malang 65145, Indonesia
Agroforestry Research Group, Faculty of Agriculture, Brawijaya University, Jl. Veteran no 1, Malang 65145, Indonesia

**James M. Roshetko**
World Agroforestry Centre, ICRAF Southeast Asia, Bogor 16001, Indonesia

**Widianto Widianto**
Faculty of Agriculture, Brawijaya University, Jl Veteran, Malang 65145, Indonesia

**Gildas Geraud Comlan Assogba**
Plant Production Systems, Wageningen University and Research, 6700 AK Wageningen, The Netherlands

**Rika Ratna Sari**
Plant Production Systems, Wageningen University and Research, 6700 AK Wageningen, The Netherlands
Agroforestry Research Group, Faculty of Agriculture, Brawijaya University, Jl. Veteran no 1, Malang 65145, Indonesia

**Didik Suprayogo**
Agroforestry Research Group, Faculty of Agriculture, Brawijaya University, Jl. Veteran no 1, Malang 65145, Indonesia

**Erika Speelman and Arend Ligtenberg**
Laboratory of Geo-Information Science and Remote Sensing, Environmental Sciences, Wageningen University & Research, 6708 PB Wageningen, The Netherlands

**Gert Jan Hofstede**
Information Technology Group, Social Sciences, Wageningen University & Research, 6706 KN Wageningen, The Netherlands
Centre for Applied Risk Management (UARM), North-West University, Potchefstroom 2520, South Africa

**Margaret Githinji and Paulina Rosero**
Information Technology Group, Social Sciences, Wageningen University & Research, 6706 KN Wageningen, The Netherlands

**Ai Farida**
Applied Climatology, IPB University, Bogor 16680, Indonesia

**Ali Yansyah Abdurrahim**
Research Centre for Population, Indonesian Institute of Sciences (LIPI), Jakarta 12190, Indonesia
Rural Sociology, IPB University, Bogor 16680, Indonesia

**Soeryo Adiwibowo**
Rural Sociology, IPB University, Bogor 16680, Indonesia

**Andrew Miccolis**
World Agroforestry (ICRAF), 66095-903 Belém/PA, Brazil
Forest Ecology and Forest Management Group, Wageningen University & Research, 6708 PB Wageningen, The Netherlands

**Lisa Tanika and Marielos Peña-Claros**
Forest Ecology and Forest Management Group, Wageningen University & Research, 6708 PB Wageningen, The Netherlands

**Arief Lukman Hakim**
Agroforestry Research Group, Faculty of Agriculture, Brawijaya University, Jl. Veteran no 1, Malang 65145, Indonesia
Forest Ecology and Forest Management Group, Wageningen University & Research, 6708 PB Wageningen, The Netherlands
Faculty of Science and Technology, Universitas Islam Raden Rahmat, Malang 65163, Indonesia

**Charles Nduhiu Wamucii and Adriaan J. Teuling**
Hydrology and Quantitative Water Management Group, Wageningen University & Research, 6700 AA Wageningen, The Netherlands

**Elisabeth Lagneaux**
Plant Production Systems, Wageningen University and Research, 6700 AK Wageningen, The Netherlands
iES Landau, Institute for Environmental Sciences, University of Koblenz-Landau, 76829 Landau, Germany

**Federico Andreotti**
Laboratory of Geo-Information Science and Remote Sensing, Environmental Sciences, Wageningen University & Research, 6708 PB Wageningen, The Netherlands
CIRAD, UPR GREEN, F-34398 Montpellier, France

**Pieter van Oel**
Water Resources Management, Wageningen University & Research, 6700AA Wageningen, The Netherlands

**George Kimbowa**
Water Resources Management, Wageningen University & Research, 6700AA Wageningen, The Netherlands
Faculty of Engineering, Busitema University, 236 Tororo, Uganda

**Lisa Best**
Laboratory of Geo-Information Science and Remote Sensing, Environmental Sciences, Wageningen University & Research, 6708 PB Wageningen, The Netherlands
Tropenbos Suriname, University Campus (CELOS Building), Paramaribo, Suriname

**Usha Satnarain**
Environmental Policy, Department of Social Sciences, Wageningen University & Research, 6706 KN Wageningen, The Netherlands

**Catherine Muthuri**
World Agroforestry (ICRAF), 00100 Nairobi, Kenya

**Edi Purwanto**
Tropenbos Indonesia, Bogor 16163, Indonesia

**Danaë Rozendaal**
Plant Production Systems, Wageningen University and Research, 6700 AK Wageningen, The Netherlands
Centre for Crop Systems Analysis, Wageningen University & Research, 6700 AK Wageningen, The Netherlands

**Stephanie Chizmar, Miguel Castillo, Raul Rivera, Erin Sills, Robert Abt, Rajan Parajuli and Frederick Cubbage**
Department of Forestry and Environmental Resources, North Carolina State University, 2800 Faucette Dr., Raleigh, NC 27695, USA

**Dante Pizarro**
Livestock and Climate Change Center, Universidad Nacional Agraria La Molina, Lima 15000, Peru

**Hector Vasquez and Wilmer Bernal**
Departamento de Zootecnia, Universidad Nacional Toribio Rodriguez de Mendoza de Amazonas, Chachapoyas 477002, Peru

**Sikstus Gusli**
Department of Soil Science, Faculty of Agriculture, Hasanuddin University, Makassar 9245, Indonesia
Natural Resource Research and Development Center, Hasanuddin University, Makassar 9245, Indonesia

**Sri Sumeni, Riyami Sabodin, Ikram Hadi Muqfi and Mustakim Nur**
Department of Soil Science, Faculty of Agriculture, Hasanuddin University, Makassar 9245, Indonesia

**Daniel Useng**
Natural Resource Research and Development Center, Hasanuddin University, Makassar 9245, Indonesia
Department of Agricultural Engineering, Hasanuddin University, Makassar 9245, Indonesia

**Mai Phuong Nguyen, Van Thanh Pham and Elisabeth Simelton**
World Agroforestry (ICRAF), Hanoi 100000, Vietnam

**Duong Dinh Nguyen**
Institute of Geography, Vietnam Academy of Science and Technology, Hanoi 100000, Vietnam

**Peter Steward and Todd Rosenstock**
World Agroforestry (ICRAF), Nairobi 00100, Kenya

**Hoang Anh Le**
Department of Science, Technology and Environment, Ministry of Agriculture and Rural Development, Hanoi 100000, Vietnam

**Van Hung Do**
World Agroforestry (ICRAF) Vietnam, 249A Thuy Khue Street, Thuy Khue Ward, Tay Ho District, Hanoi, Vietnam
Department of Crop Production Ecology, Swedish University of Agricultural Sciences (SLU), 750 07 Uppsala, Sweden

**Nguyen La, Van Thach Nguyen, Huu Thuong Pham and Rachmat Mulia**
World Agroforestry (ICRAF) Vietnam, 249A Thuy Khue Street, Thuy Khue Ward, Tay Ho District, Hanoi, Vietnam

**Göran Bergkvist**
Department of Crop Production Ecology, Swedish University of Agricultural Sciences (SLU), 750 07 Uppsala, Sweden

**A. Sigrun Dahlin**
Department of Soil and Environment, Swedish University of Agricultural Sciences (SLU), 750 07 Uppsala, Sweden

**Ingrid Öborn**
Department of Crop Production Ecology, Swedish University of Agricultural Sciences (SLU), 750 07 Uppsala, Sweden
World Agroforestry Centre (ICRAF) Headquarters, UN Avenue, Nairobi 00100, Kenya

# Index

www.ingramcontent.com/pod-product-compliance
Lightning Source LLC
LaVergne TN
LVHW061510220926
848485LV00017B/777